Modern CNS Drug Discovery

Rudy Schreiber

Editor

Modern CNS Drug Discovery

Reinventing the Treatment of Psychiatric
and Neurological Disorders

 Springer

Editor
Rudy Schreiber
Department of Neuropsychology and Psychopharmacology,
Faculty of Psychology and Neuroscience
Maastricht University
Maastricht, The Netherlands

ISBN 978-3-030-62353-1   ISBN 978-3-030-62351-7   (eBook)
https://doi.org/10.1007/978-3-030-62351-7

Preface

Merging the Old and the New in Modern CNS Drug Discovery
This book is dedicated to my mentor and *drug hunter* Jean De Vry.

His passion and talent was to find new drugs, and, man, he was good at it. Perhaps no surprise as he learned the trade from Dr. Paul, who is one of the all-time greats in drug discovery. Dr. Paul Janssen (1926–2003) founded a company of the same name (now Johnson & Johnson) and was involved in the discovery of 79 new medicines, including the prototypical antipsychotic haloperidol. This book is written for everyone who feels passionate about finding new drugs for the treatment of debilitating diseases of the brain. I wish it helps you unleash your potential and move one step closer to becoming a drug hunter. Our vision for the modern drug discoverer is described in the chapter "Innovator, Entrepreneur, Leader: The Tripartite Drug Discovery Neuroscientist." We argue that besides being creative, innovative, and an expert in a technological area, drug discovery professionals also need to have entrepreneurial competences, such as proficiency in developing ideas into successful new treatments, as well as leadership competences that meet the demands of a creative workforce seeking for sense-making. Yes, it is a tall order. But this is what is needed to meet the tremendous challenges involved in finding new CNS drugs and turn the tide on an emerging health crisis. And there is reason for optimism!

Indeed, we embrace a "techno-optimist" worldview with a strong belief that we are at the cusp of translating unprecedented progress in computing and data sciences, engineering, and the neurosciences into effective treatments for life-threatening diseases. The chapters on brain imaging by *Frans van den Berg* (PET) and *Mitul Mehta* (MRI) show how technology that barely existed when I first met Jean is now routinely used in drug development, and with great impact. We can observe already some early examples of successful translation of such novel approaches, often leading to treatments that are no longer based on the small molecule weight molecules that historically provided the majority of medicines. Other approaches are gaining ground, such as the recently approved biologics and gene therapy to treat spinal muscular atrophy (Spinraza, Zolgensma, and Evreysdi) and new CGRP antibodies against migraine (Erenumab, Fremanezumab, and Galcanezumab). Another emerging trend is the treatment with brain stimulation, and this is described in the two chapters by *Tom de Graaf* and his colleagues.

But I am getting ahead of myself. Every journey begins with the first step, and for drug discovery, this is the generation of a good new idea, which is an obvious but often underappreciated fact. Nowadays, this often translates into targeting a gene that has been associated with a disease. *Arjan Blokland* provides a critical appraisal of how we identify novel molecular targets, especially the Genome Wide Association Studies (GWAS) that are often used. Next, *Larry Melnick* eloquently outlines the principles of neurogenomics and the application to neuropsychiatric diseases with complex heredity. *Ziva Korda* and her colleagues give a comprehensive overview of the use of epigenetics in drug discovery, a relatively novel discipline in the CNS field that is intriguing but also complex and challenging. Especially in (early) drug discovery, animal models play a critical role for demonstrating proof of mechanism, proof of concept, and PK-PD relationship. *Kris Rutten* provides a comprehensive overview of the use of animal models in pain research with a focus on translational aspects. Besides being efficacious, a new drug also needs to be safe and possess a favorable metabolic profile; *Jacco Briede* expertly covers these aspects in his chapter.

Successful transition from the discovery phase to the clinical development phase is challenging, and translational neuroscience aims to bridge that gap. As mentioned previously, imaging methods play a central role therein (chapters by *van den Berg* and *Mehta*). For many decades, a true workhorse in this area has been the electrophysiology-based methods. *Anke Sambeth* provides an excellent overview of these methods and their use in drug development. We have two more examples how translational science can be successfully applied in clinical studies. *Chris Edgar* describes the use of cognitive test outcomes for clinical drug development, whereas *Paddy Janssen* approaches this from a lesser-known but interesting perspective with his chapter on sexual psychopharmacology. Finally, moving away from the conventional path of using the Diagnostic and Statistical Manual (DSM)-5 for classifying diseases, the final chapter in this part by *Bill Potter* and *Bruce Cuthbert* addresses a novel nosology for CNS diseases based on Research Domain Criteria (RDoC) that define novel biotypes.

Following generation of a discovery and translational data package, the asset moves into clinical development. *Wim Riedel* introduces this part and shares his view on the challenges in clinical development. The chapter by *Eef Theunissen* provides great insights into the *nuts and bolts* of phase 1 clinical testing. *Daniel Vargas* describes the exciting story of the development of a first in class treatment: Erenumab for migraine prophylaxis. *Kim Kuypers* describes how microdosing is used in psychedelic research. The final chapter by *Kathy McCarthy* and *Niki Gallo* describes the different steps in the registration process and current trends for CNS drug approvals by the FDA, the regulatory agency of the USA.

It has been incredibly fun to edit this book. All the authors – often (former) colleagues and/or even friends – were amazing in stepping up to the plate to deliver a quality manuscript, often accommodating my special wishes. A heartfelt and big THANK YOU! This book would not have existed without the persistence of Ina Stoeck from Springer. Ina was inspired by our Research Master in *Drug Development and Neuro-Health* and encouraged us to write a book about it. Indeed, several course coordinators contributed to this book. Many thanks to Ina for her persistence and great support during this project.

A few years ago, I met Jean during the thesis defense of his son. After a long career in drug discovery he was disillusioned, and, never afraid of making strong statements, exclaimed, *"I feel like a pastor who wakes up one day to discover that God does not exist."* Didn't I mention that drug discovery is tough? When you decide to commit yourself to becoming a drug hunter there will be times when you may feel like *"being on your own"* and that you are chasing something unreachable. I hope that this book will give you guidance and support to overcome such moments of doubt. That you will remember the successes like Erenumab, and that you will sustain your belief that *"It can be done."*

The Purkinje Pattern

One day, while examining a brain slice through my confocal microscope, I observed something unexpected – I witnessed the beauty inside the brain.

As a neuroscientist, I studied Purkinje cells, a type of neuron that resides in the cerebellum at the back of the brain. These highly branched cells are responsible for helping us learn new movements, coordinate our limbs, and maintain a balanced posture.

The cover image is a Purkinje cell from my calcium imaging experiments in the lab, captured using confocal microscopy and fluorescent dyes. To create the artistic effects, I used unconventional light sources to add a bit of flair. Purkinje cells are approximately 200 microns (0.2 millimeters) tall, which is relatively large for a neuron.

While observing these cells, I realized that they bear a striking structural resemblance to trees. In fact, this "Purkinje Pattern," which consists of larger branches subdividing into smaller branches, is present all throughout nature on both microscopic and macroscopic scales.

Look around, and you'll find examples of the Purkinje Pattern not only in tree branches, but in plant roots, coral, antlers, lightning, capillary networks, river tributaries, veins in leaves, and decision-making diagrams. What kinds of branches have you seen today?

Dana H. Simmons, Ph.D.
Dana-Simmons.com

Rudy Schreiber
Maastricht, The Netherlands
September 2020

Contents

I   Introduction

1   Innovator, Entrepreneur, Leader: The Tripartite Drug
Discovery Neuroscientist... 3
Rudy Schreiber, Mark Govers, and Arie van der Lugt

**II   Innovation in the Discovery of Novel
Therapeutic Approaches**

2   Drug Discovery in CNS: Finding a Target for What?.............. 25
Arjan Blokland

3   Neurogenomics with Application to Schizophrenia
and Other Major Neuropsychiatric Diseases with Complex
Heredity... 35
Laurence Melnick

4   Epigenetics in Drug Discovery: Achievements and
Challenges... 57
Ziva Korda, Ehsan Pishva, and Daniel L. A. van den Hove

5   Conventional Behavioral Models and Readouts in Drug
Discovery: The Importance of Improving Translation 77
Kris Rutten

6   Safety and Drug Metabolism: Toward NCE and First in
Human 93
Jacco J. Briedé

7   The Various Forms of Non-invasive Brain Stimulation and
Their Clinical Relevance 103
*Tom A. de Graaf, Alix Thomson, Felix Duecker,
and Alexander T. Sack*

8   Is Non-invasive Brain Stimulation the Low-Hanging Fruit? ... 115
Tom A. de Graaf, Shanice E. W. Janssens, and Alexander T. Sack

III   Translational Medicine and Technology

9   Electrophysiology: From Molecule to Cognition, from
Animal to Human.. 131
Anke Sambeth

10   MRI in CNS Drug Development 149
      Mitul A. Mehta

11   Positron Emission Tomography in Drug Development 165
      Frans van den Berg and Eugenii A. (Ilan) Rabiner

12   Application of Cognitive Test Outcomes for Clinical Drug
      Development 183
      Chris J. Edgar

13   Exciting Research in the Field of Sexual
      Psychopharmacology: Treating Patients with Inside
      Information 199
      Paddy Janssen

14   A Paradigm Shift from DSM-5 to Research Domain Criteria:
      Application to Translational CNS Drug Development 211
      William Potter and Bruce Cuthbert

IV   Clinical Development and Regulatory Approval

15   The Special Challenges of Developing CNS Drugs 231
      Wim Riedel

16   Phase 1 Clinical Trials in Psychopharmacology 235
      Eef Theunissen

17   Early Development of Erenumab for Migraine Prophylaxis ... 245
      Gabriel Vargas

18   Microdosing Psychedelics as a Promising New
      Pharmacotherapeutic 257
      Kim P. C. Kuypers

19   Partnering with the FDA 275
      Katie McCarthy and Niki Gallo

Correction to: Positron Emission Tomography in Drug
Development C1

Editor Biography

Rudy Schreiber obtained his B.S. in medicinal biology and his Ph.D. in neuropharmacology from the University of Groningen in the Netherlands. He started his career in central nervous system drug discovery in 1993. Subsequently, he progressed from bench scientist to senior leadership roles at global pharmaceutical companies, including Servier (France), Bayer (Germany), Roche (CA), and Sepracor (now Sunovion Pharmaceuticals, MA). From 2010 till 2012, he was the SVP of the Neurobiology Group of the contract research organization, Evotec (Germany). Rudy's drug discovery experience covers a wide range of psychiatric and neurologic indications. He also worked with many different classes of molecular targets and led projects ranging from the idea stage throughout proof of principle testing. He has published about 80 papers and 30 patents. In 2012, he founded Suadeo Drug Discovery Consulting LLC in Boston. Rudy joined Maastricht University in 2018 and is the coordinator of the 2-year research master program *Drug Development and Neuro-Health*. He is a *Homo Aquaticus* who lives on a boat in Maastricht Marina and spends times rowing on the Maas river or sailing offshore.

Contributors

Arjan Blokland   Department of Neuropsychology and Psychopharmacology, Faculty of Psychology and Neuroscience, Maastricht University, Maastricht, The Netherlands
a.blokland@maastrichtuniversity.nl

Jacco J. Briedé   Department of Toxicogenomics, Faculty of Health Medicine and Life Sciences (FHML), Maastricht University, Maastricht, The Netherlands
j.briede@maastrichtuniversity.nl

Bruce Cuthbert   National Institute of Mental Health, National Institutes of Health, Bethesda, MD, USA
bcuthber@mail.nih.gov

Tom A. de Graaf   Brain Stimulation and Cognition Section, Department of Cognitive Neuroscience, Faculty of Psychology and Neuroscience, Maastricht University, Maastricht, The Netherlands
tom.degraaf@maastrichtuniversity.nl

Felix Duecker   Brain Stimulation and Cognition Section, Department of Cognitive Neuroscience, Faculty of Psychology and Neuroscience, Maastricht University, Maastricht, The Netherlands
felix.duecker@maastrichtuniversity.nl

Chris J. Edgar   Cogstate Ltd, London, UK
cedgar@cogstate.com

Niki Gallo   Halloran Consulting Group, Boston, MA, USA
NGallo@Hallorancg.com

Mark Govers   Faculty of Health, Medicine and Life Sciences, Health Services Research, School for Public Health and Primary Care, Maastricht University, Maastricht, The Netherlands
m.govers@maastrichtuniversity.nl

Paddy Janssen   Department of Clinical Pharmacy and Toxicology, Maastricht University Medical Centre, Maastricht, The Netherlands
Department of Hospital Pharmacy, VieCuri Medical Centre, Venlo, The Netherlands
Faculty of Psychology and Neuroscience, Section Neuropsychology & Psychopharmacology, Maastricht University, Maastricht, The Netherlands
paddy.janssen@mumc.nl

Shanice E. W. Janssens   Brain Stimulation and Cognition Section, Department of Cognitive Neuroscience, Faculty of Psychology and Neuroscience, Maastricht University, Maastricht, The Netherlands
shanice.janssens@maastrichtuniversity.nl

Ziva Korda   Department of Psychiatry and Neuropsychology, School for Mental Health and Neuroscience (MHeNs), Maastricht University, Maastricht, The Netherlands
z.korda@alumni.maastrichtuniversity.nl

Kim P. C. Kuypers   Department of Neuropsychology & Psychopharmacology, Faculty of Psychology & Neuroscience, Maastricht University, Maastricht, The Netherlands
k.kuypers@maastrichtuniversity.nl

Katie McCarthy   Halloran Consulting Group, Boston, MA, USA
KMcCarthy@hallorancg.com

Mitul A. Mehta   Department of Neuroimaging, Institute of Psychiatry, Psychology & Neuroscience, King's College London, London, UK
mitul.mehta@kcl.ac.uk

Laurence Melnick Embeca Biosciences LLC, Watertown, MA, USA
lmelnick.embeca@gmail.com

Ehsan Pishva Department of Psychiatry and Neuropsychology, School for
Mental Health and Neuroscience (MHeNs), Maastricht University, Maastricht,
The Netherlands
University of Exeter Medical School, College of Medicine and Health, Exeter
University, Exeter, UK
e.pishva@maastrichtuniversity.nl

William Potter National Institute of Mental Health, National Institutes of
Health, Bethesda, MD, USA
wzpottermd@gmail.com

Eugenii A. (Ilan) Rabiner Imperial College London, Hammersmith Hospital,
London, UK
ilan.rabiner@invicro.co.uk

Wim Riedel Maastricht University, Maastricht, The Netherlands
w.riedel@maastrichtuniversity.nl

Kris Rutten Research Services Division, WuXi AppTec, Shanghai, China

Alexander T. Sack Brain Stimulation and Cognition Section, Department of
Cognitive Neuroscience, Faculty of Psychology and Neuroscience, Maastricht
University, Maastricht, The Netherlands
Department of Psychiatry and Neuropsychology, School for Mental Health
and Neuroscience (MHeNs), Brain+Nerve Centre, Maastricht University Medical Centre+ (MUMC+), Maastricht, The Netherlands
a.sack@maastrichtuniversity.nl

Anke Sambeth Department of Neuropsychology and Psychopharmacology,
Faculty of Psychology and Neuroscience, Maastricht University, Maastricht,
The Netherlands
anke.sambeth@maastrichtuniversity.nl

Rudy Schreiber Department of Neuropsychology and Psychopharmacology,
Faculty of Psychology and Neuroscience, Maastricht University, Maastricht,
The Netherlands
rudy.schreiber@maastrichtuniversity.nl

Eef Theunissen Department of Neuropsychology and Psychopharmacology,
Faculty of Psychology and Neuroscience, Maastricht University, Maastricht,
The Netherlands
e.theunissen@maastrichtuniversity.nl

Alix Thomson Brain Stimulation and Cognition Section, Department of Cognitive Neuroscience, Faculty of Psychology and Neuroscience, Maastricht University, Maastricht, The Netherlands

Department of Psychiatry and Neuropsychology, School for Mental Health and Neuroscience (MHeNs), Brain+Nerve Centre, Maastricht University Medical Centre+ (MUMC+), Maastricht, The Netherlands
alix.thomson@maastrichtuniversity.nl

Frans van den Berg   Imperial College London, Hammersmith Hospital, London, UK
Frans.VanDenBerg@invicro.co.uk

Daniel L. A. van den Hove   Department of Psychiatry and Neuropsychology, School for Mental Health and Neuroscience (MHeNs), Maastricht University, Maastricht, The Netherlands
Department of Psychiatry, Psychosomatics and Psychotherapy, University of Würzburg, Würzburg, Germany
d.vandenhove@maastrichtuniversity.nl

Arie van der Lugt   Department of Cognitive Neuroscience, Faculty of Psychology and Neuroscience, Maastricht University, Maastricht, The Netherlands
arie.vanderlugt@maastrichtuniversity.nl

Gabriel Vargas   CuraSen Therapeutics, San Mateo, CA, USA
vargas@curasen.com

Introduction

Contents

Chapter 1   Innovator, Entrepreneur, Leader:
               The Tripartite Drug Discovery
               Neuroscientist – 3
               *Rudy Schreiber, Mark Govers, and
               Arie van der Lugt*

Innovator, Entrepreneur, Leader: The Tripartite Drug Discovery Neuroscientist

Rudy Schreiber, Mark Govers, and Arie van der Lugt

Contents

1.1   The Need for a New Generation of Drug Discovery
        Neuroscientists – 4

1.2   The Student as an Innovator: *From Problem Via
        Ideas to Solutions* – 7
1.2.1   Divergent Thinking: Idea Generation – 9
1.2.2   Convergent Thinking: Idea Evaluation and Implementa-
        tion – 12

1.3   Entrepreneurial Competences: *Where Imagination
        and Logic Meet in Creating Value* – 14
1.3.1   Turning Ideas into Value Propositions – 16

1.4   Leading Creative and Innovative People: *Where
        the Rubber Hits the Road* – 18

        References – 21

The trend toward a knowledge-based society is accelerating. As a consequence, creativity, idea generation, innovation, and the translation of these ideas into action are vital capabilities for many organizations. The ability to translate faster in a coherent manner increasingly determines competitive advantage, especially in fields invested in finding solutions for highly complex problems, such as discovering and developing new treatments for psychiatric and neurologic diseases. This requires additional competences, besides being creative, innovative, and an expert in a technological area. Professionals also need to have entrepreneurial competences, such as proficiency in developing ideas into successful new treatments, and, finally, (transformational) leadership competences that meet the demands of a creative workforce seeking for sense-making. The drug discovery field places a premium on such "*tripartite experts*" who are committed to lifelong learning to develop, master, and perfect their innovative, entrepreneurial, and leading competences. Acquiring these competences gradually by "*learning on the job*" is history. To meet the challenges of our transforming times, we need to foster a receptive mindset in students. Consequently, we have to renew their education to structure and accelerate the process for becoming proficient champions in these three areas.

Learning Objectives
- Creativity. The student[1] can create novel *ideas* by experiment and play by using a combination of divergent and convergent thinking.
- Innovation. The student can work on an idea with decisions and actions which may lead to an appropriate *solution* to solve problems.
- Entrepreneurial. The student can translate potential solutions into *value propositions* regardless of the availability of favorable contingencies.

- Leading. The student can lead creative, innovative, and entrepreneurial individuals and teams seeking for feasible, valuable, and *sense-making solutions* and has the skills to get support from the organization.
- Orchestrating. The student can orchestrate teamwork with *tripartite homo medicamento inventa* specimens.

» "*For those of us who have established our careers in drug discovery, at the end of time it will be our good works that will have made a difference in the world we live and our inner sense of contributing to a noble cause, namely, the war on disease, agony and fear that needlessly reigns upon millions of human beings because of the lack of medicines. Indeed, drug discovery is a noble cause and raison d'être for transformational leadership by way of virtuous thought, word and deed. Let us take time to think, speak and act in new ways in our quest for such good works in drug discovery with a greater understanding of the words of wisdom that virtue is its own reward*" (Sawyer 2008).

1.1 The Need for a New Generation of Drug Discovery Neuroscientists

There is a huge unmet medical need for novel treatments for disorders of the central nervous system (CNS). One in every four people will develop mental or behavioral disorders at some stage in life, and virtually no cures exist. Unfortunately, CNS drug discovery[2] is perhaps the most exceptionally complex, long (up to 15 years), expensive (up to 1.5 BN), and high-risk (5% success in clinical development) endeavors one can partake in. The brain is our most complex organ with 86 billion neurons and 100 trillion synapses, which is 1000 times

---

1   We use *student* in the widest possible sense: the need for lifelong learning in modern society makes us all students for life!

2   When referring to *drug discovery* this in general covers all therapeutic approaches (e.g., biologics; medical devices, eHealth etc.) and not only small molecules.

the number of stars in our galaxy. Our understanding of how the brain operates is growing rapidly. Thanks also to a rapid pace of innovation in neuroscience technology (e.g., see chapter by de Graaf on transcranial brain stimulation). Notwithstanding such encouraging and impressive developments, finding novel treatments that will cure CNS diseases remains *very* challenging.

One root cause is the aforementioned complexity of the brain and our incomplete understanding of the pathophysiology of its disorders. On top of this, the drug discovery model has been broken for more than a decade (Scannell et al. 2012), the resulting decrease in productivity was followed by a sharp decrease in global CNS drug discovery activity in the early 10's. Fortunately, there has been renewed investment in neuroscience from pharmaceutical companies and a real promise of developing new, impactful treatments.

> Realizing the commercial potential of these investments requires a well-trained neuroscience workforce able to interact collaboratively in multidisciplinary teams focused on advancing projects through the drug discovery pipeline, including in the many areas of specialization (clinical, regulatory, commercial, partnerships, external opportunities, policy, etc. (Akil 2016)).

Training this new kind of workforce is the objective of the Maastricht Drug Development and NeuroHealth research master (DN RMa) program.[3] The mission of the DN RMa program to train a new generation of *drug hunters* dovetails with three developments: a need for academia to *provide alternative career paths*; the focus of universities on *translating science into products*; and an increased acknowledgment of the importance of *entrepreneurial competences* for the future workforce.

With regard to the first development, there has been a rapid growth in the number of

neuroscientists. In the USA alone, the number of new PhDs per year more than doubled between 2003 and 2013. Importantly, there are annually new faculty positions *for only about 10% of the new PhDs*, illustrating an urgent need for providing alternative career paths (Akil 2016). Students are very receptive to seek careers outside universities. Many do not value a non-academic career as a back-up option anymore but are interested in a range of opportunities within and outside academia. As a consequence, neuroscience training programs are responsible to provide their students with the "*the tools, skills, and knowledge to enable the trainees to make effective contributions to the workforce*" (Akil 2016).

Concerning the second development, universities increasingly require that academic research groups translate science into product ideas and "*valorize*" novel scientific insights. This development coincides with the increased reliance of the traditional powerhouses of drug discovery – big Pharma & Biotech – on academic research for the generation of ideas and performance of the preclinical stages of drug discovery. This requires the capability to translate science into product ideas and valorize novel scientific insights. Valorization is defined as "*the process of value creation from knowledge, by making it applicable and available for economic or societal utilization, and by translating it in the form of new business, products, services, or processes.*" Hence, the academic scientist in the neurosciences is often incentivized to find an idea for a novel treatment and to develop this idea through the discovery stages and perhaps even early human testing. A main objective is to generate intellectual property (often patents), which is essential for the subsequent steps such as fundraising, spinning off a company, and/or getting into a partnership to test the asset in humans. Creativity and innovation are core competences[4] of the academic culture, making it a good environ-

---

3 ▶ https://www.maastrichtuniversity.nl/education/master/research-master-cognitive-and-clinical-neuroscience-specialisation-drug-development

4 Competences are abilities, behaviors, knowledge, and skills that impact the success of employees and organizations. They can include general skills (like communication skills), role-specific skills, and leadership skills, as well as others.

1

ment for the *generation* of ideas. *Developing ideas* to a product requires additional competences, such as the creation of a value proposition assessment (see entrepreneurial section 1.3) that describes how the product will precisely address the unmet medical need of the patient, as well as the demands of the medical professional, the payer, and the regulatory agency (see ► Chap. 19).

Finally, the development of the *entrepreneurial capacity* of European citizens and organizations is one of the key policy objectives for the EU and Member States. The European

Commission identified sense of initiative and entrepreneurship as one of the key competences necessary for a knowledge-based society. The so-called "EntreComp framework" offers a basis for the development of curricula and learning activities fostering entrepreneurial competences. Many Entrecomp competences (■ Fig. 1.1) are covered in the DN RMa curriculum. For example, the competences *working with others* and *learning through experience* are implicit to the *Problem-Based Learning* teaching method that we use at Maastricht University.

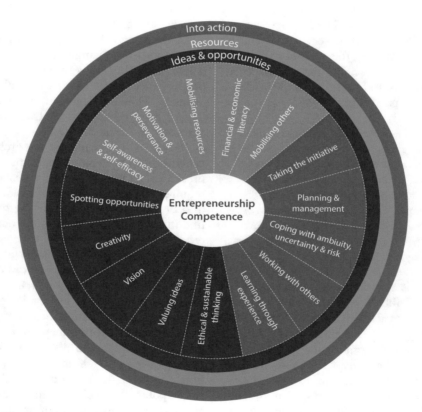

■ **Fig. 1.1** The Entrecomp framework describes entrepreneurship as a competence that can be applied to all spheres of life including the starting up of ventures. It encompasses three competence areas and 15 competences. It aims to establish a bridge between education and work building on entrepreneurship as a competence. (Figure from Bacigalupo et al. 2016)

Problem-Based Learning

Problem-Based Learning (PBL) is a teaching method in which complex real-world problems are used to stimulate student learning of concepts and principles, as opposed to direct presentation of facts and concepts (Dolmans et al. 2005; Schmidt et al. 2011). The educational principles of Maastricht University (▶ https://edlab.nl/pbl-learning-principles/) hold that learning should be collaborative, constructive, contextual, and self-directed (CCCS).

1. Contextual – PBL uses real everyday problems. Hence the learning material is more relevant and will be easier to apply on real situations.
2. Constructive – PBL is a student-centered approach in which learners construct their own knowledge and the teacher or tutor serves as a guide on the side.
3. Collaborative – PBL stimulates students to co-construct knowledge and to share ideas and knowledge.
4. Self-directed – PBL promotes self-directed learning skills among students. Examples are planning, reflection, evaluation of understanding, and managing information and resources.

Another example is the workshop *project management* that addresses the competences *planning and management*. Finally, in the workshop *valorization*, we address entrepreneurial, innovator and leadership competencies (◻ Fig. 1.2).[5] These are increasingly required in the modern drug discovery process and were added to the curriculum to complement the

---

5 The following competences are covered in the valorization workshop (and implicitly in some other courses as well): *spotting opportunities, creativity/ valuing ideas, mobilizing others, motivation, and perseverance.*

technical and research competences that have been traditionally the focus of academic curricula in the neurosciences. In the next sections, we will lay out in greater detail the competences and how together they form the essential building blocks for the twenty-first-century tripartite drug discovery neuroscientist.

1.2 The Student as an Innovator: From Problem Via Ideas to Solutions

> Learning Goal 1: *The student can create novel ideas by experiment and play by using a combination of divergent and convergent thinking.*

In the first two decades of the twenty-first century, the calls for educational institutions to pay more attention to creativity and innovation in their educational programs have become louder and louder. The European Commission even designated 2009 as *"the European Year of Creativity and Innovation."* Based on the claim that creativity and innovation are crucial for economic prosperity and both personal and social well-being, it is argued that these competences should be developed more in education. The calls for creativity and innovation are not limited to politics. Many business organizations and non-governmental organizations (NGOs) also consider creativity and innovation to be a top priority (e.g., the Pistoia Alliance to lower barriers to innovation in life science; Pistoia Alliance 2020). The creative process is at the heart of innovation and often the words are used interchangeably. Innovation implies creation: making something new. Creativity is an active process necessarily involved in innovation. Creativity is about new ideas, new ways of looking at things. The creative process leads to novel and useful solutions for given prob-

1

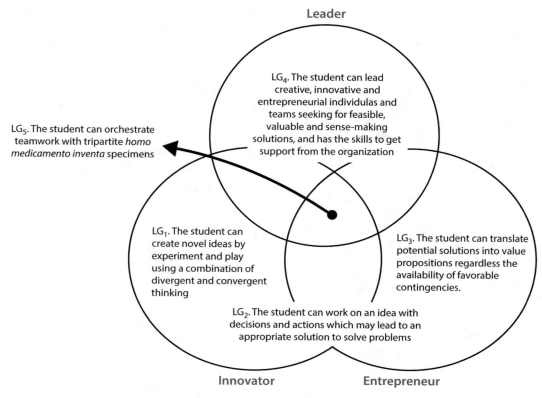

Fig. 1.2 Tripartite competences for drug discovery neuroscientists. Innovation is a creative, nonlinear process that relies on divergent and convergent thinking to generate and develop novel ideas to solve problems. It bridges to entrepreneurial thinking and acting to push these ideas forward to value propositions. Transforma-tional leadership competences (charisma, motivation, intellectual stimulation, and individualized consideration) are required to facilitate the conditions for professionals to collaborate and to orchestrate their innovative and entrepreneurial competences to valuable and sense-making outcomes. *LG* learning goal

lems (Amabile 1990). According to Kampylis and Berki (2014), creative thinking is the type of thinking that enables students to apply their imagination to generate ideas, questions, and hypotheses and to experiment with alternative ideas and solutions. Of course, these alternative ideas and solutions also need to be evaluated.

So how can we promote a more creative approach to solving problems in students? Much of early education is aimed at regulation and conformity. *First*, we train children to sit still and focus for more prolonged periods of time on the teacher's voice of reason. Natural exploratory behaviors and play are mostly limited to breaks in the scholastic program. After the acquisition of basic skills like reading, writing, and mathematics, most educational efforts are dedicated to the development of analytical skills. Students learn things

in a piecemeal fashion. They learn about the building blocks, for example the Lego bricks of life in biology, like proteins or cells, and how they can be combined using logic or some scientific law to make up a more complex structure at a higher level. In this aspect-based approach,[6] students are trained how to break down big problems in smaller problems. They are instructed how they can get from A to B via a detailed procedure which describes all of the steps that need to be taken in between. We even organize our educational system in this way (and our businesses and to a large extent our societies). It should, therefore, come as no surprise that many students (and their teachers who have been educated in the same way)

---

6   As compared with an *integral* approach.

find it very hard to deal with situations where they are required to find a solution for problems which are more ill defined and complex. These types of problems are often referred to as wicked problems (cf. Buchanan 1992). Cases where it is not clear at all how to break them down into sub-problems. To solve this type of problem, imagination and understanding are key. This insight is also expressed in this famous Einstein quote: *Logic will get you from A to B. Imagination will take you everywhere.* This creative problem-solving process is often quite chaotic. It can be a matter of going one step forward and two steps back. It is often inefficient and driven forward by waves of imaginative divergent thoughts and feelings. From this wealth of associations, the most promising ideas are selected and turned into proposed solutions which need to be put to the test.

> One of the first things students need to experience and thereby learn about creative problem-solving is that it does not progress in a linear fashion from A to B.

The idea of natural evolution can serve as a first powerful example of nonlinear problem-solving. The insect which mimics the leaf of a plant represents an appropriate solution to avoid predation over many generations (see ◘ Fig. 1.3).

Another good source of information on the nonlinear nature of creative problem-solving is provided by accounts of the creative process of highly creative individuals. The psychologist Simonton (2007) used Picasso's sketches for Guernica to establish whether this famous creative process was following a more systematic monotonic convergent pattern or a non-monotonic Darwinian pattern of blind variation and selection (see ◘ Fig. 1.4). The results of this study strongly suggest that the creative process underlying Picasso's painting Guernica is a Darwinian nonlinear process of trial and error. It is a process of blind variation because at this point it is not clear what the end result will look like and there is no explicit goal he is working toward. Of course, this does not imply that all of the variations in a creative

◘ **Fig. 1.3** Looking like a leaf is an evolved solution to the problem of being eaten by a hungry bird (with permission from the photographer; Arthur Anker)

process must be completely random. More systematic and incremental improvements based on expertise can also play an important part in creative problem-solving. If you have headed up a dead alley or taken a wrong turn, you will need to recognize at some point that this is the case. Experience and expertise are important for this diagnosis.

Dead alleys and wrong turns are not the only way in which you can get stuck in creative problem-solving. The most obvious way in which you can get stuck is simply by running out of ideas. Guilford's alternative use test (1967) is a nice way to introduce this (cf Kurzberg and Reale 1999). Simply by trying to come up with as many different ways to use a brick or a paperclip as you can, you will directly experience what it feels like to run dry in the generation of ideas.

1.2.1 Divergent Thinking: Idea Generation

Guilford (1950) was the first to propose to separate creative problem-solving into divergent thinking and convergent thinking. Divergent thinking is all about the generation of ideas. According to Torrance (1974), people's ability to engage in divergent thinking can be measured by evaluating the ideas they have generated with respect to four different aspects:
1. Fluency: the *number of ideas* generated
2. Flexibility: the *number of categories* the ideas are based on or fit in

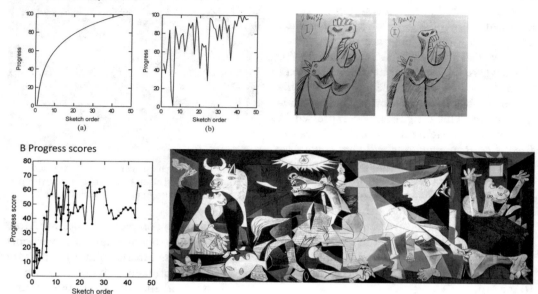

Fig. 1.4 Simonton asked people to judge the sketches Picasso made in preparation for his masterpiece Guernica in terms of how close people thought they were with respect to the final result in the painting (for example the famous horse). At the top left of the panel, the two different theoretical predictions are plotted, a monotonic development in which every step is a small step forward (on the left) and a nonlinear, more erratic pattern (on the right). The bottom plot shows the progress scores given for every sketch in the order they were produced by Picasso. On the right, two different sketches for the horse and the end result below (sketches with permission from Taylor and Francis; painting © Pablo Picasso, Guernica, 1937 with permission of c/o Pictoright Amsterdam 2020)

3. Elaboration/quality: the *depth* or filling out of the ideas
4. Originality: the degree of *novelty* of the ideas

These aspects of idea generation provide an effective structure to present some of the tools and techniques which are commonly used to enhance idea generation. Another way to look at these tools and techniques is to say that they are ways in which we can help ourselves to become unstuck in a creative problem-solving process (for a nice overview of more creative problems techniques, see Heijne and van der Meer (2019)).

Fluency: Brainstorming and Brain Writing Brainstorming is still the most popular approach to idea generation. This technique also plays important part in the Maastricht PBL approach to learning (► https://edlab.nl/pbl-learning-principles/). Two important core principles in brainstorming are "*postponing judgment*" and "*quantity breeds quality*." From these two principles, four basic rules for brainstorming were derived:

− Freely associate any ideas without boundaries
− No critical assessment or evaluation of ideas
− Quantity of ideas is the primary goal
− Encourage the combination and improvement of other ideas (sometimes referred to as piggybacking or hitch-hiking)

Brainstorming is a popular and easy way to generate more ideas. Groups should be led by an experienced facilitator. Information should be recorded, ideally without interrupting the flow. Research has shown that brainstorming can not only lead to an increase in the number of solutions but also produce solutions that are classified as more creative (Stroebe et al. 2010).

Brainwriting (Geschka et al. 1976) refers to a modified version of brainstorming that involves each person being given a sheet of paper that clearly identifies the problem and contains a grid, usually with three columns. Each person is then required to write three ideas on his/her sheet and then either pass the sheet to the person on their right or return it to a central pile. This process continues, with each person building on previously written ideas, until a specified time limit is reached or the sections of the grid are all full. Evidence shows that the sharing of written ideas using the brainwriting tool can lead to better idea generation compared to verbal discussions (Isaksen et al. 2011).

Flexibility/Originality: Creative confrontation In divergent thinking, it is important to create ideas beyond the obvious. Creative confrontation refers to a whole family of techniques which are aimed at promoting flexibility in idea generation, i.e., the ability to come up with different types of ideas. The basic assumption in *Creative Confrontation* is that the introduction of a different frame of thought or perspective can lead to new ideas. Koestler set the theoretical stage for the development of these techniques (Koestler 1964) by describing the confrontation of two or more different frames, or in his words matrices, of thought as the fundamental principle of any creative act. However, the practical use of these methods was already pioneered earlier by using sequences of analogies (Gordon 1961). A decade later, de Bono coined the term *lateral thinking* to refer to similar techniques (de Bono 1970). Many different techniques are available, ranging from random words to visual stimulation, from analogies to simply going on a walk. What all of these techniques have in common is that they somehow try to open up the problem and solution space (cf. Gassmann and Zeschky 2008). In the *Design Thinking* approach (IDEO-LLC, 2012), empathy is an important way of promoting flexibility by moving the perspective from the designer to the user (Brown and Katz 2009).

For creative problem-solving in a team, diversity and the availability of different perspectives appear to be a bit of a double-edged sword (Webber and Donahue 2001). Earlier research has demonstrated that group diversity can enhance creativity and innovation (Webber 2002) and also improve task completion time (Jassawalla and Sahittal 2002). However, it has also been shown that team diversity can reduce group cohesion, increase conflict and member stereotyping, and lower team performance (Bunderson and Sutcliffe 2002; Johnson and Johnson 2000). These findings underline the importance of good team dynamics. Experiential learning approaches can help boost creativity in teams by counteracting these threats to team performance by enhancing team trust, providing a safe environment, and making members aware of the importance of different team roles for optimal team performance (Kolb 1984). Perspective-taking in teams involves attempting to understand the viewpoint, feelings, and thoughts of another person (Parker et al. 2008), and it has been shown to be important in team creativity (Hoever et al. 2012). A technique like the Six Thinking Hats (De Bono 2009) provides an individual analog for perspective-taking in a team, which involves putting on imaginary "hats." Each hat treats a problem from a particular viewpoint: the "white" hat, for instance, focuses on the acquisition of facts or information (see also ◘ Fig. 1.5).

Perspective-taking is thought to boost creativity by expanding the 'conceptual space' (Vernon et al. 2016). New perspectives can lead to different ways of looking at and thinking about a problem or solution that might otherwise have been overlooked. Mednick (1962) suggested that highly creative individuals have a 'shallow' hierarchy of concepts (where concepts related to a target are more easily accessible), whereas low-creativity individuals have a "steep" hierarchy (where less-related concepts to the target are overwhelmed by stereotypically related concepts). According to Mednick, the shallow hierarchy helps people to come up with more flexible ideas. Similar spatial metaphors for creative problem-solving can be found in Martindale's connectionist approach to creative problem-solving (Martindale 1995) and the Klondike space model (Perkins 1994) which compares creative problem-solving to gold digging.

1

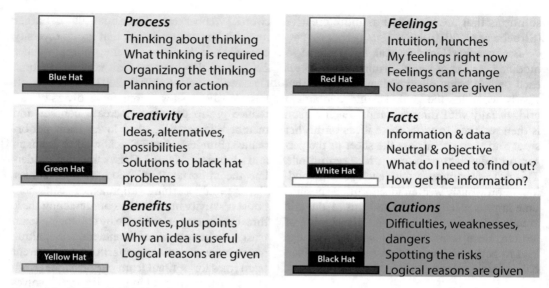

	Process		Feelings
Blue Hat	Thinking about thinking What thinking is required Organizing the thinking Planning for action	Red Hat	Intuition, hunches My feelings right now Feelings can change No reasons are given
	Creativity		Facts
Green Hat	Ideas, alternatives, possibilities Solutions to black hat problems	White Hat	Information & data Neutral & objective What do I need to find out? How get the information?
	Benefits		Cautions
Yellow Hat	Positives, plus points Why an idea is useful Logical reasons are given	Black Hat	Difficulties, weaknesses, dangers Spotting the risks Logical reasons are given

Fig. 1.5 De Bono's Thinking Hats: a technique for inviting different perspectives (De Bono 2009)

Perkins identifies four critical challenges every creative system (individual or group, animal or machine) has to meet: the rarity problem: nuggets (feasible problem solutions, viable designs, etc.) are rare; the isolation problem: promising regions in solution space are separated by wide distances; the oasis problem: the comfort with existing solutions discourages ventures that leave the known, recognized paths; and finally, the plateau problem: there is no gradient indicating the direction to rewarding regions. Having a good idea is like finding gold in the desert of non-solutions and non-designs.

Good ideas are not enough. Innovation and creative thinking also involve hard work. Persistence and perseverance are necessary to carry things through, as many good ideas never get developed further and bad ideas should not be transformed into bad products or bad processes. Creativity is not just a matter of anything goes (cf Baer 2016). This common misunderstanding of creative thinking is also reflected in the frequent false opposition of scientific critical thinking and artistic creative thinking. In line with Treffinger and Isaksen c.s. (Treffinger et al. 2006), we argue that critical thinking and creative thinking are complementary and need to work together in harmony to solve complex problems. In his

1878 book, *Human, All Too Human: A Book for Free Spirits*, Nietzsche observed: *[Artists have a vested interest in our believing in the flash of revelation, the so-called inspiration… shining down from heavens as a ray of grace.]* In reality, *the imagination of [the good artist] or thinker produces continuously good, mediocre or bad things, but his judgment, trained and sharpened to a fine point, rejects, selects, connects… All great [artists and] thinkers are great workers, indefatigable not only in inventing, but also in rejecting, sifting, transforming, ordering. This is where convergent thinking comes in.*

Following divergent thinking, the ideas and information will need to be organized using convergent thinking, i.e., putting the various ideas back together in some structured way and selecting those ideas that are judged to merit pursuing further. Creative problem-solving can be seen as the constant interplay of divergent and convergent thinking.

1.2.2 Convergent Thinking: Idea Evaluation and Implementation

Learning Goal 2: *The student can work on an idea with decisions and actions which may lead to an appropriate solution to solve problems.*

How can you transform an idea into an action? When an individual or a group recognizes some ideas as interesting and worth following up on, they may need support in strengthening those options, elaborating on them, and deepening them. In the quest for the highest quality ideas and solutions, they need to be developed further. Based on rational analysis and evaluation, effective choices need to be made to prepare for successful implementation. Exciting and intriguing new ideas are not necessarily useful or workable without extended effort and productive thinking. A first transformation needs to be made from exploration to potential exploitation. Idea evaluation involves examining promising options closely to determine what steps will need to be taken. If there are a few promising options, all of which might be implemented, the principal focus will be on refining or developing options, making them as strong as possible.

> If there are several promising options, not all of them can (or may need to) be implemented, the task may focus more on ranking options or on setting effective priorities. This is where the hard work comes in. *This is also the stage where domain-specific knowledge and expertise provides crucial criteria and instruments for diagnosing the best options (Baer 2015).*

When many new and promising options exist, the main aim will be to limit the number of choices by clustering, compressing, and condensing the ideas to make them more manageable, or to evaluate a number of options very systematically using explicit criteria. For example, to strengthen or refine an option, or a cluster of ideas with a common original creative theme, you could do a *SWOT analysis* to determine the Strengths, Weaknesses, Opportunities, and Threats of a particular (cluster of) ideas. Other techniques in convergent thinking are about setting priorities, for example by rating each of several options against all the others, one pair at a time (Fox and Fox 2000), or by categorizing the ideas in Now-How_Wow (Raison 1997; see also ◘ Fig. 1.6).

All ideas have their strengths and weaknesses. Research has shown that teams tend not to recognize their most creative ideas.

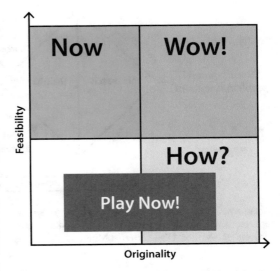

◘ **Fig. 1.6** Clustering ideas: Now-How_Wow. Blue ideas (Now) are easy to implement and often use previous examples and are low risk/quick wins. These ideas have high acceptability. Green ideas (Wow!) are exciting, innovative, and offer potential breakthroughs. They make a distinction and can be implemented. Yellow ideas (How?) have a visionary nature and consist of future ideas and dreams. These are challenging and consist of the green ideas of the future. From ▶ https://gamestorming.com/how-now-wow-matrix/. The How-Wow-Now Matrix is adapted from the work done by the Center for Development of Creative Thinking (COCD). Information about the COCD Matrix was published in the book, "*Creativity Today*" authored by Ramon Vullings, Igor Byttebier, and Godelieve Spaas (With permissiom from COCD)

There seems to be a strong tendency to select feasible and desirable ideas, at the cost of originality (Rietzschel et al. 2010; Putnam and Paulus 2009). It appears that people perceive originality and feasibility to be incompatible. To overcome this feasibility bias, Isaksen and Treffinger and colleagues (Isaksen et al. 2000) have proposed a number of guidelines for convergent thinking. The principle of *affirmative judgment* is the first guideline. It states that a review of the strengths of an idea should precede the discussion of any limitations. Affirmative judgment limits the detrimental effects caused by negative thinking by requiring members to treat limitations as resolvable and not as immutable facts. Other guidelines for convergence included taking a deliberate and open approach, giving preference to novelty, and staying focused

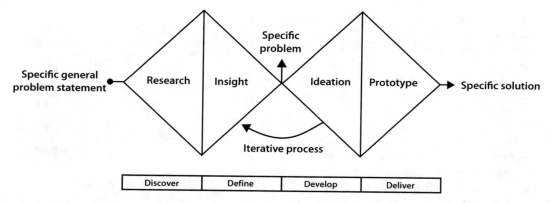

■ **Fig. 1.7** The design process (Source: ▶ https:// commons.wikimedia.org/wiki/File:Double-diamond-process.jpg. Author: Olga Carreras Montoto. This file is licensed under the Creative Commons Attribution 4.0 International license)

on the problem statement. The choice of an appropriate convergent thinking technique appears to be determined most by the number of options to be vetted. Hits, hotspots, and highlighting is a popular selection and compression technique for managing large quantities of options (Isaksen and Treffinger 1985). Matrices, pairwise comparisons, and sorting tasks are more suitable for situations where fewer alternatives are available. When the most promising idea or cluster of ideas has been selected, other people need to be convinced of this particular solution to the original problem as well. This is where leadership and persuasive communication skills also come into play. Leadership (see 1.4) is also crucial in the very first and necessary step in creative problem-solving: understanding the problem (challenge).

> Actually, innovation and creative problem-solving do not start with an idea, but they start with a problem and often this problem is quite ill defined. At this stage, a concerted and systematic effort needs to be made to define the problem.

This problem definition involves creating opportunities, exploring existing data, and framing the problem in a way that triggers curiosity and maximizes the initial motivation to solve a problem (Isaksen et al. 2011). As we will see later, there is an important role for the leader in the problem definition process. This important

first stage is also clearly visualized in the popular double diamond depiction of the creative design process (see ■ Fig. 1.7). The diamond shapes represent divergent and convergent thinking and represent points of iteration for research, observation, learning, prototyping, and testing (Design Council 2020a, b).

However insightful this view of the design process as a sequence of divergent and convergent thinking may be, it remains essentially an empty framework. It gives no concrete indications on how to actually carry out these phase and how to turn creative ideas into appropriate solutions. The main challenge in *the development of better innovator competences* is to know when to apply divergent or convergent thinking and to use as many iterations of creative diamonds as needed to get from a general problem to a good idea and from a good idea to a solution. Importantly, this process will help manage the danger of falling in love with one's own ideas at the potential cost of the development of better ideas.

1.3 Entrepreneurial Competences: *Where Imagination and Logic Meet in Creating Value*

> Learning Goal 3: *The student can translate potential solutions into value propositions regardless of the availability of favorable contingencies.*

Convergent and divergent thinking and acting in innovation and creative problem-solving bridge to entrepreneurial thinking and acting, where imagination and logic meet in creating value (Banathy 1996).

▶ Traditionally, entrepreneurial is thought of as entrepreneurship referring to starting a new business, scaling for profit, and creating business capital (Villacci 2019). This definition is quite narrow. It leaves out the core notion that entrepreneurship is about an entrepreneurial way of thinking and acting.

Such thinking and acting is opportunity driven and holistic in approach for the purpose of creating economic, cultural, and/or social value for others. For the purpose of creating value, it boils down to the willingness of trying out new, unusual ways of acting not hindered by the (un)availability of resources and the level of (un)certainty. The ability to rapidly sense, act, and mobilize, even under uncertain conditions, is a distinctive feature for an entrepreneurial mindset (McGrath and MacMillan 2000; Ireland et al. 2003; Haynie et al. 2010); see also ◧ Fig. 1.1 for the Entre-Comp framework.

▶ An entrepreneurial mindset is instrumental for successfully turning novel ideas into value. It "adds," in other words, the competences to bring innovations into practice regardless of the availability of favorable contingencies. Being able to effectively deal with these contingencies differentiates a true entrepreneurial spirit.

Such an entrepreneurial journey is iterative and goes through a process of profoundly exploring a problem (divergent thinking) and then taking focused action (convergent thinking). The "double diamond" we used before to illustrate the entire process from problem to idea to solution connects both ways of thinking into a coherent approach. Inspired by this design process, several approaches emerged to support the entrepreneurial professional, like for instance designing value propositions

(Osterwalder et al. 2014).[7] A value proposition is actually embedded in a business model, which Osterwalder and Pigneur (2010) describe as *"the rationale of how an organization creates, delivers and captures value."*

▶ Value Proposition Design (VPD) was introduced to stress that a business model should ground on a feasible and realistic value proposition: *"the benefits customers can expect from your product or service"* (Osterwalder et al. 2014).

It turned out that failing business models often build on wobbly value propositions; therefore, VPD was developed to *"manage the messy and nonlinear process of value proposition design and reduce risk by systematically applying adequate tools and processes"* (Osterwalder et al. 2014). Before explaining VPD in detail, we point out to two essential starting points which should be integrated in practice.

▶ First, it is essential to start with understanding what the problem actually is about and for whom it is a problem, before to start developing ideas with solution capabilities. It is crucial to avoid thinking about solutions as long as possible. After all, thinking about solutions makes one blind for understanding the problem – as it is human nature to fall in love with one's own ideas.

To quote Einstein: *"If I had an hour to solve a problem I'd spend 55 minutes thinking about the problem and five minutes thinking about solutions"*. A second important point is that designing comes down to testing assumptions and from these learnings refine the design. The learnings can also show that the direction in which a design is developing should be reconsidered. The willingness to change the course in which a design is developing

---

7 Several approaches are available, which advocate similar objectives, processes, and tool boxes, like lean startup (Ries 2011). In practice, several approaches are combined.

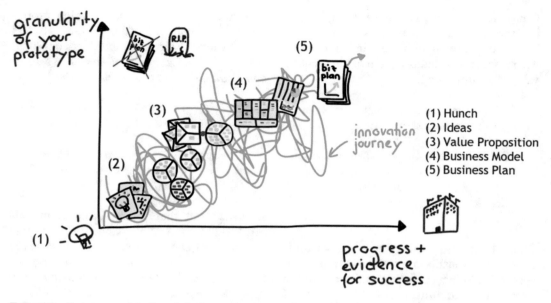

Fig. 1.8 Entrepreneurial view of the innovation journey. (Adapted from ► https://www.strategyzer.com/blog/posts/2017/2/16/prototype-learn-iterate)

and to give up some of the ideas one has been working on is crucial to avoid design fixation (Crilly 2018). To emphasize this, Ries (2011) introduces and promotes the concept of *pivoting*. It is "*a structured course correction designed to test a new fundamental hypothesis*" (Ries 2011), explaining that "*we keep one foot rooted in what we've learned so far, while making a fundamental change in strategy in order to seek greater validated learning*" (Ries 2011). Pivoting may feel like going backward, but it is essential for staying open-minded to find solutions that will solve problems from the perspective of the problem-holder, the (future) users of the product and service.

Figure 1.8 visualizes the non-linearity of the innovation journey (see also Fig. 1.4) of designing ideas and testing assumptions. Pivoting is visualized by course corrections and by saying good-bye to certain ideas (R.I.P.). By means of designing, testing, and pivoting, the prototype has been refined and evidence has been collected to increase success.

1.3.1 Turning Ideas into Value Propositions

In the previous paragraph it has been explained how ideas can be generated. It marks the starting point of each innovation and entrepreneurial journey. Turning first ideas – hunches – into value proposition is a next decisive step in the journey. Here the VDP approach proves its added value (see also Fig. 1.8). VPD consists of two spheres: the customer profile visualized by the circle and the value map visualized by the square (Osterwalder et al. 2014). The *customer profile* describes a specific customer profile in a structured and detailed way. It is broken down into *jobs, pains,* and *gains. Jobs* describe what a customer is trying to get done in their work and in their lives, as expressed in their own words. *Pains* describe bad outcomes, risks, and obstacles related to jobs, as experienced by the customer. *Gains* describe the outcomes customers want to achieve or concrete benefits they are seeking. *The value map* describes the features of a specific value proposition in a

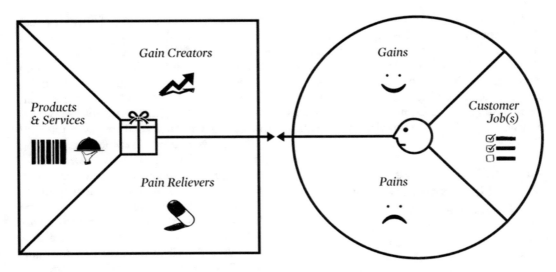

Fig. 1.9 Value proposition design. (Taken from Osterwalder et al. 2014, with permission from J. Wiley)

structured and detailed way. It is broken down into *products and services, pain relievers,* and *gain creators. Products and services* explain the core of the offering to the targeted customer. *Pain relievers* describe how the products and services alleviate the customer's pains. *Gain creators* describe how products and services create customer gains. A value proposition design has the potential to deliver value for customers if the value map and customer profile fit, i.e., when products and services produce pain relievers and gain creators that match the pains and gains that are important for the targeted customer. In drug discovery, the VPD for an asset will inform the so-called "*Target Product Profile (TPP)*"[8] that is defined for the clinical

development and the commercial opportunity for a new medicine that addresses the unmet medical needs of patients (☐ Fig. 1.9).

After a feasible and realistic value proposition is designed and tested, the development of a business model and a realization plan are next. For successfully realizing and delivering solutions – whether or not linked to developing new businesses – business and management competences are asked for. Without proper business and management thinking and acting, brilliant solutions will still fail to solve actual problems. We will not address this further as it goes beyond the scope of this chapter; for suggestions, see, for instance, Kawasaki (2015), Ries (2011), and Kerzner (2017).

When talking about innovations in an entrepreneurial context, the phenomenon of ambidexterity should be highlighted. It concerns the ability to simultaneously pursue both *incremental innovations* – focused on improving the present – and *discontinuous innovations* – focused on exploring new opportunities (O'Reilly and Tushman 2013; Tushman and O'Reilly 1996). In drug discovery, these two types of innovations lead to *best-in-class* and *first-in-class* therapeutics, respectively (Schulze and Ringel 2013). Students and professionals in the drug discovery field should be aware that both exploitation and exploration can co-exist

---

8   A TPP is a strategic development process tool that provides a summary of the product under development, as well as its desired characteristics and features and the features that provide a competitive advantage. A TPP is aimed to ensure that the company embarks on an efficient development program to reach the desired commercial outcome. It thus provides a structure for the scientific, technical, clinical, and market information and gives all stakeholders a clear vision of the product objectives to help guide research and development decisions. It is a dynamic document that is reviewed and adjusted as needed throughout the development process (cited from presentation by Pierandrea Maglia and Massimo Bani at Maastricht University).

and more quickly alternate. In our vision of the entrepreneurial mindset, we need drug discovery neuroscientists who are committed and trained to discover and develop first-in-class drugs, although it is appreciated that – especially in commercial settings – there will often be a risk-balanced portfolio with both projects focusing on first and best-in-class treatments.

1.4 Leading Creative and Innovative People: *Where the Rubber Hits the Road*

> Learning Goal 4: *The student can lead creative, innovative, and entrepreneurial individuals and teams seeking for feasible, valuable, and sense-making solutions and has the skills to get support from the organization.*

> Learning Goal 5: *The student can orchestrate teamwork with tripartite homo medicamento inventa specimens.*

Successfully managing an enterprise in a high-risk, complex, dynamic, and rapidly moving field such as CNS drug discovery requires extraordinary leadership competences. Many veterans of this industry would argue that this has been the Achilles' heel of the field and a major factor in its downfall in the early 2000s. The consolidation in the pharmaceutical industry and the concomitant scaling up and industrialization of its R&D infrastructure and *"management reductionism"*[9] reduced the innovative culture of the R&D organizations

and the creative risk-taking of their scientists (Garnier 2008; Thong 2015). So the need to build and foster an innovative culture is acknowledged, although this actually has not yet led to a meaningful cultural shift in the (CNS) drug discovery industry. A new generation of well-trained drug discovery neuroscientists will be needed to lead a return to an innovative culture. These leaders have significant impact on all steps in *creative work,*[10] that is, problem definition, idea generation, idea development, and idea implementation.

These new leaders will have to master a specific set of leadership competences to support a creative workforce going through these steps, as has been described excellently by Mumford and colleagues in 2002. These competences are dictated by the traits of creative workers and the characteristics of their environment. These workers are characterized by professionalism, expertise, a desire for autonomy, high intrinsic and achieve motivation, and high focus. Their innovative environment relies on expertise that often develops slowly, requires a need to collaborate, is marked by high levels of uncertainty and ambiguity, and is resources intensive. Many of these factors contribute to an inherent conflict with the organization that needs to be managed.

> The leadership skills and competences that we will focus on are (1) technical and creative problem-solving skills, (2) providing structure, (3) planning and sense-making, (4) strategic thinking, (5) organizational expertise, and (6) persuasion and social intelligence.

First, as creative work occurs on complex and ill-defined problems, the imposition of *structure* (2) represents a key component of the creative process. Accordingly, problem definition, or problem construction, activities are important aspects of creative thought (see also last citation in section 1.2.2). Leaders can help with the problem definition and

---

9   This term has been described by Robert Thong as follows. "[Companies] are expected to deliver predictable revenue and profit streams …This expectation leads … to regarding …the overall R&D organization as "machines" that can be "engineered" to convert certain specified inputs into predictable outputs, with productivity enhanced through the control of a small number of key management "levers". I refer to this logic as *"management reductionism"*– indeed a far cry from the creativity and innovation culture that is essential to find solutions for complex problems!"

10   Defined as *"work which occurs when the tasks involve complex and ill-defined problems and performance requires the generation of novel, effective solutions."*

understanding process by bridging with the organizational needs and goals. Therefore, they must have a clear understanding of the organization's strategy and be able to frame requests for support in terms of broader strategic objectives (5).

Second, technical expertise and *creative problem-solving skills* (1) are essential for leading creative people because they provide a basis for structuring as well as the credibility needed to exercise influence. As mentioned by Mumford et al. (2002), *"Given the strong professional identity of creative people, it may prove difficult for leaders lacking technical expertise and creative problem-solving skills to: (a) adequately represent the group, (b) communicate effectively with group members, (c) appraise the needs and concerns of followers, (d) develop and mentor junior staff, and (e) assess the implications of group members' interactions".* The powerful influences on group performance by these competences involve *social influence* (6), more specifically, the before mentioned work focus, achievement motivation, and autonomy of creative people frame a situation where expertise is the most powerful form of influence at the disposal of a leader (Mumford et al. 2002). Cognitive influences are also important for the evaluation of ideas and to provide feedback to followers or other managers, which is almost impossible when leaders lack expertise and creative problem-solving skills.

---

11 Four distinguishing characteristics of transformational leaders (*"the four i's"*) make this leadership style especially effective in an environment driven by creativity and innovation: idealized influence or charisma, inspirational motivation, intellectual stimulation, and individualized consideration (Bryant 2003). *Idealized influence* can be seen as a leader acting as a "role model" and encouraging followers to have pride, faith, and respect in themselves, their leaders, and their organizations. *Inspirational motivation* involves behavior to motivate and inspire followers by providing a shared meaning and a challenge to those followers. *Intellectual stimulation* is the frequency with which leaders encourage employees to be innovative in their problem-solving and solutions. *Individualized consideration* means acting as a coach and mentor to help followers and students to reach their full potential.

Third, to help people define the problems, vision-based or *translational leadership* skills can influence creativity and innovation (3, 6). For example, one study employed a brainstorming task where transformational and transactional leadership[11] were manipulated with regard to task instructions. It was found that *fluency* and *flexibility* (see ▶ Sect. 1.3.1) were higher in the transformational condition. Vision and direction should be framed in terms of more concrete production missions to define goals and clarify paths to goal attainment (Jung 2001 cited in Mumford et al. 2002).

This goal setting ties into a fourth leadership competency: *planning and sense-making (3)*. When multiple parties are working together in the development of a complex, creative product or technology, a lack of structure can emerge. A leader's sense-making activities help to create a *shared mental model* about the causes and consequences of actions that significantly influences performance. Such sense-making activities help to ensure coordination and joint problem-solving by people from different backgrounds or with different expertise (see also ▶ Sect. 1.4.1). Especially when projects move from generation to development, increases in diversity and complexity of activities will put a higher demand on both organizational expertise and sense-making activities.

Both a leader's initiation of structure/ planning and goal clarity were rated high by creative workers under conditions of uncertainty (Arvey et al. 1976; Weick 1995; cited in Mumford et al. 2002). Successful planning, goal setting and resource allocation require an in-depth understanding of both the product, process, or technology and the organization, its capabilities, and its markets. Thus, the effective leadership of creative efforts will require good *organizational expertise* (5) as well as significant technical expertise (1).

Fifth, successful leadership of creative efforts will require *persuasion* and *social intelligence (6)*. The need for persuasion is attributable to the people being led – autonomy, professionalism, and critical nature can make it challenging to persuade creative people – and the importance of resources in creative work. Creative work requires substantial resources that are invested in risky efforts. Together with negative attitudes toward creativity, discounting the value of new ideas, and other negative forces such as the previously mentioned feasibility bias (Rietzschel et al. 2010; Putnam and Paulus 2009), this can lead to the premature rejection of viable ideas. Therefore, the leaders persuasive skills and ability to "sell" new ideas are of utter importance. Persuasion can also be indirect as leaders need social perceptiveness, flexibility, wisdom, and social appraisal skills, in other words, social intelligence (Mumford et al. 2002).

In conclusion, leaders of creative people have several important roles such as supporting problem definition and providing structure, evaluating their ideas, integrating their ideas with the needs of the organization, creating conditions where people can generate ideas in the first place, and at times being a demanding critic of an idea and its potential. This requires an integrative style that allows the leader to orchestrate expertise, people, and relationships to foster the generation, structuring and promotion of novel ideas (Mumford et al. 2002).

How then to prepare our aspirant drug hunters for their future as leaders of creative efforts? There are comprehensive sets of research-based leadership-competency development resources available (for example, the Korn Ferry Leadership Architect™ Global Competency Framework[12]). One of the applications is for organizations to configure a competency model that aligns with their business objectives and talent strategy in order to achieve their business goals. One example

how it works. One can back-translate the most relevant competences from this framework to those that we identified for the tripartite leader; look what the less-skilled competences are and follow the proposed tips for improvement. For example, the cluster *creating the new and different (C)* is obviously a big hitter for an innovative and entrepreneurial environment, and especially the competence *cultivates innovation* (#19; For Your Improvement 2014, p.178). Suppose self-assessment reveals that one is less skilled and *"presents ideas that are ordinary, conventional, and from the past"*. From the 14 tips that are offered under the *Tips for Improvement* section, one could look at tip #11: *"Want creative ideas while brainstorming? Use multiple tools and techniques."* And, perhaps no surprise, the reader will run into many of the tools that we introduced in this chapter's section *"The Student as an Innovator."* Now that the circle is closed we can proceed to the conclusions.

Conclusions

Traditionally, drug discovery neuroscientists have been focusing on developing potential solutions from a neuroscientific perspective. The new-generation drug-discovery neuroscientist has to raise her/his game and should develop into a tripartite drug discovery neuroscientist who is able to orchestrate innovation and entrepreneurial and leadership activities to lead teams to generate good ideas and turn these ideas into valuable solutions. In this chapter, we explained the competences of this new-discovery neuroscientist and introduced several methods and techniques to support this new role.

---

12 In a nutshell, this framework comprises the following: (1) four *factors* – these are groups of competences that form a cohesive theme. These competencies share some thematic similarities. Factors can be derived from statistics or content analysis. (2) Twelve *clusters* – statistically supported groupings of related competences that represent a broader scope of skills and behaviors that contribute to success in the skill. (3) 38 *competences* – skills and behaviors required for success that can be observed. (4) Ten *career stallers and stoppers* – these are behaviors generally considered problematic or harmful to career success.

References

Akil H (2016) Neuroscience training for the 21st century. Neuron 90:917–926

Amabile TM (1990) Within you, without you: the social psychology of creativity, and beyond. In: Runco MA, Albert RS (eds) Sage focus editions, Vol. 115. Theories of creativity. Sage Publications, Newbury Park, pp 61–91

Arvey RD, Dewhirst HD, Boling JC (1976) Relationships between goal clarity, participation in goal setting, and personality characteristics on job satisfaction in a scientific organization. J Appl Psycho 61:103–105

Bacigalupo M, Kampylis P, Punie Y, Van den Brande G (2016) EntreComp: the entrepreneurship competence framework. Luxembourg: Publication Office of the European Union; EUR 27939 EN; https://doi.org/10.2791/593884

Banathy, B. H. Designing Social Systems in a Changing World: A Journey Toward a Creating Society. Plenum, 1996

Baer J (2015) The importance of domain specific expertise in creativity. Roeper Rev 37:165–178

Baer J (2016) Creativity and the common core need each other. In: Ambrose D, Sternberg RJ (eds) Creative intelligence in the 21st century: grappling with enormous problems and huge opportunities. Sense, Rotterdam, pp 175–190

Barnfield-h and Lombardo-M (2014). FYI: For Your Improvement - Competencies Development Guide, 6th Edition, Korn-Ferry.

Brown T, Katz B (2009) Change by design: how design thinking transforms organizations and inspires innovation. Harper Business, New York

Bryant SE (2003) The role of transformational and transactional leadership in creating, sharing and exploiting organizational knowledge. J Leadersh Organ Stud 9:32–44

Buchanan R (1992) Wicked problems in design thinking. Des Issues 8:5–21

Bunderson JS, Sutcliffe KM (2002) Comparing alternative conceptualizations of functional diversity in management teams: process and performance effects. Acad Manag J 45(5):875–893. https://doi.org/10.2307/3069319

Crilly N (2018) 'Fixation' and 'the pivot': balancing persistence with flexibility in design and entrepreneurship. Int J Design Creat Innov 6(1–2):52–65

De Bono E (1970) Lateral thinking: creativity step by step. Harper & Row

De Bono E (2009) Six thinking hats. Penguin Books, London

DesignCouncil (2020a, June 10) What is the framework for innovation? Design Council's evolved Double Diamond. https://www.designcouncil.org.uk/news-opinion/what-framework-innovation-design-councils-evolved-double-diamond

DesignCouncil (2020b, June 10) The Double Diamond: a universally accepted depiction of the design process. https://www.designcouncil.org.uk/news-opinion/double-diamond-universally-accepted-depiction-design-process

Dolmans DH, De Grave W, Wolfhagen IH, Van Der Vleuten CP (2005) Problem-based learning: future challenges for educational practice and research. Med Educ 39(7):732–741

Fox JM, Fox RL (2000) Exploring the nature of creativity. Kendall/Hunt, Dubuque

Garnier JP (2008) Rebuilding the R&D engine. Harvard Business Rev 86:68–70

Gassmann O, Zeschky M (2008) Opening up the solution space: the role of analogical thinking for breakthrough product innovation. Creat Innov Manag 17:97–106. https://doi.org/10.1111/j.1467-8691.2008.00475.x

Geschka H, Schaude GR, Schlicksupp H (1976) Modern techniques for solving problems. Int Stud Manag Organ 6:45–63

Gordon WJJ (1961) Synetics: the development of creative capacity. Harper and Row, New York

Guilford JP (1950) Creativity. Am Psychol 5:444–454

Guilford JP (1967) The nature of human intelligence. McGraw-Hill, New York

Haynie JM, Shepherd D, Mosakowski E, Earley PC (2010) A situated metacognitive model of the entrepreneurial mindset. J Bus Ventur 25(2):217–229

Heijne, K & van der Meer, H. (2019). Road Map for Creative Problem Solving Techniques: Organizing and Facilitating Group Sessions. Amsterdam: Boom

Hoever IJ, van Knippenberg D, van Ginkel WP, Barkema HG (2012) Fostering team creativity: perspective taking as key to unlocking diversity's potential. J Appl Psychol 97:982–996. https://doi.org/10.1037/a0029159

IDEO LLC (2012) Design thinking for educators toolkit, 2nd edn. IDEO, New York

Ireland RD, Hitt MA, Sirmon DG (2003) A model of strategic entrepreneurship: the construct and its dimensions. J Manag 29:963–990

Isaksen SG, Treffinger DJ (1985) Creative problem solving: the basic course. Bearly Limited, Buffalo

Isaksen SG, Dorval KB, Treffinger DJ (2000) Creative approaches to problem solving, 2nd edn. Kendall/Hunt, Dubuque

Isaksen SG, Dorval KB, Treffinger DJ (2011) Creative approaches to problem solving: a framework for innovation and change, 3rd edn. SAGE, London

Jassawalla AR, Sahittal HC (2002) Building collaborative new product processes: why instituting team is not enough. SAM Adv Manag J 68:27–36

Johnson DW, Johnson FP (2000) Joining together: group theory and group skills, 7th edn. Allyn & Bacon, Boston

Kampylis P, Berki E (2014) Nurturing creative thinking. Int Acad Education 6

Kawasaki G (2015) The Art of the Start 2.0: The Time-Tested. Battle-Hardened Guide for Anyone Starting Anything

Kerzner H (2017) Project management: a systems approach to planning, scheduling, and controlling. John Wiley & Sons

Koestler A (1964) The act of creation. Macmillan

Kolb D (1984) Experiential learning: experience as the source of learning and development. Prentice-Hall, Englewood Cliffs

Kurtzberg RL, Reale A (1999) Using Torrence's problem identification techniques to increase fluency and flexibility in the classroom. J Creat Behav 33(3):202–207

Martindale C (1995) Creativity and connectionism. In: Smith SM, Ward TB, Finke RA (eds) The creative cognition approach. The MIT Press, pp 249–268

McGrath RG, MacMillan IC (2000) The entrepreneurial mindset: strategies for continuously creating opportunity in an age of uncertainty (Vol. 284). Harvard Business Press, Boston

Mednick S (1962) The associative basis of the creative process. Psychol Rev 69(3):220–232. https://doi.org/10.1037/h0048850

Mumford MD, Scott GM, Gaddid B, Strange JM (2002) Leading creative people: orchestrating expertise and relationships. The Leadership Quarterly 13:705–750

O'Reilly CA III, Tushman ML (2013) Organizational ambidexterity: past, present, and future. Acad Manag Perspect 27(4):324–338

Osterwalder A, Pigneur Y (2010) Business model generation: a handbook for visionaries, game changers, and challengers. John Wiley & Sons, Hoboken

Osterwalder A, Pigneur Y, Bernarda G, Smith A (2014) Value proposition design: how to create products and services customers want. John Wiley & Sons, Hoboken

Parker SK, Atkins PW, Axteil CM (2008) Building better work places through individual perspective taking: a fresh look at a fundamental human process. In: Hodgkinson G, Ford K (eds) International review of industrial and organizational psychology, vol 32. Wiley, Chichester, pp 149–196

Perkins DN (1994) Creativity: beyond the Darwinian paradigm. In: Boden MA (ed) Dimensions of creativity. The MIT Press, Cambridge, MA, pp 119–142

Pistoia Alliance (2020). https://www.pistoiaalliance.org/

Putman VL, Paulus PB (2009) Brainstorming, brainstorming rules and decision making. J Creat Behav 43:23–39

Raison M (1997) COCD-box: Eindproduct deskundigheidstraining. Center for Development of Creative Thinking, Belgium

Ries E (2011) The lean startup: how today's entrepreneurs use continuous innovation to create radically successful businesses. Crown Business, New York

Rietzschel EF, Nijstad BA, Stroebe W (2010) The selection of creative ideas after individual idea generation: choosing between creativity and impact. Br J Psychol 101:47–68

Sawyer TK (2008) Transformational leadership in drug discovery by way of virtuous thought, word and deed. Chem Biol Drug Dis 71:507–510

Scannell JW, Blanckley A, Boldon H, Warrington B (2012) Diagnosing the decline in pharmaceutical R&D efficiency. Nat Rev Drug Discov 11:191

Schmidt HG, Rotgans JI, Yew EH (2011) The process of problem-based learning: what works and why. Med Educ 45(8):792–806. https://doi.org/10.1111/j.1365-2923.2011.04035.x

Schulze U, Ringel R (2013) What matters most in commercial succes: first-in-class or best-in-class? Nat Rev Drug Dis 12:419–420

Simonton DK (2007) The creative process in Picasso's Guernica sketches: monotonic improvements versus nonmonotonic variants. Creat Res J 19(4):329–344. https://doi.org/10.1080/10400410701753291

Stroebe W, Nijstad BA, Rietzschel EF (2010) Beyond productivity loss in brainstorming groups: the evolution of a question. In: Zanna MP, Olson JM (eds) Advances in experimental social psychology, vol 43. Academic Press, San Diego, pp 157–203. https://doi.org/(...)0065-2601(10)43004-X

Thong R (2015) Root causes of the pharmaceutical R&D productivity crisis. https://scitechstrategy.com/2015/03/31/root-causes-of-the-pharmaceutical-rd-productivity-crisis/

Torrance EP (1974) Norms technical manual: Torrance tests of creative thinking. Ginn and Co, Lexington

Treffinger DJ, Isaksen SG, Dorval KB (2006) Creative problem solving: an introduction, 4th edn. Prufrock Press, Waco

Tushman ML, O'Reilly CA (1996) The ambidex- trous organization: managing evolutionary and rev- olutionary change. Calif Manag Rev 38:1–23

Vernon D, Hocking I, Tyler CT (2016) An evidence-based review of creative problem solving tools: a practitioners resource. Hum Resource Dev Rev 15(2):230–259. https://doi.org/10.1177/1534484316641512

Villacci J (2019, August 29) What is the definition of entrepreneurship? https://entrepreneurship.babson.edu/entrepreneurship-definition

Webber SS (2002) Leadership and trust: facilitating cross-functional team success. J Manag Dev 21(3):201–214

Webber SS, Donahue LM (2001) Impact of highly and less job-related diversity on work group cohesion and performance: a meta-analysis. J Manag 27(2):141–162

Weick KE (1995) Sensemaking in organizations. Sage, Thousand Oaks

https://innovationenglish.sites.ku.dk/model/double-diamond-2/

https://states-of-change.org/stories/proof-of-concept-prototype-pilot-mvp-whats-in-a-name

https://www.maastrichtuniversity.nl/education/master/research-master-cognitive-and-clinical-neuroscience-specialisation-drug-development

https://www.weforum.org/agenda/2015/02/how-to-maintain-the-entrepreneur-mindset/

Innovation in the Discovery of Novel Therapeutic Approaches

Contents

Chapter 2   Drug Discovery in CNS: Finding a Target for
What? – 25
Arjan Blokland

Chapter 3   Neurogenomics with Application to
Schizophrenia and Other Major
Neuropsychiatric Diseases with
Complex Heredity – 35
Laurence Melnick

Chapter 4   Epigenetics in Drug Discovery:
Achievements and Challenges – 57
*Ziva Korda, Ehsan Pishva, and
Daniel L. A. van den Hove*

Chapter 5   Conventional Behavioral Models and
Readouts in Drug Discovery: The
Importance of Improving Translation – 77
Kris Rutten ⓘD

Chapter 6   Safety and Drug Metabolism: Toward NCE
and First in Human – 93
Jacco J. Briedé

Chapter 7   The Various Forms of Non-invasive Brain
             Stimulation and Their Clinical
             Relevance – 103
             Tom A. de Graaf, Alix Thomson,
             Felix Duecker, and Alexander T. Sack

Chapter 8   Is Non-invasive Brain Stimulation the Low-
             Hanging Fruit? – 115
             Tom A. de Graaf, Shanice E. W. Janssens,
             and Alexander T. Sack

Drug Discovery in CNS: Finding a Target for What?

Arjan Blokland

Contents

2.1   Finding the Target: Some Examples – 26

2.2   Target Identification – 28
2.2.1   GWAS – 28
2.2.2   Will GWAS Help Us to Find Novel Targets? – 29

2.3   What Can We Treat? – 30
2.3.1   The Disease? – 30
2.3.2   Symptoms? – 30
2.3.3   Systems Biology: Diseaseome – 31

          References – 32

2

In the last decades, not much progress has been made in finding new treatments for CNS diseases. The well-established concepts of target identification and validation are at the core for drug discovery and development programs. It is argued that this is the most critical step in finding new treatments. Drug target finding has been pursued using Genome-Wide Association Study (GWAS) methods, but this has not been proven successful for CNS diseases yet. Various potential issues with the GWAS approach are listed and may dampen the validity of using GWAS for target finding. Two recent alternative methods for target finding are discussed. One approach is related to the Research Domain Criteria (RDoC) initiative in which different functional domains are connected with neurodevelopmental and biological mechanisms in a matrix format. Thereby, drug targets could be identified for these brain functions. Another approach mentioned here is based on the "diseaseome" (network medicine), which is a data-driven approach using molecular biology and genetic information to find treatment-based mechanism. Although these two approaches may seem promising, no promising clinical treatments are available yet. Target identification still remains the most important and challenging step in drug discovery and development for CNS diseases.

🎓 **Learning Objectives**
- Understanding why target identification and validation are essential for drug development. Current CNS drugs are developed on the basis of very general neurotransmitter deficits in CNS diseases. Some clinical effective treatments were not developed on basis of an existing scientific rationale.
- Understanding potential novel approaches in finding drug targets (GWAS, RDoC, Diseaseome/Network medicine), and that this needed to change the idea of treating diseases, treating symptoms, or underlying mechanisms.

2.1 Finding the Target: Some Examples

The most crucial step in drug discovery is finding the right target for a disease. It is at the heart of all the following steps that are taken in developing a drug. The choice is leading in choosing in vitro and in vivo test models and efforts for further target optimization that eventually leads to the proposal of a drug candidate. The subsequent clinical development also relies on the drug target assumptions and how it can treat a disease. In case a novel drug fails in drug development, it cannot be determined whether the drug failed in one of the stages of drug discovery or drug development, or whether the target was not valid. Hence, the author considers target selection as the most critical step in finding new drugs.

Therefore, finding the right target requires careful attention and should be based on sound scientific knowledge of the disease and what is the underlying cause of the disease. It is interesting to note that some treatments of CNS diseases have not been based on drug development but were found via interesting routes (serendipity). For example, the use of lithium in psychiatry is known since the middle of the nineteenth century (Shorter 2009). The account that the first physician (William Hammond) gave in 1871 for using lithium in mania was that "… to diminish the amount of blood in the cerebral vessels, and to calm any nervous excitement that may be present." Cleary, this is not based on current ideas of drug development. Nevertheless, lithium is still being used as a mood stabilizer, but the mechanism of action (target) for its clinical effect is not clear.

Another recent interesting example is the development of ketamine as an antidepressant (Wei et al. 2020). Although there was a major focus on the monoaminergic hypothesis of depression, some researchers argued that the NMDA receptor plays a significant role in neuroplasticity, and thereby a target for depression. Interestingly, they used an antagonist to increase neuroplasticity. This is counterintuitive since neuroplasticity is achieved with NMDA receptor activation (Malenka and Nicoll 1993). Some recent ideas have been developed how ketamine may lead to its clinical effect (e.g., Collo and Merlo Pich 2018; Fukumoto et al. 2019), which may link also to other mechanisms of action or metabolites (Zanos and Gould 2018). In 1990, animal studies showed that NMDA antagonists were effective in animal models of depression (Trullas and Skolnick 1990).

Based on these findings (supported by additional studies), and the ability to test this NMDA antagonist in humans, a clinical study confirmed the antidepressant effect of ketamine (Berman et al. 2000). Actually, the most interesting (unexpected?) effect was that there was a rapid relief of depressive symptoms. This could not predicted on basis of the animal models since the behavioral effects were similar to the effects of other antidepressant drugs (like SSRIs). Although there seemed to be rationale for developing NMDA antagonists for depression, the actual mechanism of action and the rapid onset of action in humans cannot be fully explained yet. There are some other examples to show how serendipity played a role in finding novel drug treatments (Ban 2006).

There are also examples where drugs were developed on basis of a scientific rationale. For example, the development of SSRIs was based on the observation that monoamines were depleted in depression (Hillhouse and Porter 2015). Using a drug discovery and development program, these drugs were developed and were found to be effective in the treatment of depression. In a similar manner, l-dopa was developed for lower dopamine levels in Parkinson's disease (Hornykiewicz 2010), and acetylcholinesterase inhibitors were developed for the decreased levels of acetylcholine levels in Alzheimer's disease (Pepeu and Giovannini 2009). Although, these drugs were developed at least 40 years ago, they are still clinically used and are relatively effective in many patients with an acceptable adverse side-effect profile. Interestingly, the main improvements that have been made over the years are related to the formulation to obtain more steady plasma levels for these drugs.

With some exceptions (e.g., memantine for Alzheimer's disease, and clozapine in schizophrenia, ritalin for ADHD, and some new drugs in MS), the development of novel CNS drugs has been found to be limited. Although there has been a tremendous increase in knowledge in how the brain, neurons, and signaling pathways work over the last decades, this did not bring newer and better drugs for the major CNS disease. One example of this is the enormous amount of scientific progress in the field of Alzheimer's disease and the development of anti-amyloid drugs. Although there has been a tremendous academic effort and many drug development projects in the pharmaceutical industry, the development of these potential disease-modifying drugs has been very disappointing thus far (e.g., Oxford et al. 2020).

Case Study

I would like to use the amyloid-based treatment as an example why drug development may have failed for this drug. One issue could be that amyloid may not be the critical protein that causes the neurodegeneration in Alzheimer's disease. Clearly, there is overwhelming data showing that amyloid is neurotoxic (Selkoe and Hardy 2016), but there are some doubts whether this can fully explain the neurodegeneration in Alzheimer's disease (e.g., Dourlen et al. 2019; Makin 2018). The question that can be asked here is whether the target selection for Alzheimer's disease has been right. Is amyloid at the core of Alzheimer's disease? The failures of the anti-amyloid drugs may suggest that this is the case. However, it has been argued that the treatment with these drugs came too late to treat the disease (Jack et al. 2013). The anti-amyloid treatment should start before the amyloid starts to induce neurodegeneration. An analogy that

2

may make this clearer: "You do not treat a heart attack with cholesterol lowering drug, although high cholesterol levels may lead to a heart attack" (quote by Prof Harald Schmidt).

Thus, if we still assume that amyloid is the target for Alzheimer's disease, we should start treatment at an earlier stage, before amyloid starts having neurotoxic effects. This implies that we need to know who will develop Alzheimer's disease on an individual basis. This is also referred to as an early biomarker. Currently, there is a great effort finding biomarkers for Alzheimer's disease, which would identify people who will develop the disease (e.g., Zetterberg and Bendlin

2021). Based on specific and valid biomarkers, people could then be treated with the anti-amyloid drug. Interestingly, a recent study suggested that P-tau217 could be considered as an early biomarker for Alzheimer's disease (Palmqvist et al. 2020). It is unclear whether this hints to a role of tau or amyloid in the early development of Alzheimer's disease. This is of course of utmost importance to find out since it will determine which target should be chosen for drug development. If the rationale for the drug target is not invalid, this effort may be deemed to be useless. This again highlights the capital importance of selecting the right target for a disease.

2.2 Target Identification

There are several ways in which targets can be selected. A first approach was mentioned above: examine neurobiological changes in a disease and find a target that can reverse this change. This is of course the most sensible strategy to identify a drug target. But when studying a disease, there will be many differences when you compare this with a healthy normal person. For example, when you would consider finding a target for depression, you may choose from a wide pool of potential relevant drug targets. When searching PubMed with the key words "drug target depression review" (about 1500 hits, August 2020), you will find a wide variety of potential relevant drug targets for this disease. The major question is of course which target should be chosen for a drug discovery program? One strategy is to find as much evidence that the target is relevant for the disease. This can be done with screening the literature and compiling the evidence in favor of this drug target. One set of data is usually based on animal studies in which various animal models are used to evaluate drugs that affect the specific target (e.g., NMDA receptor for depression). If the available data show that different drugs that act on the same specific target reverse the depressive symptoms in different animal models of depression, this increases the notion that the

target is relevant for depression. Clearly, there should also be evidence from human studies that indicate that the target is relevant. This can be done with imaging studies, blood- and CSF samples, genetic analyses, pharmacological challenge studies, and immortal pluripotent stem cells. These efforts should be carefully evaluated before selecting a target for a drug discovery program.

- Studying the biology of a disease and finding differences with healthy controls.
- Finding the most relevant relation between biology and disease characteristics.
- Identifying the target for new molecules to modulate the target.

2.2.1 GWAS

For some time (starting around 1995), genetic analyses have been used to identify targets in different CNS diseases (This topic is also discussed in more detail in the ▶ Chap. 3). The most fundamental data needed for these analyses come from genome-wide association studies (GWAS), usually containing more than 10,000 patients (Breen et al. 2016). The idea is that GWAS can define loci, containing different genes, which are affected in a disease. This could lead to the discovery of new targets or may lead to novel model systems to be used to screen new compounds (e.g., Wei et al. 2019). In general, these GWAS data can further be analyzed in combination with other databases such as pathway analysis (e.g., REACTOME), brain systems biology information (e.g., BrainSeq), and drug target lists (e.g., ChEMBL). This combined information should be a powerful tool for screening new targets for different CNS diseases. Of note, some additional steps are recommended before defining drug targets based on GWAS (e.g., Plenge 2019; Wendland and Ehlers 2016). When a specific target has been identified and a drug developed, the GWAS information can also identify patients that should be most sensitive to the new drug. This is also referred to as precision medicine or personalized medicine (e.g., Tam et al. 2019). Further details of this approach can be found via the website of the Psychiatric Genomics Consortium (▶ https://www.med. unc.edu/pgc/).

2.2.2 Will GWAS Help Us to Find Novel Targets?

There are some issues with this GWAS analysis (for a critical appraisal, see Breen et al. 2016; Tam et al. 2019). A first point is that epigenetics may play an important role in many diseases (See ▶ Chap. 4). Therefore, in addition to using GWAS, epigenome-wide association studies (EWAS) could also provide valuable information to identify novel targets (Sweatt 2009). It should be noted that the epigenetic changes that are related to a specific CNS disorder may determine which method (GWAS or EWAS) could be more successful in identifying a drug target. A second potential issue is related to the finding that GWAS analysis only revealed very few existing drug targets (Cao and Moult 2014). This suggests that targets for which effective drugs were developed were not picked up by GWAS analysis. Although this is concerning, this may be related to a current limited availability of data in various databases at that time. Over the

last years, the databases have grown exponentially and may be better in identifying existing- and novel drug targets. A third point is related to the loci that show up in the so-called Manhattan plots only explain a very small amount of the variation in disease. However, integrating the data in a pathway- and network-informed interpretation could increase the effect size of the relevant loci involved in disease (Gaspar et al. 2019). This could possibly increase the explained variance of a target, making it more relevant.

A fourth potential issue is that GWAS may not take into account the developmental windows during which brain diseases develop and may be related to specific cell types (Fernando et al. 2020). Combining GWAS data with studies using specific human-induced pluripotent stem cells and using CRISPR-CAS methods to induce disease-specific models may lead to valid target identification (Fernando et al. 2020). A fifth issue is related to the importance of a locus for a disease. From a GWAS analysis, it cannot be concluded whether the locus is causally linked to a disease (Tam et al. 2019). It is not fully clear how a causal link can be established using these methods.

A sixth issue may also be related with the complexity of CNS diseases. We take schizophrenia as an example in order to make this point. Schizophrenia has an enormous variation in the clinical expression of a disease and has primarily been characterized on basis of negative and positive symptoms. However, it appears to be very difficult to clinically diagnose these patients (e.g., Fountoulakis et al. 2019). Moreover, there seems to be an overlap between diagnoses of different psychiatric disorders. For example, the boundary between schizophrenia spectrum disorders and autism spectrum disorder (e.g., Sunwoo et al. 2020) or bipolar disorder (e.g., Yamada et al. 2020) appears to be quite blurry. Although this clinical perspective may be far away from a drug target and the use of GWAS, it should be clear that the GWAS do not take this huge variation of clinical features into account. GWAS databases do not consider the variation of clinical severity or possible misdiagnosis of the disease into consideration. Of note, a similar heterogeneity and overlap in clinical features can also be found for affective disorders and dementias.

As mentioned above, GWAS and EWAS need a high volume of data in order to make some predictions on valid targets. These datasets include the heterogeneity of a disease and may even include wrongly diagnosed patients. This may explain, at least to some extent, why only a marginal level of variation can be explained by the individual loci. Conversely, when a drug target is identified and a drug is developed for schizophrenia, it can be questioned whether this drug will work in all schizophrenia patients. It was suggested that personalized medicine could be used to overcome this issue, but it will be a challenge on which basis these patients should be selected. We would need very good biomarkers to determine which patients should receive a specific treatment.

> GWAS have been used as a tool to understand and explain brain diseases at a molecular level. Large databases and complex data analyses are required for making sensible inferences and proposing a drug target. There are different potential pitfalls when using this method.

2.3   What Can We Treat?

2.3.1   The Disease?

In the previous section, I mentioned the problem of the complexity of clinical diagnoses of brain diseases. Brain diseases are associated with a spectrum of clinical features, which can be differentially affected in patients. If GWAS identifies a valid target for schizophrenia and a drug would be developed, it can be questioned whether this will treat schizophrenia (I am still using the example of schizophrenia, but this also applies to other CSN diseases). For example, would it treat all positive- and negative symptoms in patients? Probably, this will be a very unlikely outcome

that one drug will turn a schizophrenic patient into a healthy person. It could be argued that drugs could be developed for different targets, which could affect pathways that are critically involved in schizophrenia (e.g., Kondej et al. 2018). Whether this will be a fruitful way to find drug targets to treat a disease will need to come from future drug discovery and development programs.

2.3.2   Symptoms?

There have been some other ideas about how to treat schizophrenia. For example, it has been suggested that cognition is a core symptom of schizophrenia (a.k.a. CIAS, cognitive impairment associated with schizophrenia) and that treatment of the cognitive symptoms could be very beneficial for these patients (e.g., Sinkeviciute et al. 2018). This could be regarded independent of the clinical heterogeneity. So, instead of treating "schizophrenia" one could also focus on one of the symptoms of the disease, which may be easier to diagnose and to assess. Moreover, considering cognition as a symptom may be interesting since this may be diagnosed across different brain diseases. Thus, treating cognitive symptoms may be independent of a disease, and one treatment may have beneficial effects in different brain diseases. For example, the alpha-7 nicotinic receptor could be considered as a target CIAS and dementia (Lewis et al. 2017; Yang et al. 2017). It should be noted that cognition is still a collection of different brain functions, but this could be further specified (e.g., working memory).

This approach of looking at symptoms rather than looking at psychiatric disease diagnosis has also been by Research Domain Criteria (RDoC) initiative, which started in 2009 (see website ▶ https://www.nimh.nih.gov/research/research-funded-by-nimh/rdoc/index.shtml). This approach is also briefly mentioned in the ▶ Chap. 14. In essence, the aim of RDoC is to "understand the nature of mental health and illness in terms of varying degrees of dysfunction in general psychological/biological systems." The aim is to break down the different domains of human functioning and to study these in a neurodevelopmental and environmental context. Within this matrix, they define various levels going from genes, molecules all the way up to patient self-reports. Instead of diagnosing mental illnesses on various symptoms with all the issues of heterogeneity and comorbidities, the RDoC approach would overcome these issues. Moreover, having a matrix with different functional domains, which are connected with neurodevelopmental and biological mechanisms, can provide valid drug targets for the symptoms defined in each behavioral domain. It should be noted that this is not explicitly stated by the RDoC consortium, and drug development is not the major aim of this initiative.

It should be noted that if this approach seems viable in finding drug targets for treating symptoms, and this would change the way we approach diseases. Instead of treating diseases as currently classified, symptoms should be defined that could apply for different diseases. Thus, the FDA only approves novel drugs for diseases and not for symptoms. Whether this will change will depend on successful projects that show that we can improve brain disease at a clinical level on basis of this approach.

2.3.3 Systems Biology: Diseaseome

A final topic that I would like to bring up in this chapter is related to systems biology. This is a relative new approach in finding novel drug targets. This discipline emerged from the developments in the field of genome-scale molecular biology and molecular genetics, which allows to obtain a better understanding of human biology in health and disease (Goh and Choi 2012). In combination with the increasing knowledge about cellular networks, molecular or regulatory mechanisms, protein interactions, or gene–disease interactions, researchers were able to build a diseaseome (Goh and Choi 2012). This approach leads to a clustering of different diseases based on shared molecular or regulatory mechanisms. One example for drug discovery in stroke has been proposed on basis of a common pathomechanism underlying vascular, neurological, and metabolic disease phenotypes (Langhauser et al. 2018). Using this network medicine approach, the cyclic guanosine monophosphate (cGMP) signaling pathway was identified as a potential shared common mechanism underlying stroke. Some promising data were obtained in animal studies using this approach (Casas et al. 2019), but the validity of this approach needs to be substantiated in further clinical studies.

One important assumption of this method is that diseases should not be based on symptoms or clinical phenotypes but on a pathomechanism. The rationale for this is that you cannot treat a phenotype or symptom but that you can modulate and normalize a mechanism. This requires a somewhat different approach for defining drug targets for CNS diseases, especially for psychiatric indications. Some examples of this approach can be found for Alzheimer's disease (Chatterjee and Roy 2016), Parkinson's disease (Midic et al. 2009), and multiple sclerosis (Khankhanian et al. 2016). Another outcome of using this diseaseome approach is that comorbidities can be revealed on the basis of shared mechanisms. For example, using these methods a research consortium showed that depression showed comorbidities with anxiety and somatic disorders such as obesity, irritable bowel syndrome, fibromyalgia, and migraine (Marx et al. 2017). These findings should indicate possible molecular-level mechanisms and offer new drug targets for treating depression and these other diseases.

> CNS diseases are clinically very heterogeneous and maybe we cannot treat a clinically defined disease. Defining symptoms, preferably based on neurobiological mechanisms, may be considered as a more relevant manner to define drug targets. Symptom treatment can be applied in different diseases.

2

Conclusion

The development of new drugs based on novel targets in CNS diseases has not been very successful in the last decades. Novel tools enabled us to unravel the genetics of human diseases but have not been very helpful in finding novel targets for CNS diseases. One approach is to understand the genetic aberrations in disease states that should be leading in finding novel drug targets. However, this may be associated with some limitations related to this method. It remains to be demonstrated whether RDoC or the diseaseome approaches will provide new targets for symptoms and biological mechanisms. One consequence of these approaches is that we should change our views on current CNS diseases, especially psychiatric diseases: not treating a disease, but treating symptoms with a defined neurobiological mechanism.

References

Ban TA (2006) The role of serendipity in drug discovery. Dialogues Clin Neurosci 8(3):335–344

Berman RM, Cappiello A, Anand A, Oren DA, Heninger GR, Charney DS, Krystal JH (2000) Antidepressant effects of ketamine in depressed patients. Biol Psychiatry 47(4):351–354. https://doi.org/10.1016/s0006-3223(99)00230-9

Breen G, Li Q, Roth BL, O'Donnell P, Didriksen M, Dolmetsch R et al (2016) Translating genome-wide association findings into new therapeutics for psychiatry. Nat Neurosci 19(11):1392–1396. https://doi.org/10.1038/nn.4411

Cao C, Moult J (2014) GWAS and drug targets. BMC Genomics 15(Suppl 4):S5. https://doi.org/10.1186/1471-2164-15-S4-S5

Casas AI, Hassan AA, Larsen SJ, Gomez-Rangel V, Elbatreek M, Kleikers PWM et al (2019) From single drug targets to synergistic network pharmacology in ischemic stroke. Proc Natl Acad Sci U S A 116(14):7129–7136. https://doi.org/10.1073/pnas.1820799116

Chatterjee P, Roy D (2016) Insight into the epigenetics of Alzheimer's disease: a computational study from human interactome. Curr Alzheimer Res 13(12):1385–1396. https://doi.org/10.2174/1567205013666160803151101

Collo G, Merlo Pich E (2018) Ketamine enhances structural plasticity in human dopaminergic neurons: possible relevance for treatment-resistant depression. Neural Regen Res 13(4):645–646. https://doi.org/10.4103/1673-5374.230288

Dourlen P, Kilinc D, Malmanche N, Chapuis J, Lambert JC (2019) The new genetic landscape of Alzheimer's disease: from amyloid cascade to genetically driven synaptic failure hypothesis? Acta Neuropathol 138(2):221–236. https://doi.org/10.1007/s00401-019-02004-0

Fernando MB, Ahfeldt T, Brennand KJ (2020) Modeling the complex genetic architectures of brain disease. Nat Genet 52(4):363–369. https://doi.org/10.1038/s41588-020-0596-3

Fountoulakis KN, Dragioti E, Theofilidis AT, Wikilund T, Atmatzidis X, Nimatoudis I et al (2019) Staging of schizophrenia with the use of PANSS: an international multi-center study. Int J Neuropsychopharmacol 22(11):681–697. https://doi.org/10.1093/ijnp/pyz053

Fukumoto K, Fogaca MV, Liu RJ, Duman C, Kato T, Li XY, Duman RS (2019) Activity-dependent brain-derived neurotrophic factor signaling is required for the antidepressant actions of (2R,6R)-hydroxynorketamine. Proc Natl Acad Sci U S A 116(1):297–302. https://doi.org/10.1073/pnas.1814709116

Gaspar HA, Gerring Z, Hubel C, Major Depressive Disorder Working Group of the Psychiatric Genomics, C, Middeldorp CM, Derks EM, Breen G (2019) Using genetic drug-target networks to develop new drug hypotheses for major depressive disorder. Transl Psychiatry 9(1):117. https://doi.org/10.1038/s41398-019-0451-4

Goh KI, Choi IG (2012) Exploring the human diseasome: the human disease network. Brief Funct Genomics 11(6):533–542. https://doi.org/10.1093/bfgp/els032

Hillhouse TM, Porter JH (2015) A brief history of the development of antidepressant drugs: from monoamines to glutamate. Exp Clin Psychopharmacol 23(1):1–21. https://doi.org/10.1037/a0038550

Hornykiewicz O (2010) A brief history of levodopa. J Neurol 257(Suppl 2):S249–S252. https://doi.org/10.1007/s00415-010-5741-y

Jack CR Jr, Knopman DS, Jagust WJ, Petersen RC, Weiner MW, Aisen PS et al (2013) Tracking pathophysiological processes in Alzheimer's disease: an updated hypothetical model of dynamic biomarkers. Lancet Neurol 12(2):207–216. https://doi.org/10.1016/S1474-4422(12)70291-0

Khankhanian P, Cozen W, Himmelstein DS, Madireddy L, Din L, van den Berg A et al (2016) Meta-analysis of genome-wide association studies reveals genetic overlap between Hodgkin lymphoma and multiple sclerosis. Int J Epidemiol 45(3):728–740. https://doi.org/10.1093/ije/dyv364

Kondej M, Stepnicki P, Kaczor AA (2018) Multi-target approach for drug discovery against schizophrenia. Int J Mol Sci 19(10):3105. https://doi.org/10.3390/ijms19103105

Langhauser F, Casas AI, Dao VT, Guney E, Menche J, Geuss E et al (2018) A diseasome cluster-based drug repurposing of soluble guanylate cyclase activators from smooth muscle relaxation to direct neuroprotection. NPJ Syst Biol Appl 4:8. https://doi.org/10.1038/s41540-017-0039-7

Lewis AS, van Schalkwyk GI, Bloch MH (2017) Alpha-7 nicotinic agonists for cognitive deficits in neuropsychiatric disorders: a translational meta-analysis of rodent and human studies. Prog Neuro-Psychopharmacol Biol Psychiatry 75:45–53. https://doi.org/10.1016/j.pnpbp.2017.01.001

Makin S (2018) The amyloid hypothesis on trial. Nature 559(7715):S4–S7. https://doi.org/10.1038/d41586-018-05719-4

Malenka RC, Nicoll RA (1993) NMDA-receptor-dependent synaptic plasticity: multiple forms and mechanisms. Trends Neurosci 16(12):521–527. https://doi.org/10.1016/0166-2236(93)90197-t

Marx P, Antal P, Bolgar B, Bagdy G, Deakin B, Juhasz G (2017) Comorbidities in the diseasome are more apparent than real: what Bayesian filtering reveals about the comorbidities of depression. PLoS Comput Biol 13(6):e1005487. https://doi.org/10.1371/journal.pcbi.1005487

Midic U, Oldfield CJ, Dunker AK, Obradovic Z, Uversky VN (2009) Protein disorder in the human diseasome: unfoldomics of human genetic diseases. BMC Genomics 10(Suppl 1):S12. https://doi.org/10.1186/1471-2164-10-S1-S12

Oxford AE, Stewart ES, Rohn TT (2020) Clinical trials in Alzheimer's disease: a hurdle in the path of remedy. Int J Alzheimers Dis 2020:5380346. https://doi.org/10.1155/2020/5380346

Palmqvist S, Janelidze S, Quiroz YT, Zetterberg H, Lopera F, Stomrud E et al (2020) Discriminative accuracy of plasma Phospho-tau217 for Alzheimer disease vs other neurodegenerative disorders. JAMA 324(8):772–781. https://doi.org/10.1001/jama.2020.12134

Pepeu G, Giovannini MG (2009) Cholinesterase inhibitors and beyond. Curr Alzheimer Res 6(2):86–96. https://doi.org/10.2174/156720509787602861

Plenge RM (2019) Priority index for human genetics and drug discovery. Nat Genet 51(7):1073–1075. https://doi.org/10.1038/s41588-019-0460-5

Selkoe DJ, Hardy J (2016) The amyloid hypothesis of Alzheimer's disease at 25 years. EMBO Mol Med 8(6):595–608. https://doi.org/10.15252/emmm.201606210

Shorter E (2009) The history of lithium therapy. Bipolar Disord 11(Suppl 2):4–9. https://doi.org/10.1111/j.1399-5618.2009.00706.x

Sinkeviciute I, Begemann M, Prikken M, Oranje B, Johnsen E, Lei WU et al (2018) Efficacy of different types of cognitive enhancers for patients with

schizophrenia: a meta-analysis. NPJ Schizophr 4(1):22. https://doi.org/10.1038/s41537-018-0064-6

Sunwoo M, O'Connell J, Brown E, Lin A, Wood SJ, McGorry P, O'Donoghue B (2020) Prevalence and outcomes of young people with concurrent autism spectrum disorder and first episode of psychosis. Schizophr Res 216:310–315. https://doi.org/10.1016/j.schres.2019.11.037

Sweatt JD (2009) Experience-dependent epigenetic modifications in the central nervous system. Biol Psychiatry 65(3):191–197. https://doi.org/10.1016/j.biopsych.2008.09.002

Tam V, Patel N, Turcotte M, Bosse Y, Pare G, Meyre D (2019) Benefits and limitations of genome-wide association studies. Nat Rev Genet 20(8):467–484. https://doi.org/10.1038/s41576-019-0127-1

Trullas R, Skolnick P (1990) Functional antagonists at the Nmda receptor complex exhibit antidepressant actions. Eur J Pharmacol 185(1):1–10. https://doi.org/10.1016/0014-2999(90)90204-J

Wei Y, de Lange SC, Scholtens LH, Watanabe K, Ardesch DJ, Jansen PR et al (2019) Genetic mapping and evolutionary analysis of human-expanded cognitive networks. Nat Commun 10(1):4839. https://doi.org/10.1038/s41467-019-12764-8

Wei Y, Chang L, Hashimoto K (2020) A historical review of antidepressant effects of ketamine and its enantiomers. Pharmacol Biochem Behav 190:172870. https://doi.org/10.1016/j.pbb.2020.172870

Wendland JR, Ehlers MD (2016) Translating neurogenomics into new medicines. Biol Psychiatry 79(8):650–656. https://doi.org/10.1016/j.biopsych.2015.04.027

Yamada Y, Matsumoto M, Iijima K, Sumiyoshi T (2020) Specificity and continuity of schizophrenia and bipolar disorder: relation to biomarkers. Curr Pharm Des 26(2):191–200. https://doi.org/10.2174/1381612825666191216153508

Yang T, Xiao T, Sun Q, Wang K (2017) The current agonists and positive allosteric modulators of alpha7 nAChR for CNS indications in clinical trials. Acta Pharm Sin B 7(6):611–622. https://doi.org/10.1016/j.apsb.2017.09.001

Zanos P, Gould TD (2018) Mechanisms of ketamine action as an antidepressant. Mol Psychiatry 23(4):801–811. https://doi.org/10.1038/mp.2017.255

Zetterberg H, Bendlin BB (2021) Biomarkers for Alzheimer's disease-preparing for a new era of disease-modifying therapies. Mol Psychiatry 26(1):296–308. https://doi.org/10.1038/s41380-020-0721-9

Neurogenomics with Application to Schizophrenia and Other Major Neuropsychiatric Diseases with Complex Heredity

Laurence Melnick

Contents

3.1   Introduction – 37
3.1.1   Drug Discovery with Incomplete Understanding of Disease or Drug Mechanism – 37
3.1.2   The Neuropsychiatric Pharmacology for SCZ, BPD, MDD, and ASD Largely Consists of Variations of Old Drugs – 38
3.1.3   The Heritable Component of Neuropsychiatric Disorders – 39

3.2   Variations in the Human Genome That Are Most Commonly Queried for Associations with Neuropsychiatric Disease – 39
3.2.1   Single Nucleotide Polymorphisms (SNPs) and Genome-Wide Association Studies (GWAS) – 40
3.2.2   From SNP to Inferred Function Using Expression Quantitative Trait Loci (eQTL) – 42
3.2.3   Protein Coding Variations and Whole Genome Exome Sequencing (WES) – 43
3.2.4   Whole Genome Sequencing (WGS) – 43

© The Author(s), under exclusive license to Springer Nature Switzerland AG 2021
R. Schreiber (ed.), *Modern CNS Drug Discovery*, https://doi.org/10.1007/978-3-030-62351-7_3

3.2.5   Structural Variations (SV) Including Copy Number
         Variations (CNVs) – 43
3.2.6   Genetic Variants with Large or Small Influences on
         Neuropsychiatric Disease – 43

3.3   Three Large Genome-Wide Studies (GWAS)
         of Schizophrenia – 44
3.3.1   Two Major Observations Critical for Drug
         Discovery – 44
3.3.2   A Large Consortium for Meta-Analysis – 45
3.3.3   GWAS and Polygenic Contributions to Disease – 45
3.3.4   Findings from Additional Large GWAS Studies – 46

3.4   Genetic Findings That Cross Current Diagnostic
         Categories – 47
3.4.1   Pooled Data for Five Diseases Shows Common
         Risk Loci – 47
3.4.2   Shared Genetic Risks for Different Diagnoses Are
         Specific to Disorders – 47
3.4.3   As with SNP Risk Alleles, Certain De Novo Highly
         Influential Risk Alleles Also Cross Diagnostic
         Boundaries – 47

3.5   Genomic Studies and Drug Discovery – 48
3.5.1   Concerns for Application of GWAS Data Toward
         Neuropsychiatric Drug Discovery – 48
3.5.2   A Genomics-Identified Target for Drug Discovery,
         the Complement Encoding C4 Gene Structural Variants,
         and a Confluence of Observations – 50
3.5.3   Current Efforts Toward Revealing the Fundamental
         Biology, Informing Clinical Practice, and Delivering New
         Therapeutic Targets – 50

3.6   On Order from Complexity – 54

         References – 54

There is a pressing need for novel pharmaceutical interventions for the major neuropsychiatric diseases schizophrenia (SCZ), bipolar disorder (BPD), major depressive disorder (MDD) and autism spectrum disorder (ASD). The heritable component of these disorders provides an avenue for understanding disease mechanism leading to novel pharmaceutical interventions. Progress toward this understanding is described. Variations in the genome include single nucleotide polymorphisms (SNPs) generally but not always of small effect size. Only a small subset of SNPs are within protein coding sequences. Structural variants (SVs) including copy number repeats (CNVs) can have larger influence on phenotype. These genetic variations are queried for influence on diseases using genome-wide association studies (GWASs) to reveal a largely polygenic influence on disease. This consists of multiple small contributions from common polymorphisms. Mutations with large effect size, for example from protein coding changes or chromosomal structural variations, often include de novo mutations. These also influence neuropsychiatric disease. Importantly, the genetic findings cross current diagnostic boundaries. The genomic studies seek improved mechanistic understanding leading to improved diagnostic and clinical practice and importantly provision of a basis for discovery of novel pharmaceuticals. Efforts toward these goals are discussed.

☺ Learning Objectives
- The state of novel drug discovery for typical forms of schizophrenia, bipolar disorder, major depressive disorder, and autism spectrum disorder
- The complex heritable of neuropsychiatric disease
- Variations in human genomes that are the objects of genomic studies
- Most commonly used methods for large neuropsychiatric genome-wide studies
  - Single nucleotide polymorphism (SNP) analysis
  - Whole exome sequencing
  - Analysis of structural variants including copy number variants

- Large genome-wide association study (GWAS) findings for schizophrenia indicating:
  - The largely polygenic genetic contributions toward schizophrenia and other major neuropsychiatric diseases
  - Common genetic contributions across disease categories
- Current diagnostic practice and how genomic studies can inform efforts toward mechanism-based diagnosis
- Genomic studies in support of drug discovery; successes and concerns

3.1 Introduction

The complexity of the brain is reflected in the diversity of neuropsychiatric disorders. Even using current methods, there is considerable heterogeneity within the diagnostic categories schizophrenia (SCZ), bipolar disorder (BPD), major depressive disorder (MDD) and autism spectrum disorder (ASD). This is due to incomplete understanding of the mechanistic bases of these disorders. Despite enormous strides in the basic science of neurobiology, diagnosis is still based on disease phenotype, not mechanism. The heritable component of these disorders provides an avenue for achieving an understanding of mechanism. This is foreseen to enable both mechanism-based drug discovery and, importantly, more precise diagnosis for both improved genomic studies and clinical trials.

3.1.1 Drug Discovery with Incomplete Understanding of Disease or Drug Mechanism

Let us begin with the following questions. For typical forms of schizophrenia (SCZ), bipolar disorder (BPD), major depressive disorder MDD, and autism spectrum disorder (ASD), do we have:
- A mechanism-based diagnostic?
- A reliably predictive animal disease model?

- A genetics-based method, based on disease mechanism, not tolerance, metabolism, or other non-efficacy-related drug properties, for differentiating the effectiveness of psychiatric medications for a specific disorder? For understanding the mechanism-based difference between responders and non-responders?
- A specific novel target or biochemical pathway singled out from the several genetically identified genetic risk factors or pathways to provide a mechanistic basis for discovery of novel effective drugs for SCZ/BPD/MDD/ASD?

At this time, it is difficult to answer affirmatively for the first three questions. For the fourth question, the genetics of these diseases is multifactorial. The diversity of biological functions associated with genetic risk factors, as well as the presence of genetic risk variants in healthy as well as affected individuals, makes choice of targets difficult.

Neuropsychiatric genomics has been highly successful at identifying genes, pathways, and circuits that are associated with risk for disease and in some cases have motivated drug discovery toward specific targets. Continued progress will lead to new avenues for drug discovery.

A key aim of this document is the discussion of the efforts of neuropsychiatric genomics to move the answers to the above questions to the affirmative, offering mechanism-based diagnoses and novel disease targets for drug discovery.

3.1.2 The Neuropsychiatric Pharmacology for SCZ, BPD, MDD, and ASD Largely Consists of Variations of Old Drugs

There has been a conspicuous novel-mechanism based for the major psychiatric disorders schizophrenia (SCZ), bipolar disorder (BPD), and autism spectrum disorder (ASD), an exception being the repurposing of ketamine for major depressive disorder (MDD) (◻ Table 3.1). For example, lithium, the dominant treatment for BPD was reported to be effective for the treatment of mania in 1949 (Cade 1949; Shorter 2009). As another example, all currently approved antipsychotic medications share the property of blockage of dopamine D2 receptors, (Patel et al. 2014) a mechanism documented by Delay and Deniker in 1952 (Ban 2007). At

◻ **Table 3.1** Origin of pioneer pharmaceuticals for neuropsychiatric disease (Barondes 1993; Brunton et al. 2018)

Pioneer pharmaceutical and application	Origin	Original or proposed use	Discovery of CNS effects	Psychiatric therapeutic mechanism
Reserpine[a] (schizophrenia)	Hindu medicinal herb	High blood pressure	Traditional use suggests CNS	Monoamine storage inhibition
Chlorpromazine[a] (schizophrenia)	Synthesized as H1 antihistamine	Pre-surgery sedation	Tranquilizing suggests SCZ agitation application	Dopamine D2 receptor antagonist
Iproniazid[a] (depression)	Developed as antibacterial	Tuberculosis	Tuberculosis patients showed anti-depression	Monoamine oxidase inhibitor
Imipramine[a] (depression)	Chlorpromazine analog for SCZ	SCZ but ineffective	Anti-depression	Tricyclic reuptake inhibitor

[a]Serendipitous discovery while looking for something else

the time of this writing, the hope for a role of neuropsychiatric genomics of providing approved drugs based on novel mechanism has largely not been met.

3.1.3 The Heritable Component of Neuropsychiatric Disorders

Evidence from family studies including twin studies and, importantly, adoptive twin studies establishes a familial component of SCZ etiology (Ingraham and Kety 2000; Sullivan et al. 2003; Gejman et al. 2010). For example, in the year 1971, Seymour S. Kety, David Rosenthal, Paul H. Wender, and Fini Schulsinger published the results of twin studies demonstrating a clear inherited component of schizophrenia (Kety et al. 1971).

» ...a significantly higher than usual prevalence of schizophrenia-related illness was found among the biological relatives of adopted schizophrenics, but not among their adoptive relatives. ...The findings support a genetic transmission of vulnerability to schizophrenia.

For monozygotic twins, where one is diagnosed with schizophrenia, the probability of the other being affected is approximately 40%.

Two fundamental insights emerge from these studies. First, there is a large heritable component of the disease and second, importantly, this genetic influence by itself is insufficient to cause the disorder.

Heritability of neuropsychiatric disease has been found to hold true to a degree for several psychiatric disorders (◘ Table 3.2). These observations motivated the search for the contributing genes. In defiance of early expectations, the results have been far from straightforward.

Examples from a wealth of literature describe the methods and results of genomic analysis (Corvin et al. 2010; Geschwind and Flint 2015), application to drug discovery (Breen et al. 2016; Sullivan et al. 2018), and interpretation of findings including criticisms of methods (Craddock and Owen 2010). For schizophrenia (SCZ), bipolar disorder (BPD), and major depressive disorder (MDD), the pharmaceutical industry

◘ **Table 3.2** Estimated heritability of neuropsychiatric disease based on twin/family studies

Disorder	Variance explained (%)	95% Confidence intervals
Schizophrenia	76	69–83
Autism spectrum disorder	56	37–82
Bipolar disorder	58	42–64
Major depressive disorder	32	19–40

Derived from Geschwind and Flint, supplementary table S1 (Geschwind and Flint 2015)

has provided therapeutics that are profoundly both effective and wanting.

In the following sections, we review the genomic variations in human populations that are queried in neuropsychological genomic studies. This is followed by a survey of findings largely centered on a highly powered study by the Schizophrenia Working Group of the Psychiatric Genomics Consortium (2014), referred to as PGC 2014, and subsequent additional studies. We then consider what is needed for application of the genomic findings toward a source of novel therapeutic targets.

3.2 Variations in the Human Genome That Are Most Commonly Queried for Associations with Neuropsychiatric Disease

┌ Definition ─────────────────

Allele: Allele is one of the variant forms of a gene at a particular locus on a chromosome. Different alleles can produce variation in inherited characteristics.

3

> **Definition**
>
> *Polymorphism:* Polymorphisms are one of two or more variants of a particular DNA sequence. The most common type of polymorphism involves variation at a single base pair. Polymorphisms can also be much larger in size and involve long stretches of DNA (*National Human Genomics Research institute Talking Glossary* n.d.).

> **Definition**
>
> *Single Nucleotide Polymorphism (SNP):* SNPs are common, but minute, variations that occur in human DNA at a frequency of 1 in every 1000 bases. a SNP is a single base-pair site within the *genome* at which more than one of the four possible base pairs is commonly found in natural populations. Over 10 million SNP sites have been identified and mapped on the sequence of the genome, providing the densest possible map of genetic differences (*National Human Genomics Research institute Talking Glossary* n.d.).

> **Definition**
>
> *Copy number variation (CNV)* is when the number of copies of a particular gene varies from one individual to the next. Following the completion of the Human Genome Project, it became apparent that the genome experiences gains and losses of genetic material. The extent to which copy number variation contributes to human disease is not yet known. It has long been recognized that some cancers are associated with elevated copy numbers of particular genes (*National Human Genomics Research institute Talking Glossary* n.d.).

> **Definition**
>
> *Genome-wide association study (GWAS)* is an approach used in genetics research to associate specific genetic variations with particular diseases. The method involves

scanning the genomes from many different people and looking for genetic markers that can be used to predict the presence of a disease. Once such genetic markers are identified, they can be used to understand how genes contribute to the disease and develop better prevention and treatment strategies (*National Human Genomics Research institute Talking Glossary* n.d.).

Genome-wide association studies (GWAS) generally focus on three types of genomic variants (◻ Fig. 3.1). These genomic variations are (1) single nucleotide polymorphisms (SNPs), (2) changes in protein coding sequences, generally also single nucleotide changes, studied using whole exome sequencing (WES), and (3) larger structural variants (SVs), generally but not always greater than 1000 bases. Copy number variants (CNVs) are the most studied of the structural variants. The three variants are described in ◻ Fig. 3.1. SNPs in general identify genetic loci associated with small influences on disease phenotype. In contrast, protein coding variations that confer loss, gain or change of function (LoF, GoF, CoF), or chromosomal structural variations can have stronger, more penetrant influences on disease. The small, incremental contributions of SNPs are usually associated with typical forms of disease. The larger influences from protein changes or structural variations are often associated with atypical disease phenotypes.

3.2.1 Single Nucleotide Polymorphisms (SNPs) and Genome-Wide Association Studies (GWAS) (◻ Fig. 3.1a)

SNPs are single nucleotide polymorphisms that are common in the population. These are distributed throughout the genome and are used to identify regions that are associated with phenotypes. Most SNPs have no intrinsic influence on phenotype. For SNPs to be useful

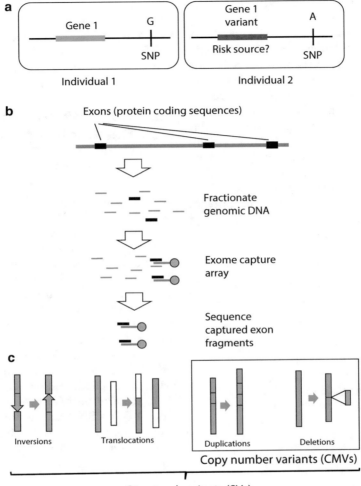

Fig. 3.1 Three types of genomic variations that comprise the major objects of current genome-wide neuropsychiatric genetics studies

a Single nucleotide polymorphisms (SNPs)
- Polymorphisms generally common in the population
- Overwhelming majority in non-coding regions (protein coding exons comprise only 1–2% of the human genome)
- Generally small contributions per SNP to phenotype (low penetrance)
- Evolutionally conserved
- SNP microarray chips allow whole genome analysis for low cost per individual
- Associated risk genes inferred from proximity to SNPs, eQTLs, or other sources

b Protein coding variations queried using whole exome sequencing (WES) studies
- Can be silent, missense for possible gain or loss of function, nonsense for truncation of protein, and possible complete loss of function. Can influence splice patterns

- Can be highly penetrant and therefore informative of mechanism; however, mutations of strong influence on phenotype are generally associated with atypical forms of disease
- Often occur de novo

c Structural variations (SV), including copy number variations (CNVs)
- Generally >1000 bases but can be much smaller, distributed throughout genome (Malhotra and Sebat 2012)
- Can be de novo mutation in the individual or in recent ancestry
- Possibe role in rapid evolution of human brain (O'Bleness et al. 2012)
- Effect size limited by embryonic lethality
- "Load" of CNVs greater in SCZ compared with controls and compared with BPD (Craddock and Owen 2010)
- SCZ risk can also confer risk for multiple neuropsychiatric phenotypes, e.g. autism and mental retardation (Li et al. 2016)

3

for population studies, each SNP variant must be represented above a defined frequency in the population (and in this sense is common). Only 1–2% of the human genome consists of protein coding sequence and most SNPs are outside of these. In general, SNPs themselves do not have functional significance; rather, they identify regions that can have regulatory or other influences. Exceptions to this include some SNPs that are within protein-coding regions. The COMT gene provides an example for which, in certain studies, SNP variants in protein-coding sequences have been associated with behavioral phenotypes (Zubieta et al. 2003; Goghari and Sponheim 2008; Voisey et al. 2011).

Definition

Expression quantitative trait locus (eQTL): An eQTL is a locus that explains a fraction of the genetic variance of a gene expression phenotype. Standard eQTL analysis involves a direct association test between markers of genetic variation with gene expression levels typically measured in tens or hundreds of individuals. This association analysis can be performed proximally or distally to the gene. One of the major advantages of eQTL mapping using the GWAS approach is that it per-

mits the identification of new functional loci without requiring any previous knowledge about specific *cis* or *trans* regulatory regions (Nica and Dermitzakis 2013).

3.2.2 From SNP to Inferred Function Using Expression Quantitative Trait Loci (eQTL) (◘ Fig. 3.2)

To a varying degree, the influence of a SNP identified risk locus can be inferred. In the most fortuitous case, rare SNPs are within regions that encode protein sequences or functional RNAs. For other SNPs, proximity to a protein-coding sequence by itself is not definitive; for example, the SNP could be within a region of undocumented function. Expression quantitative trait locus (eQTL) analysis (Nica and Dermitzakis 2013) provides a basis for associating a SNP with its influence on transcription. eQTLs associate the allelic configuration of a SNP to the transcription of a gene or set of genes. These eQTLs/SNP associations can be tissue specific and for neuropsychiatric genetics, postmortem brain tissue can be particularly informative. Associations of SNPs with

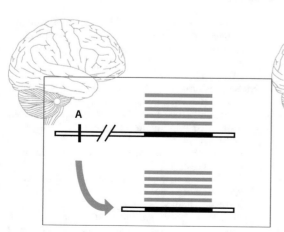
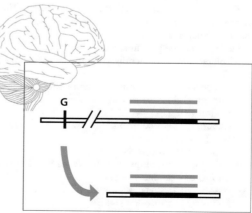
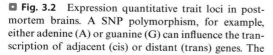

◘ **Fig. 3.2** Expression quantitative trait loci in postmortem brains. A SNP polymorphism, for example, either adenine (A) or guanine (G) can influence the transcription of adjacent (cis) or distant (trans) genes. The association of SNP variants with different tissue-specific transcription patterns, for example in postmortem brain samples, can result in inferred function

gene expression can be cis for nearby genes or trans for distant genes. eQTL studies can be computationally intensive including the integration of information from transcriptional microarrays including tens of thousands of genes with genomic SNP analysis array data.

3.2.3 Protein Coding Variations and Whole Genome Exome Sequencing (WES) (■ Fig. 3.1b)

Ng et al. (2009) provide methods for determining the sequences of the exome, protein coding, and exon regions of the genome. The discovered variations in inferred proteins can have neutral to various degrees of influence. Important information is derived from complete loss or function, for example from nonsense mutations resulting in truncated encoded proteins. Other changes in function such as incremental loss or gain of function might derive from missense mutations. These variations that change protein expression or function often occur de novo and for cases of complete loss or extreme change of function, generally are associated with atypical disease.

3.2.4 Whole Genome Sequencing (WGS)

As the cost of DNA sequencing continues to decrease, whole genome sequencing is finding application for certain neuropsychiatric genetic studies. The largest WGS study to date (Halvorsen et al. 2020) included WGS data from 1,162 Swedish schizophrenia cases and 936 ancestry-matched population controls. This high-resolution study discovered SCZ associations with rare variants that alter the physical configuration of chromatin in ways that influence gene expression (Szabo et al. 2019).

3.2.5 Structural Variations (SV) Including Copy Number Variations (CNVs) (■ Fig. 3.1c)

A variety of genomic structural variations are associated with neuropsychiatric disease (Malhotra and Sebat 2012). The term generally indicates changes of greater than 1000 bases but SVs can be much smaller. Perhaps surprisingly, structural variations can typically comprise the same approximately 0.1% variation load on the genome as provided by SNPs. This is in light of the larger size of each SV.

3.2.6 Genetic Variants with Large or Small Influences on Neuropsychiatric Disease

As discussed, the individual SNP contribution to neuropsychiatric disease or to any phenotype is generally below a threshold for unambiguous determination. In contrast, variants that influence protein structure/function or chromosome structure can by themselves have strong, highly penetrant influences. (The term penetrance indicates the degree of genetic influence.) In general, highly penetrant genetic variations are found in atypical disease cases and are often de novo in origin. This is in contrast to disease-associated SNP alleles that are generally stable over genomic evolution with each SNP risk variant having small effect.

> A range of genomic variations provide a basis for searching genetic associations with neuropsychiatric disease. Single nucleotide polymorphisms provide a low resource per subject method for genomic scanning to provide associations, each of which generally provides small influence on disease. Other genomic variations such as changes in protein coding or chromosome structure can each exert greater disease influence; however, in general, this influence is associated with atypical forms of disease.

3.3 Three Large Genome-Wide Studies (GWAS) of Schizophrenia

Genome-wide association studies (GWAS), also termed whole genome association studies (WGAS), are hypothesis-free methods for identifying, across the entire genome, associations between genetic regions and traits including diseases. These studies include both SNPs and other genetic variants that can be associated with a trait. Here we focus on GWAS studies for SCZ, perhaps the most ambitious challenge for the neuropsychiatric community. These highly resourced studies are powered by large case (disease phenotype) and control sample sizes in a major effort to determine the genetic underpinnings and fundamental mechanisms of the disease. We discuss three large genome-wide mega-analyses listed in ◘ Table 3.3.

3.3.1 Two Major Observations Critical for Drug Discovery

First, a major contribution to disease phenotype is polygenic. Small contributions from many common polymorphisms comprise a major portion of the genetic contribution to SCZ. These polygenic contributions are in most cases polymorphisms found in the general population but enriched in affected individuals. Each of these contributions by itself

◘ **Table 3.3** Three large GWAS schizophrenia studies discussed in this chapter

Study	Cases/controls	Risk alleles above threshold for genome-wide significance	Major conclusions
Schizophrenia Working Group of the Psychiatric Genomics Consortium 2014 (PGC 2014) (Note 1)	36,989/113,075	108	Largest published GWAS study Demonstrated the power of increased sample sizes for discovery in SCZ genetics research Polygenic nature of genetic influence on SCZ finds risk alleles in a diversity of biological pathways
Pardiñas et al. 2018 (Note 2)	Additional 11,260/24,542	50 novel	Novel GWAS findings and meta-analysis with previous GWAS Risk alleles enriched in evolutionarily stable genetic regions
Lam et al. 2019 (Note 3)	Asian GWAS study 22,778/35,362	44 novel	European and Asian GWAS studies compared and combined Combined populations add power to study Different allele frequencies in different populations influence GWAS determinations Polygenic risk scores have diminished predictive power across European/Asian populations

1. *Biological insights from 108 schizophrenia-associated genetic loci* (Schizophrenia Working Group of the Psychiatric Genomics Consortium 2014)

2. *Common schizophrenia alleles are enriched in mutation-intolerant genes and in regions under strong background selection* (Pardiñas et al. 2018)

3. Comparative genetic architectures of schizophrenia in East Asian and European populations (Lam et al. 2019)

is generally below the threshold of producing a consistently definable phenotype. The polymorphisms are associated with a wide range of biological functions.

The second important finding, consistent with a range of studies, is the identification of risk alleles for SCZ that are common to different categories of neuropsychological disorder, most notably BPD (◼ Table 3.5).

3.3.2 A Large Consortium for Meta-Analysis

The Schizophrenia Working Group of the Psychiatric Genomics Consortium (PGC) was assembled "to combine all available schizophrenia samples with published or unpublished GWAS genotypes into a single, systematic analysis." The three goals of the consortium for neuropsychiatric disease are to (a) reveal the fundamental biology, (b) inform clinical practice, and (c) deliver new therapeutic targets (Sullivan et al. 2018).

Results of these analyses are discussed below. The first discussed meta-analysis, the largest published study to date (Schizophrenia Working Group of the Psychiatric Genomics Consortium 2014), included 36,989 cases and 113,075 controls. Also included were 1,235 parent-affected offspring trios (parents, affected child, and possibly other family members). The study identified 108 significant SCZ-associated genetic loci.

3.3.3 GWAS and Polygenic Contributions to Disease

To quote from the major 2014 publication of the Schizophrenia Working Group of the Psychiatric Genomics Consortium (2014), "Although risk variants range in frequency from common to extremely rare, estimates suggest half to a third of the genetic risk of schizophrenia is indexed by common alleles genotyped by current genome-wide association study (GWAS) arrays." Polygenicity is supported by rigorous analysis designed to rule out as a source of bias population stratification (differences in allele frequencies between sub-populations generally due to non-random mating, e.g., different ancestry between case and control groups). The results are summarized in ◼ Table 3.4. In brief, unfortunate combinations of multiple polymorphisms that in general are prevalent in the unaffected population contribute to neurological disorder. Effects generally occur through altering gene expression. Only 10 of the risk allele SNPs were located within protein coding (exonic) sequences. Risk loci are enriched for expression quantitative trait loci (eQTL). Nearly 75% are associated with (but rarely within) protein-coding genes and 40% with a single gene, and a further 8% are within 20 kb of a gene. Additional findings include the following:

1. *The risk alleles cover genes involved in a wide range of pathways (◼ Table 3.4).* Much of this work is consistent with dopaminergic and other SCZ disease models; however, the diversity of pathways precludes, at this time, a definitive basis for discrimination among disease models.

2. *For schizophrenia, GWAS identified risk SNPs coincide with more severe de novo mutations found in other studies of neuropsychiatric disease.* For example, the SNP identified SCZ risk loci *DPYD, ESAM, RP1, ZDHHCS,* and *CCDC39* correspond to previously identified loss of function SCZ risk loci for these genes. Also, SCZ SNP risk loci were found in genes for which loss of function or change of function is associated with intellectual disability and autism spectrum disorder. These observations support the disease relevance of the SNP-based discoveries.

3. *SCZ SNP identified risk alleles are enriched for active enhancers, particularly within the brain and immune system.* SCZ associations are strongly enriched at enhancers active in immune tissues, including B-lymphocyte lineages involved in acquired immunity (CD19 and CD20 lines). Previous epidemiological studies suggested immune dysregulation in schizophrenia (Khandaker et al. 2015). A gene that regulates the complement gene C4 (Section 3.5.2), of current interest for drug discovery, was found among the 108 risk alleles identified in the study.

3

Table 3.4 Findings from the SCZ PGC 2014 consortium study (Schizophrenia Working Group of the Psychiatric Genomics Consortium 2014)

General findings	Effects generally through altering gene expression. Risk loci are enriched for expression quantitative trait loci (eQTL) 75% are inferred to be associated with (but generally not within) protein-coding genes (40%, a single gene) A further 8% are within 20 kb of a gene.
Associations relevant to hypotheses of cause and treatment of schizophrenia	DRD2 (the target of all effective antipsychotic drugs) Many genes (e.g., GRM3, GRIN2A, SRR, GRIA1) are involved in glutamatergic neurotransmission and synaptic plasticity Voltage-gated calcium channel subunits CACNA1C, CACNB2 and CACNA1I are associated with risk.
The brain and immunity	SCZ associations strongly enriched at enhancers active in immune tissues Previous epidemiological studies suggested immune dysregulation in schizophrenia (Khandaker et al. 2015). Complement C4 level-regulating gene associated with one of the risk SNPs found in this study
Pathways involved	G protein coupled receptor signaling Glutamatergic neurotransmission Neuronal calcium signaling Neurodeveopment Synaptic function and plasticity Immune function

3.3.4 Findings from Additional Large GWAS Studies (◻ Table 3.3)

Pardinas et al. (2018) added 11,260 cases and 24,542 controls to the large PGC 2014 GWAS study described above. The meta-analysis in combination with previous GWAS studies resulted in a determination of 145 alleles significantly associated with SCZ risk. The additional GWAS data from the Pardinas study was highly correlated with data from the previous PGC 2014 study (Schizophrenia Working Group of the Psychiatric Genomics Consortium 2014).

This study highlights six functional gene sets, each independently associated with SCZ, including targets of FMRP (fragile X mental retardation protein), abnormal behavior, 5-HT2C receptor complex, abnormal nervous system electrophysiology, voltage-gated calcium channel complexes, and abnormal long-term potentiation.

In addition, this study queried a set of 3,230 genes that were evolutionarily intoler-

ant to loss of function (LoF) mutations. This set of genes was found to be enriched for common SNPs associated with SCZ risk. Similar associations were not found for the other polygenic diseases, ALZ, type 2 diabetes, and neuroticism. The authors suggest various explanations for the evolutionary maintenance of these SCZ associated alleles. One argument, balanced selection, is not inconsistent with the risk alleles providing selective advantage in certain genetic or environmental contexts (Section 3.5.1).

Lam et al. (2019) documents a large Asian population-based SCZ GWAS study. The meta-analysis of the combined Asian population GWAS with European GWAS studies enhanced the power of both of these studies resulting in the determination of additional SCZ risk allele findings for a total of 208. Whereas there is a general similarity between populations for most of the SCZ risk allele findings, certain incongruities were found. European and Asian populations displayed different frequencies for certain polymorphisms. Notably, polymorphisms in the MHC locus comprise the highest frequency source of discovered risk for European populations (Schizophrenia Working Group of the Psychiatric Genomics Consortium 2014). In contrast, this region does not meet the threshold of significance for the Asian population. This appears to result from the lower frequency of the relevant SNP variants in the Asian population, not the absence of disease association of these variants. Additionally, and not surprisingly, the predictive power of polygenic risk scores, an approximate method for gauging risk based on GWAS data (Section 3.5.3.2), is significantly weakened when applied across European and Asian populations.

> Genome-wide association studies (GWAS) including tens of thousands of diseased (termed case) and control subjects have resulted in the identification of over 200 significant variations associated with schizophrenia. For the overwhelming majority of these associations, each variation contributes a very small disease influence. This points to the polygenic, i.e., multiple small influences, nature of the genetic influence on schizophrenia. There is a degree of confluence between findings from SNP-based studies and findings from protein coding loss of function studies. In general, GWAS from European ancestry coincides with the findings form Asian ancestry, with certain differences likely due to different allele frequencies in the two populations.

3.4 Genetic Findings That Cross Current Diagnostic Categories

In addition to the genetic risk for SCZ that is disease-specific, several genetic risk associations are also risk associations for other neuropsychiatric diseases. This information may provide insights toward providing disease mechanism-based distinctions among diagnostic categories. Examples are discussed below.

3.4.1 Pooled Data for Five Diseases Shows Common Risk Loci

In one interesting study, instead of searching for risk loci from defined disease groups, data was cross-analyzed from five diagnostically distinct disorders: autism spectrum disorder, attention deficit-hyperactivity disorder, bipolar disorder, major depressive disorder, and schizophrenia (Cross-Disorder Group of the Psychiatric Genomics Consortium 2013). Four loci were found ($p < 5 \times 10 - 8$) to contribute risk across the pool of disorders including two found to be within non-coding regions of the L-type calcium channel subunits, $CACNA1C$ and $CACNB2$. Referring to data from specific disease-based studies, the $CACNA1C$ channel was identified as a risk factor for both BPD and SCZ. The $CACNB2$ channel was found for all five disorders.

3.4.2 Shared Genetic Risks for Different Diagnoses Are Specific to Disorders

Maier et al. (2015) used GWAS data to derive genetic correlations between pairs of disorders (◘ Table 3.5). These findings indicate, on the genetics level, relationships between disorders including possible shared disease mechanisms that are not obvious from current phenotype-based observations.

3.4.3 As with SNP Risk Alleles, Certain De Novo Highly Influential Risk Alleles Also Cross Diagnostic Boundaries

Each SNP-identified variant exerts a small risk influence and is generally stable over generations. In contrast, variations that effect protein structure such as loss of function (LoF) or extreme change of function (CoF) can have limited heritability. Nevertheless these extreme variants provide a portion of disease risk in the population, often as de

☐ **Table 3.5** Pairwise genetic correlations

	Genetic correlation	SE
BPD/SCZ	0.590	0.048
MDD/SCZ	0.365	0.047
MDD/BPD	0.371	0.06
ASD/SCZ	0.194	0.071
ASD/BPD	0.084	0.089
ASD/MDD	0.054	0.089
ADHD/SCZ	0.055	0.046

Adapted from Maier et al. (2015)
SE standard error

novo mutations. Li et al. (2016) analyzed the combined results of multiple trio studies (unaffected parents and affected (proband) offspring and possibly other family members). They found that de novo mutations that were predicted to have extreme effect, loss of function, or change of critical function were found for autism spectrum disorder (ASD), epileptic encephalopathy (EE), intellectual disability (ID), and schizophrenia (SCZ). Additionally, a subset of these genes were found for all four disorders. Strikingly, many of these extreme de novo mutations were found to recur independently, pointing to their importance in disease etiology. Of particular interest was the recurrence of extreme mutations of *SCN2A*, encoding a voltage-gated sodium channel alpha subunit. This is reported to be one of the most frequently found de novo mutations associated with all four neuropsychiatric disorders. The contribution of extreme de novo mutations was found to be more prevalent in epileptic encephalopathy (EE) and intellectual disability (ID) compared with schizophrenia, possibly indicating a more heterogeneous disease population for EE and ID than for SCZ.

❯ Certain risk alleles for neuropsychiatric disease present risk for more than one syndrome, for example schizophrenia and bipolar disorder. This is true for SNPs and other variations including protein-coding influences. These findings motivate consideration of possible common mechanisms

for otherwise distinctly diagnosed diseases. It is hoped that genetics findings will increase mechanistic understanding and improve diagnostic precision with consequences for further genetics studies, clinical trials, and mechanism-based drug discovery.

3.5 Genomic Studies and Drug Discovery

Drug discovery for the brain is difficult. Any drug discovery program is arduous and resource intensive with unique aspects and uncertainties specific to each application. This is especially true for the discovery of novel brain therapeutics. A typical targeted (as opposed to phenotype based) discovery program can include identification of a single gene influence on disease, studies involving an isolated therapeutic target protein, cell culture models, and predictive animal models. These are not available for SCZ, BPD, MDD, and ASD drug discovery. In this section, we describe concerns of GWAS applications for drug discovery and movement toward overcoming these concerns. We also describe a promising example for drug discovery based on GWAS data leading to a confluence of observations toward an attractive drug discovery target.

3.5.1 Concerns for Application of GWAS Data Toward Neuropsychiatric Drug Discovery (☐ Fig. 3.3)

— *The absence of a mechanism-based understanding of neuropsychiatric disease.* Current diagnostic practice for SCZ, BPD, MDD, and ASD is based on phenotype, not on mechanism (☐ Fig. 3.3). Neuropsychiatric genetics identifies contributions to disease from a range of mechanistic sources. This diversity of influences makes it difficult to single out a single biological pathway or conceptual model to instruct intervention.

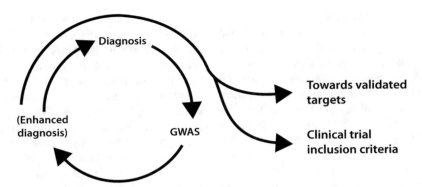

Fig. 3.3 Inclusion criteria and design of genomic and subsequent clinical studies for neuropsychiatric disease. Clinical heterogeneity results from our current paucity of disease mechanism–based classification or diagnosis. This impacts both genomic studies and clinical trials of neuropsychiatric drugs. Information from ongoing genome-wide association studies is anticipated to enhance diagnostic distinctions. The outcomes will inform inclusion criteria for further GWAS studies and, in addition, inform drug target decisions and refine inclusion criteria for clinical trials

- *Genetic findings blur the boundaries of the current diagnostic categories in the Statistical Manual of Mental Disorders (DSM) and the International Classification of Diseases (ICD)* (Craddock and Owen 2010; Lilienfeld and Treadway 2016). As discussed, genetic studies discovered common genetic risk for different neuropsychiatric diseases. We do not know to what degree these boundaries need to be redefined as additional knowledge of disease mechanism becomes available. An extreme possibility is that multiple different mechanistic sources could result in highly similar and consequently inadequately diagnosed disease phenotypes.
- *Inclusion criteria for clinical trials.* Inclusion for SCZ, BPD, ASD, and MDD is currently based on phenotypic rather than currently unknown mechanistic considerations. There is no assurance that the clinical cases included in trials are suffering from results of the same disease mechanism.
- *Inclusion criteria for GWAS studies.* The concerns for phenotype-based inclusion criteria for drug trials also apply to large-scale GWAS studies. For genomic population studies, there is a trade-off between precision or uniform diagnosis as opposed to maximization of sample size. Phenotypic imprecision has been argued to lower the power of studies to a greater degree than the advantage of performing larger studies (Consortium Cross-Disorder Phenotype Group of the Psychiatric GWAS 2009; Manchia et al. 2013; Liang and Greenwood 2015; Webber 2017). A notable exception where BPD GWAS inclusion was based on lithium response (Song et al. 2016) (Section 3.5.3.9).
- *Small effect size of genomics identified risk alleles.* One concern for drug discovery based on GWAS information is the small effect size of specific allele contributions to polygenic disease. This concern is mitigated by examples of non-neuropsychiatric pharmaceuticals. Sullivan et al. (2018) argue that a small contribution to disease of a specific risk allele does not rule out utility as a therapeutic target. They refer to *HMGCR*, which encodes the rate-limiting step for cholesterol biosynthesis (Teslovich et al. 2010) and is the target of the (LDL cholesterol) lowering statin pharmaceuticals. Coronary artery disease-based GWAS found specific SNPs to be associated with disease. These risk alleles are associated with only small changes in LDL cholesterol levels. Whereas the LDL genetic influence is small in degree, statin treatment results in large and therapeutically effective influence (Liang and Greenwood 2015). This example provides encouragement for utility of GWAS application for novel neuropsychiatric therapeutic targets.

- *The human condition; does a portion of the genetics of mental illness overlap with a genetics of creativity and extreme abilities?* It is beyond the scope of this document to delve deeply into the relationship between mental illness, a detriment to evolutionary fitness and creativity (Keller and Visscher 2015; Power et al. 2015) or other extremes of abilities (Szöke et al. 2019) that are favorable to fitness. Considering the evolutionary preservation of the SNPs that are associated with disease risk, it would not be surprising that certain risk alleles could, within certain contexts, provide fitness benefit. As more insight is gained, this consideration could influence decisions for genetics-based pharmaceutical targeting.

Toward a genetics-informed mechanism-based disease understanding and consequent diagnosis. Current efforts to leverage genomics to understand the biological basis of neuropsychiatric disease are discussed below. First, we examine propitious aspects of a currently investigated target of drug discovery that was largely identified by genomic studies. Then, we discuss a range of efforts toward genomics-enabled drug discovery.

3.5.2 A Genomics-Identified Target for Drug Discovery, the Complement Encoding C4 Gene Structural Variants, and a Confluence of Observations

Encouragement for genetic supply of targets for therapeutics derives from the complement C4 gene example (Breen et al. 2016). The PCG 2014 study described above (Schizophrenia Working Group of the Psychiatric Genomics Consortium 2014) identified the major histocompatibility complex (MHC) as containing the highest frequency of loci associated with SCZ. This locus includes the gene-encoding complement C4 that is currently a target for therapeutic development. Importantly, the attractiveness as a therapeutic target derives from a confluence of observations (Sekar

et al. 2016). These include the following: (1) common, structurally distinct C4 alleles that affect expression of C4A and C4B in the brain; (2) each allele is associated with schizophrenia risk in approximate proportion to its effect on C4A expression in postmortem brain samples; (3) C4 is expressed by neurons, localized to dendrites, axons, and synapses, and secreted; and (4) C4 promotes synapse elimination during the developmentally timed maturation of a neuronal circuit. Interestingly, of the 108 risk alleles identified by the PGC 2014 study (Schizophrenia Working Group of the Psychiatric Genomics Consortium 2014), one association, *SERPING1* (complement component 1 inhibitor), is a regulator of C4 expression. Also of interest, the MHC locus does not display significant SCZ risk loci in Asian studies. This results from different allele representation in European and Asian populations (Lam et al. 2019). This is also a reminder that even the most frequent European C4 risk alleles account for only a small fraction of risk in Europe, even less in Asia, and most affected individuals in either population do not display the C4-associated risk alleles. Nevertheless, complement C4 variants provide an attractive target for drug discovery stemming from a wide range of studies and observations that build upon genomics findings.

3.5.3 Current Efforts Toward Revealing the Fundamental Biology, Informing Clinical Practice, and Delivering New Therapeutic Targets

The following is a discussion of considerations and efforts to move the significant findings of neuropsychiatric genetics toward enhanced mechanistic understandings and therapeutic opportunities.

3.5.3.1 Research Domain Criteria (RDoC)

The concepts for RDoC were established in 2009 as an effort to change current psychiatric diagnosis to a biologically based system (Insel et al. 2010; Cuthbert and Insel 2013;

See ▶ Chap. 14). RDoC classification rests on the following three assumptions briefly paraphrased from Insel et al. (2010): (1) Mental disorders can be addressed as disorders of brain circuits. (2) Disease dysfunction in neural circuits can be identified with the tools of clinical neuroscience. (3) The data from genetics and clinical neuroscience will yield biosignatures that will augment clinical symptoms and signs for clinical management. In the context of this writing, the RDoC anticipated "biosignatures" are steps toward improved mechanistic understanding and importantly diagnosis for treatment decisions as well as inclusion criteria for genetics studies, clinical trials, and steps toward target selection for drug discovery.

Definition

Polygenic risk score (PRS) is a summary of genome-wide genetic data that gives an indication of genetic liability for a trait. This provides an approximate measure of an individual's common variant genetic propensity for a given disorder and, at a population level, can show a degree of predictive power for case-control status (Wray et al. 2007).

3.5.3.2 Polygenic Risk Scores

A polygenic risk score (PRS) is a summary of genome-wide genetic data that gives an indication of genetic liability for a trait. Predictive power is based on the aggregate of a large number of alleles, each of which by itself can have very small disease influence. PRS can be predictive for individuals showing no other indication of predisposition. For the large PGC 2014 SCZ study (Schizophrenia Working Group of the Psychiatric Genomics Consortium 2014), PRS explains 18% of variance between case and controls. PRS also discriminates between SCZ and other psychiatric diseases (Vassos et al. 2017). Although valuable for identifying a portion of the genetic contribution to neuropsychiatric disease in an individual, PRS does not provide neuropsychiatric diagnosis. It is notable that PRS scores are less powered when applied across European and Asian populations (Lam et al.

2019) which contain different frequencies of common polymorphisms including SCZ risk-associated polymorphisms.

3.5.3.3 Thoughts on Polygenicity

To recap some basic observations, polygenic risks are the aggregate of generally common polymorphisms in unfortunate combinations and unfortunate genetic or environmental contexts. Recalling that monozygotic twins can include affected and non-affected siblings, how should we consider genetics-based treatment of the affected but not the non-affected with the same genotype? Another point to consider is the possibility of context dependent benefits from certain risk alleles as discussed in Section 3.3.1. In light of these observations, it is reasonable to posit that, for many cases, rather than the SNP-identified risk genes themselves being targets for drug discovery, these multiple small contributions can point to pathways or circuits for which pathway limiting targets can be identified.

3.5.3.4 Endophenotypes as Components of Disease Syndromes That Can Display Effects in Non-disease Individuals

Polygenic effects can be considered as additive toward a liability threshold model where sufficient risk influence results in diagnosed disease (Gottesman and Gould 2003). Efforts toward determining sub-threshold phenotypes (endophenotypes) are hoped to get closer to understanding single gene influences revealing the elements of disease syndromes. Examples of endophenotypes under study for SCZ include pre-pulse inhibition, mismatch negativity, oculomotor anti-saccade, letter-number sequencing, and continuous performance tests (Greenwood et al. 2019).

Definition

Epistasis is a circumstance where the expression of one gene is affected by the expression of one or more independently inherited genes (*National Human Genomics Research institute Talking Glossary* n.d.).

3.5.3.5 Epistatic Interactions Among Risk Alleles

Here we use the term epistasis to indicate deviation from additivity of influence from genetic variants. The additive nature of polygenic risk scores hides qualitative interactions among risk alleles. These interactions might be synergistic or antagonistic between identified risk alleles and also between risk alleles and statistically not identified genetic variants. A query of the 108 risk loci identified in the PGC 2014 study (Schizophrenia Working Group of the Psychiatric Genomics Consortium 2014) found no epistatic, i.e., non-additive, interactions. This of course does not rule out epistatic interactions between risk loci and "hidden" risk loci that do not meet a statistical cutoff or that by themselves have no phenotypic contribution. Webber (2017) discusses the difficulties and potential benefits of identification of epistatic interactions for neuropsychiatric genomics. In general, direct detection of epistatic interactions is beyond the power of current GWAS studies. Hypothesis-based animal model studies have been successful. A successful discovery of synergistic influences of risk alleles using human-induced pluripotent stem cells is discussed (Section 3.5.3.7) (Schrode et al. 2019). In contrast to PRS scores based on additive influences, discovery of non-additive or synergistic influences points to functional relationships between genes, a step toward understanding disease mechanism.

> **Definition**
>
> *Induced pluripotent stem cells (iPSC) and* Induced pluripotent stem cells are adult cells that have been genetically reprogrammed to an embryonic stem cell–like state by being forced to express genes and factors important for maintaining the defining properties of embryonic stem cells (*Stem Cell Information Home Page. In Stem Cell Information [World Wide Web site]* 2016).

3.5.3.6 Human-Induced Pluripotent Stem Cells (hiPSCs) for Modeling Neuropsychiatric Disease

Several studies have used human-induced pluripotent stem cells derived from the fibroblasts of SCZ cases and controls. The fibroblasts are reprogramed into neurons or neuron progenitor cells (NPCs) and queried for phenotype such as transcription and protein profiles (Brennand et al. 2011; Hoffman et al. 2017). Findings for these disorder-specific hiPSCs include SCZ-associated alterations in cAMP and WNT pathway gene expression (Brennand et al. 2011) and miRNA dysregulation (Topol et al. 2016; Schrode et al. 2019).

3.5.3.7 Application of CRISPR Cas9-Directed Mutagenesis to Disorder-Specific hiPSCs

CRISPR Cas9-directed mutations in disorder-specific and control hiPSCs-derived neurons resulted in discovery of SCZ-associated, developmentally regulated chromosomal conformation alterations (Rajarajan et al. 2018). Importantly, a study using CRISPR to alter hiPSC-derived neurons resulted in the discovery of synergistic relationships among SCZ-associated alleles (Schrode et al. 2019).

3.5.3.8 Human-Induced Pluripotent Stem Cells (hiPSC) from Affected Individuals with Highly Penetrant Genetic Influences (◘ Fig. 3.4)

In contrast to the common polygenic contributions to disease, certain neuropsychiatric syndromes can be attributed to the pleiotropic influence of a single well-defined and highly penetrant genetic variation. Phelan-McDermid syndrome (PMDS), a neurodevelopmental disorder, is caused by heterozygous deletions of chromosome 22q13.3. The genes included in the deletion include *SHANK3* encoding a protein in the post-synaptic density (PSD). Among the phenotypes of this disorder are intellectual disability and increased risk of autism spectrum disorder. Fibroblasts

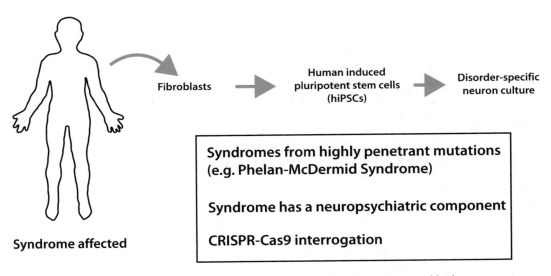

Fibroblasts

Human induced pluripotent stem cells (hiPSCs)

Disorder-specific neuron culture

Syndromes from highly penetrant mutations (e.g. Phelan-McDermid Syndrome)

Syndrome has a neuropsychiatric component

CRISPR-Cas9 interrogation

Syndrome affected

■ **Fig. 3.4** Induced pluripotent stem cells from complex syndromes that have a neuropsychiatric component

from affected individuals provide a source for induced pluripotent stem cells which are induced to form neurons displaying the genetic influence of the PMDS deletion. These disease-specific neurons displayed decreased Shank3 expression and specific synaptic transmission deficits in excitatory neurons. Furthermore, the deficits could be reversed by increasing Shank3 expression (Shcheglovitov et al. 2013). Other hiPSC studies with highly penetrant genetic alterations include Timothy syndrome (Paşca et al. 2011), a frameshift in the *DISC1* genetic region (Wen et al. 2014) and a 22q11.2 deletion (Lin et al. 2016).

3.5.3.9 Drug Response Has the Potential to Provide Mechanism-Based Diagnostic Distinctions

Failure to respond to pharmaceuticals derives from general or disease-specific sources. General failures can stem from intolerable side effects and consequent non-compliance or individual differences in drug absorption, distribution, metabolism, or excretion (ADME). Readily available tests can identify many of these general reasons, for example polymorphisms in the liver enzymes that metabolize drugs. It can be difficult to differentiate between general and disease mechanism-based drug failures, but this distinction is impor-

tant in that drug response can provide potential probes for disease mechanism. Using a broadly illustrative example, consider SCZ and BPD. Despite the degree of genetic risk similarity between SCZ and BPD (■ Table 3.5), response to lithium, for most cases, differentiates the two syndromes. Importantly, the different responses to lithium would appear to be based on underlying disease mechanisms. This is despite our current ignorance of what the specific mechanisms are.

Inclusion Criteria for a Genetic Study Based on Lithium Response

The utility of using drug response for diagnosis is illustrated by the work of Song et al. (2016). In this study, lithium responsiveness was used as a criterion for case inclusion for a BPD GWAS. For the lithium-responsive group, the risk allele within an intron of the *SESTD1* gene was identified. Importantly, inclusion criteria based on drug response were shown to increase the GWAS power leading to a genetic discovery. In addition, differentiation between lithium-responsive and non-responsive BPD derives from electrophysiological studies of BPD patient-derived iPSCs (Stern et al. 2018; Walss-Bass and Fries 2019). These studies indicate differences in neuronal excitability and spiking properties between Li responders and non-responders.

Amphetamine Response and SCZ/ADHD Risk Genetics

The work of Hart et al. (2014) provides an example of drug response association with disease predisposition. Alleles were identified showing association with sensitivity to the euphoric response to d-amphetamine. A subset of these alleles were additionally associated with low risk for SCZ or ADHD. The d-amphetamine response involves dopamine pathways consistent with models of SCZ and ADHD. Put another way, failure of specific response to amphetamine is associated with increased risk of disorder. Consistent with SCZ and ADHD disease models, this drug response could provide an early example of a probe for mechanism-based diagnosis.

Drugs as Probes for Mechanism and Mechanism-Based Diagnosis

The Song et al. and Hart et al. studies could foreshadow expanded use of enhanced diagnosis based on drug response. The degree that this can be extended to other psychiatric disorders remains to be explored. The vast bodies of data on responsiveness to currently used neuropsychiatric drugs may provide a potential source for developing mechanism-based diagnostic categories based on differential drug response.

3.6   On Order from Complexity

Just as the periodic table of the elements unified the disparate observations of the chemical world, might a unifying organization from neuropsychiatric genetics provide a coherent, simplified model or set of models for neuropsychiatric disease? Examples of organization of neuropsychiatric GWAS findings into global models are presented by Craddock and Owen (2010) and Geschwind and Flint (2015). This is the continuing challenge for this generation of neuropsychiatric geneticists.

> A range of efforts are being employed toward the discovery of novel pharmaceuticals for neuropsychiatric diseases with complex heredity. Genetic findings can lead to hypotheses and further discovery pointing to attractive discovery targets, for example the elements of the C4 complement system. Several efforts described above address the current complexity and our incomplete knowledge toward finding new avenues for novel effective neuropsychiatric pharmaceuticals.

> ## Conclusion

Discovery of the major classes of neuropsychological pioneer drugs was based on phenotype (◘ Table 3.1), and their therapeutic mechanisms are still only partially understood.

Neuropsychiatric genomics continues to expand the power of its studies and reveals the genetic underpinning of typical psychiatric disorders largely, but not entirely, as small contributions from a complexity of risk loci influencing a multiplicity of biological pathways. Mutations in protein-coding regions and genetic structural mutations also contribute to disease risk and point to possible disease mechanisms.

A striking revelation is the degree of shared genetic risk among different diagnostic categories that were previously considered to be distinct.

As the genomic studies continue, it is likely that specific diseases, now under a single diagnosis, will be distinguished by different mechanisms, leading to specific targets for interventions and more productive clinical trials using mechanism-based inclusion criteria. As new information informs better understanding, the complexity of mental illness will yield to simplifying and enabling concepts.

References

Ban TA (2007) Fifty years chlorpromazine: a historical perspective. Neuropsychiatr Dis Treat 3(4):495–500

Barondes S (1993) Molecules and mental illness. Scientific American Library HPHLP, New York

Breen G et al (2016) Translating genome-wide association findings into new therapeutics for psychiatry. Nat Neurosci 19(11):1392–1396. https://doi.org/10.1038/nn.4411

Brennand K et al (2011) Modeling schizophrenia using hiPSC neurons. Nature 473(7346):221–225. https://doi.org/10.1038/nature09915

Brunton L, Hilal-Dandan R, Knollmann B (2018) Goodman & Gilman's: the pharmacological basis of therapeutics, 13e, 13th edn. McGraw Hill, New York

Cade J (1949) Lithium salts in the treatment of psychotic excitement. Med J Aust 2(10):349–351

Charlesworth B (2012) The role of background selection in shaping patterns of molecular evolution and variation: evidence from variability on the Drosophila X chromosome. Genetics 191(1):233–246. https://doi. org/10.1534/genetics.111.138073

Consortium Cross-Disorder Phenotype Group of the Psychiatric GWAS (2009) Dissecting the phenotype in genome-wide association studies of psychiatric illness. Br J Psychiatry 195(2):97–99

Corvin A, Craddock N, Sullivan P (2010) Genome-wide association studies: a primer. Psychol Med 40(7):1063–1077. https://doi.org/10.1017/S0033291709991723

Craddock N, Owen MJ (2010) The Kraepelinian dichotomy – going, going... But still not gone. Br J Psychiatry 196(2):92–95. https://doi.org/10.1192/bjp.bp.109.073429

Cross-Disorder Group of the Psychiatric Genomics Consortium (2013) Identification of risk loci with shared effects on five major psychiatric disorders: a genome-wide analysis. Lancet 381(9875):1371–1379. https://doi.org/10.1016/S0140-6736(12)62129-1. Identification

Cuthbert BN, Insel TR (2013) Toward the future of psychiatric diagnosis: the seven pillars of RDoC. BMC Med 11:126. https://doi.org/10.1186/1741-7015-11-126

Gejman PV, Sanders AR, Duan J (2010) The role of genetics in the etiology of schizophrenia. Psychiatr Clin N Am 33(1):35–66. https://doi.org/10.1016/j.psc.2009.12.003

Geschwind DH, Flint J (2015) Genetics and genomics of psychiatric disease. Science 349(6255):1489–1494. https://doi.org/10.1126/science.aaa8954.Genetics

Goghari VM, Sponheim SR (2008) Differential association of the COMT Val158Met polymorphism with clinical phenotypes in schizophrenia and bipolar disorder. Schizophr Res 103(1–3):186–191. https://doi.org/10.1016/j.schres.2008.05.015

Gottesman IGT, Gould TD (2003) The Endophenotype concept in psychiatry: etymology and strategic intentions. Am J Psychiatry 160(4):636–645

Greenwood TA, Shutes-David A, Tsuang D (2019) Endophenotypes in schizophrenia: digging deeper to identify genetic mechanisms. J Psychiatry Brain Sci 4(2):1–26. https://doi.org/10.20900/jpbs.20190005

Halvorsen M et al (2020) Increased burden of ultra-rare structural variants localizing to boundaries of topologically associated domains in schizophrenia. Nature Commun. Springer US 11(1):1842. https://doi.org/10.1038/s41467-020-15707-w

Hart AB et al (2014) Genetic variation associated with euphorigenic effects of d-amphetamine is associated with diminished risk for schizophrenia and attention deficit hyperactivity disorder. Proc Natl Acad Sci U S A 111(16):5968–5973. https://doi.org/10.1073/pnas.1318810111

Hoffman GE et al (2017) Transcriptional signatures of schizophrenia in hiPSC-derived NPCs and neurons are concordant with post-mortem adult brains. Nat Commun. Springer US 8(1):2225. https://doi.org/10.1038/s41467-017-02330-5

Ingraham LJ, Kety SS (2000) Adoption studies of schizophrenia. Am J Med Genet 97(1):18–22

Insel T et al (2010) Research domain criteria (RDoC): toward a new classification framework for research on mental disorders. Am J Psychiatr 7(July):748–751. https://doi.org/10.1176/appi.ajp.2010.09091379

Keller MC, Visscher PM (2015) Genetic variation links creativity to psychiatric disorders. Nat Neurosci 18(7):928–929. https://doi.org/10.1038/nn.4047

Kety SS et al (1971) Mental illness in the biological and adoptive families of adopted schizophrenics. Am J Psychiatry 128(3):302–306

Khandaker GM et al (2015) Inflammation and immunity in schizophrenia: implications for pathophysiology and treatment. Lancet Psychiatry 2(3):258–270. https://doi.org/10.1016/S2215-0366(14)00122-9

Lam M et al (2019) Comparative genetic architectures of schizophrenia in East Asian and European populations. Nat Genet 51:1670–1678. https://doi.org/10.1016/j.euroneuro.2018.08.019

Li J, Cai T, Jiang Y, Chen H, He X, Chen C, Li X, Shao Q, Ran X, Li Z, Xia K, Liu C, Sun ZS, Jinyu Wu J (2016) Genes with de novo mutations are shared by four neuropsychiatric disorders discovered from NPdenovo database. Mol Psychiatry 21(2):290–297

Liang SG, Greenwood TA (2015) The impact of clinical heterogeneity in schizophrenia on genomic analyses. Schizophr Res 161:490–495

Lilienfeld SO, Treadway MT (2016) Clashing diagnostic approaches: DSM-ICD versus RDoC. Annual review of clinical psychology. Annu Rev Clin Psychol 12:435–463. https://doi.org/10.1146/annurev-clinpsy-021815-093122.Clashing

Lin M et al (2016) Integrative transcriptome network analysis of iPSC-derived neurons from schizophrenia and schizoaffective disorder patients with 22q11.2 deletion. BMC Syst Biol 10(1):1–20. https://doi.org/10.1186/s12918-016-0366-0

Maier R et al (2015) Joint analysis of psychiatric disorders increases accuracy of risk prediction for schizophrenia, bipolar disorder, and major depressive disorder. Am J Hum Genet 96(2):283–294. https://doi.org/10.1016/j.ajhg.2014.12.006

Malhotra D, Sebat J (2012) CNVs: harbingers of a rare variant revolution in psychiatric genetics. Cell (Elsevier Inc.) 148(6):1223–1241. https://doi.org/10.1016/j.cell.2012.02.039

Manchia M et al (2013) The impact of phenotypic and genetic heterogeneity on results of genome wide association studies of complex diseases. PLoS One 8(10):1–7. https://doi.org/10.1371/journal.pone.0076295

National Human Genomics Research institute Talking Glossary (n.d.). Available at: https://www.genome.gov/genetics-glossary. Accessed: 20 April 2020

Ng SB et al (2009) Targeted capture and massively parallel sequencing of 12 human exomes. Nature

461(7261):272–276. https://doi.org/10.1038/nature 08250

Nica AC, Dermitzakis ET (2013) Expression quantitative trait loci: resent and future. Philos Trans R Soc Lond B Biol Sci 368(1620):20120362. https://doi.org/10.1098/rstb.2012.0362

O'Bleness M, Searles VB, Varki A, Gagneux P, Sikela JM (2012) Evolution of genetic and genomic features unique to the human lineage. Nat Rev Genet 13(12):853–866. https://doi.org/10.1038/nrg3336. Evolution

Pardiñas AF et al (2018) Common schizophrenia alleles are enriched in mutation-intolerant genes and in regions under strong background selection. Nat Genet 50(3):381–389. https://doi.org/10.1038/s41588-018-0059-2

Paşca SP et al (2011) Using iPSC-derived neurons to uncover cellular phenotypes associated with Timothy syndrome. Nat Med 17(12):1657–1662. https://doi.org/10.1038/nm.2576

Patel KR et al (2014) Schizophrenia: overview and treatment options. P T 39(9):638–645

Power RA, Steinberg S, Bjornsdottir G, Rietveld CA, Abdellaoui A, Nivard MM, Johannesson M, Galesloot TE, Hottenga JJ, Willemsen G, Cesarini D, Benjamin DJ, Magnusson PK, Ullén F, Tiemeier H, Hofman A, van Rooij FJ, Walters G, Sigurdsson E, Tho K (2015) Polygenic risk scores for schizophrenia and bipolar disorder predict creativity. Nat Neurosci 18(7):953–955

Rajarajan P et al (2018) Neuron-specific signatures in the chromosomal connectome associated with schizophrenia risk. Science 362(6420):eaat4311. https://doi.org/10.1126/science.aat4311

Schizophrenia Working Group of the Psychiatric Genomics Consortium (2014) Biological insights from 108 schizophrenia-associated genetic loci. Nature 511(7510):421–427. https://doi.org/10.1038/nature13595

Schrode N et al (2019) Synergistic effects of common schizophrenia risk variants. Nat Genet 51(10):1475–1485. https://doi.org/10.1038/s41588-019-0497-5

Sekar A et al (2016) Schizophrenia risk from complex variation of complement component 4. Nature 530(7589):177–183. https://doi.org/10.1038/nature16549

Shcheglovitov A et al (2013) SHANK3 and IGF1 restore synaptic deficits in neurons from 22q13 deletion syndrome patients. Nature. Nature Publishing Group, a division of Macmillan Publishers Limited. All Rights Reserved 503(7475):267–271. Available at: https://doi.org/10.1038/nature12618

Shorter E (2009) The history of lithium therapy. Bipolar Disord 11(Suppl 2):4–9

Song J et al (2016) Genome-wide association study identifies SESTD1 as a novel risk gene for lithium-responsive bipolar disorder. Mol Psychiatry 21(9):1290–1297. https://doi.org/10.1038/mp.2015.165

Stem Cell Information Home Page. In Stem Cell Information [World Wide Web site] (2016).

Available at: https://stemcells.nih.gov/info/basics/6.htm. Accessed: 20 April 2020

Stern S et al (2018) Neurons derived from patients with bipolar disorder divide into intrinsically different sub-populations of neurons, predicting the patients' responsiveness to lithium. Mol Psychiatry (Nature Publishing Group) 23(6):1453–1465. https://doi.org/10.1038/mp.2016.260

Sullivan PF, Kendler KS, Neale MC (2003) Schizophrenia as a complex trait: evidence from a meta-analysis of twin studies. Arch Gen Psychiatry 60(12):1187–1192. https://doi.org/10.1001/archpsyc.60.12.1187

Sullivan PF et al (2018) Psychiatric genomics: an update and an agenda. Am J Psychiatr 175(1):15–27. https://doi.org/10.1176/appi.ajp.2017.17030283

Szabo Q, Bantignies F, Cavalli G (2019) Principles of genome folding into topologically associating domains. Sci Adv 5(4):eaaw1668. https://doi.org/10.1126/sciadv.aaw1668

Szöke A, Pignon B, Schürhoff F (2019) Schizophrenia risk factors in exceptional achievers: a re-analysis of a 60-year-old database. Sci Rep 9(1):1294. https://doi.org/10.1038/s41598-018-37484-9

Teslovich TM et al (2010) Biological, clinical and population relevance of 95 loci for blood lipids. Nature 466(7307):707–713. https://doi.org/10.1038/nature09270

Topol A et al (2016) Dysregulation of miRNA-9 in a subset of schizophrenia patient-derived neural progenitor cells. Cell Rep (ElsevierCompany) 15(5):1024–1036. https://doi.org/10.1016/j.celrep.2016.03.090

Vassos E et al (2017) An examination of polygenic score risk prediction in individuals with first-episode psychosis. Biol Psychiatry (Elsevier) 81(6):470–477. https://doi.org/10.1016/j.biopsych.2016.06.028

Voisey J et al (2011) A novel SNP in COMT is associated with alcohol dependence but not opiate or nicotine dependence: a case control study. Behav Brain Funct 7(1):51. https://doi.org/10.1186/1744-9081-7-51

Walss-Bass C, Fries GR (2019) Are lithium effects dependent on genetic/epigenetic architecture? Neuropsychopharmacology (Nature Publishing Group) 44(1):228. https://doi.org/10.1038/s41386-018-0194-6

Webber C (2017) Epistasis in Neuropsychiatric Disorders. Trends Genet (Elsevier Ltd) 33(4):256–265. https://doi.org/10.1016/j.tig.2017.01.009

Wen Z et al (2014) Synaptic dysregulation in a human iPS cell model of mental disorders. Nature 515(7527):414–418. https://doi.org/10.1038/nature13716

Wray NR, Goddard ME, Visscher PM (2007) Prediction of individual genetic risk to disease from genome-wide association studies. Genome Res 17(10):1520–1528. https://doi.org/10.1101/gr.6665407

Zubieta J-K et al (2003) COMT genotype affects μ-opioid neurotransmitter responses to a pain stressor. Science 299(5610):1240–1243. https://doi.org/10.1126/science.1078546

Epigenetics in Drug Discovery: Achievements and Challenges

Ziva Korda, Ehsan Pishva, and Daniel L. A. van den Hove

Contents

4.1   Introduction – 58

4.2   Epigenetic Mechanisms in AD – 60
4.2.1   Histone Modifications – 60
4.2.2   DNA Methylation – 60
4.2.3   Non-coding RNAs – 62

4.3   Epigenetic-Based Drug Discovery for AD – 63
4.3.1   HDAC Inhibitors as a Potential Treatment for AD – 64
4.3.2   DNMT Inhibitors as a Potential Treatment for AD – 67
4.3.3   Challenges – 68

        References – 70

The epigenome comprises a range of covalent modifications of DNA and histone proteins that establish chromatin structure and function. These processes regulate gene expression without altering the DNA sequence and play an important role in regulating, e.g. brain function and associated behaviours. Epigenetic changes are acquired throughout life in response to the environment exposed to and thus present a link between one's environment and genetic landscape. Various epigenetic processes are thought to play a role in several psychiatric and neurodegenerative disorders including Alzheimer's disease (AD). As such, the epigenome and its disease-associated signatures present a compelling drug target for the treatment of these disorders. Although studies of mental disease-related epigenetic changes are still in their infancy and face many challenges, neuroepigenetic research has contributed significantly to the AD field over the last decade. Research on epigenome-wide changes bears great potential in nominating epigenetics-related drug targets and inclusion of epigenetic modalities in, e.g. multiomics approaches may provide more robust findings, which could be ultimately translated into potential drug development applications.

Learning Objectives
- The basic molecular and cellular principles of epigenetics
- Epigenetic dysregulation in Alzheimer's disease
- The value of understanding disrupted DNA and histone modifications in Alzheimer's disease for drug discovery
- Challenges in epigenetic-based drug discovery

4.1 Introduction

Definition
Epigenetics is a field of science that investigates regulatory mechanisms of gene expression, which are accomplished without altering the DNA sequence.

Epigenetics is a field of science that investigates regulatory mechanisms of gene expression. Epigenetic changes are accomplished without altering the DNA sequence and are therefore reversible, as opposed to genetic alterations, which usually cause permanent changes to the nucleotide sequence. Epigenetic changes can be inherited through mitosis, while some of them have even been found to exert transgenerational effects (Lardenoije et al. 2015). Epigenetic mechanisms comprise DNA modifications such as DNA methylation, histone modifications as well as non-coding RNAs (Castanho and Lunnon 2019). Generally, epigenetic mechanisms control gene expression by altering the accessibility of the DNA at the respective genomic locus and hence repressing or promoting its transcription. Controlling gene expression is crucial for our development and survival, considering that all of the cells in our body contain the same genotype even though not every gene product is needed in every cell at all times. Synaptic plasticity and long-lasting forms of memory are also epigenetically regulated as they require stable coordinated gene expression changes (Rudenko and Tsai 2014). This regulation can act in a more programmed manner, as needed during cellular differentiation, or it can be dynamic, e.g. when involved in learning and memory processes (Lardenoije et al. 2015). As such, epigenetic changes are acquired throughout life in response to the environment exposed to. Thus, epigenetic mechanisms present a way for experiences to modify, e.g. cellular and behavioural responses.

> Epigenetic mechanisms produce reversible changes (i.e. DNA modifications, histone modifications and RNA-based interactions) to the DNA that are acquired throughout life in response to the environment and are crucial for our development and survival.

Definition

Alzheimer's disease (AD) is a progressive neurodegenerative disorder causing severe disability and decreased life expectancy. Clinical symptoms include decline in memory and other cognitive impairments, as well as various behavioural changes. Histopathological hallmarks include extracellular amyloid beta plaques and intracellular neurofibrillary tangles of hyperphosphorylated tau, which are thought to lead to cortical and subcortical atrophy.

In the last decade, epigenetic alterations emerged as one of the key players in the pathophysiology of Alzheimer's disease (AD). As such, in this chapter, AD will be used as a showcase on how addressing epigenetics can aid in drug discovery. AD, the most common type of dementia, is a progressive neurodegenerative disorder causing severe disability and decreased life expectancy (World Health Organisation 2017). Clinical symptoms include cognitive impairment, involving, e.g. memory decline, and various behavioural changes. Histopathological hallmarks of AD include extracellular amyloid beta (Aβ) plaques and intracellular neurofibrillary tangles of hyperphosphorylated tau, which are thought to lead to cortical and subcortical atrophy (Iatrou et al. 2017). The spatiotemporal pattern of brain regions being affected reflects the gradual progression of the disease, with abnormalities first present in the brainstem, spreading to the entorhinal cortex and hippocampus and finally affecting the neocortex associated with the manifestation of clinical symptoms (Roubroeks et al. 2017).

Genetic research on AD has substantially contributed to our understanding of the pathogenic mechanisms underlying the disorder. Familial AD (<65 years) is mainly driven by mutations in the amyloid precursor protein (APP), presenilin 1 and 2 (PSEN1, PSEN2) genes, whereas the apolipoprotein E (APOE) gene represents the most important risk factor for developing the more prevalent sporadic forms of AD (Bertram et al. 2010). Genome-wide association studies (GWAS) have associated several risk loci with AD. However, discovered variants display small risk effects and explain only about ~30% of the underlying phenotypic variance. Moreover, many of these risk variants are located in non-coding regions making it more difficult to understand their function (Narayan and Dragunow 2017). A multitude of lifestyle (e.g. lack of physical and social activity, poor nutrition, diabetes, cardiovascular diseases, abnormal sleep patterns, lack of cognitive stimulation, smoking) and environmental (e.g. chemical and metal exposure, pollution) risk factors together with the aforementioned studies suggest the involvement of epigenetic mechanisms in the development and course of the disease (Bartolotti and Lazarov 2016).

> A small proportion of phenotypic variance explained by genetic variants, together with lifestyle and environmental risk factors, suggests involvement of epigenetic mechanisms in development and progression of AD.

Although much is known about AD pathology, specific mechanisms driving the onset and progression of the disease remain elusive. Animal models that have largely contributed to our understanding of biological mechanisms involved in AD are mainly models of familial AD and thus neglect the characteristics of the sporadic forms (Zhang et al. 2020; Foidl and Humpel 2020). This notion may have substantially contributed to the fact that many potential drugs for the treatment of AD failed preclinical or clinical trials, and there are currently only a limited number of disease-relevant targets that could aid drug discovery. Considering that there is currently no pharmacological treatment that persistently improves or prevents AD symptoms, there is a great unmet medical need to identify novel targets for new therapeutic applications. As such, the epigenome and its disease-associated signatures present a compelling drug target for the treatment of AD.

4.2 Epigenetic Mechanisms in AD

4.2.1 Histone Modifications

> **Definition**
>
> Double-stranded DNA is wrapped around four pairs of histones forming a nucleosome. Nucleosomes are condensed into chromatin, which is further condensed into chromosomes. Genes can be transcribed only when the respective region of the DNA is accessible to the transcription machinery, i.e. in its less condensed state, referred to as euchromatin. Condensed DNA is referred to as heterochromatin.

Large strands of DNA are present within the nucleus of every cell, which, in comparison, is rather small. Since there is no space in the nucleus for the DNA to be freely floating, it is condensed in the form of tightly packed chromosomes. Chromosomes consist of condensed chromatin, i.e. heterochromatin, which in turn consists of nucleosomes. Nucleosomes consist of four pairs of core histone proteins (H2A, H2B, H3 and H4), double-stranded DNA wrapped around this complex and a linker histone H1, which is located on the outside part of the core complex and stabilises the wrapped DNA (Lawrence et al. 2016). In order for a gene to be transcribed, the respective region of the condense DNA needs to be opened up, i.e. a state referred to as euchromatin, to be accessible to the transcription machinery. DNA (de)condensation is regulated by various histone modifications at the level of their N-terminal tails, such as acetylation, methylation, phosphorylation, ubiquitination and sumoylation (Zhao and Garcia 2015). Each modification is accomplished, maintained and balanced by specific families of enzymes. For example, histone acetylation is carried out by histone acetyltransferases (HATs) inducing increased transcription, whereas, conversely, histone deacetylases (HDACs) remove the acetyl group from histone tails, thereby repressing gene expression (Castanho and Lunnon 2019).

> Histone modifications are accomplished, maintained and balanced by specific families of enzymes working in an opposite direction (e.g. histone acetyltransferases [HATs] and histone deacetylates [HDACs]).

Histone modifications have been implicated in learning, memory and associated neuronal and synaptic plasticity processes. However, there are only a limited number of studies investigating their role in AD. The best studied histone modification in relation to AD is histone acetylation. Histones H3 and H4 showed increased acetylation in pyramidal neurons and neocortex of AD patients, while hyperacetylation has been hypothesised to play a role in the production of Aβ or intervene with protein degradation (Gu et al. 2013; Narayan et al. 2015; Lithner et al. 2013). Similarly, histones H2 and H3 have been found to be hyperphosphorylated in the frontal cortex as well as in hippocampal astrocytes of AD patients, with histone phosphorylation thought to have the same effect on gene expression as acetylation (Rao et al. 2012; Myung et al. 2008). Furthermore, a recent large-scale EWAS found that tau pathology correlates with acetylation levels on histone H3 with a significantly larger effect in the open chromatin compartments (Klein et al. 2019).

4.2.2 DNA Methylation

> **Definition**
>
> DNA methylation (DNAm) takes place at CpG sites, i.e. where a cytosine nucleotide is followed by a guanine nucleotide in the linear sequence of bases. Demethylation can occur either actively or passively. Active demethylation happens through the process of oxidation, whereas passive demethylation takes place during cell replication due to reduced activity of DNMTs.

DNA methylation (DNAm) represents one of the most stable epigenetic modifications and owing to the relative ease of assessing it, also the most investigated one. DNAm takes place at CpG sites, i.e. at regions of the DNA where a cytosine nucleotide is followed by a guanine nucleotide in the linear sequence of bases. DNAm occurs when a methyl group is added to the fifth position of cytosine by DNA methyltransferases (DNMTs), forming 5-methylcytosine (5mC) (Hwang et al. 2017). Demethylation can occur passively or actively. Passive demethylation takes place during cell replication, and the concomitant duplication of the DNA, in the absence or with reduced activity of DNMTs, resulting in a loss of 5mC (Kohli and Zhang 2013). Active demethylation occurs through the process of oxidation by the action of ten-eleven translocation (TET) proteins, first resulting in 5-hydroxymethylcytosine (5hmC), which can be further oxidised into 5-formylcytosine (5fC) and then to 5-carbocylcytosine (5caC) (Roubroeks et al. 2017). During the last step of the demethylation cycle, 5caC is decarboxylated into unmodified cytosine. For a recent review, see Roubroeks et al. (2017).

Human studies on global DNAm in AD are largely inconclusive and point towards gene-specific regulation. Various studies have found differential DNAm in the hippocampus, temporal and frontal cortex of AD patients, although some of them found hypomethylation while others reported hypermethylation (Rao et al. 2012; Coppieters and Dragunow 2011; Mastroeni et al. 2010; Coppieters et al. 2014; Chouliaras et al. 2013; Bradley-Whitman and Lovell 2013). These contradictive findings can be at least partially explained by cell-type-specific DNAm signatures in normal aging and AD since some genes show specific differential DNAm in neurons or glia cells (Phipps et al. 2016; Gasparoni et al. 2018). Next to this, methodological differences between the various studies may represent an alternative explanation for the apparent discrepancy in findings.

Major technological advances over the last decade have prompted the methylomic profiling of brain tissue in AD. Methylome-wide association studies (MWAS) in recent years have nominated several differentially methylated regions (DMRs) and differentially methylated positions (DMPs) and associated them with genes dysregulated in AD. The first study using the Illumina Infinium HumanMethylation450k array, which is the most widely used method for MWAS to date, implicated 71 DMRs in the dorsolateral prefrontal cortex in AD and, consequently, highlighted seven dysregulated genes in the proximity of these DMRs (De Jager et al. 2014). A parallel study investigating brain region-specific DNAm in AD implicated hypermethylation of the Ankyrin 1 (*ANK1*) gene region in cortical regions but not in the cerebellum (Lunnon et al. 2014). Furthermore, it seems there is a loss of active DNA demethylation of *ANK1* in AD leading to *ANK1* hypermethylation, which is present already in the early stages of the disease (Smith et al. 2019a). Hypermethylation of *ANK1* was also found in Huntington's disease and Parkinson's disease, although differential methylation in these two neurodegenerative diseases was not observed in the regions struck by pathology (i.e. striatum and substantia nigra, respectively) (Smith et al. 2019b). *ANK1*, encoding for the ANK1 protein, which is known to be involved in binding of membrane proteins, is, to date, the most replicated finding of MWAS (De Jager et al. 2014; Lunnon et al. 2014; Smith et al. 2019a, b; Semick et al. 2019; Mastroeni et al. 2017). As such, it was the first AD-associated gene nominated from EWAS to be characterised in an animal model (Higham et al. 2019). In a Drosophila model of AD, Ankyrin 2 (*ANK2*) was chosen over *ANK1*, since human *ANK1* shows higher amino acid identity to the Drosophila's *ANK2* than to *ANK1*. Furthermore, *ANK2* is neuron-specific in Drosophila and when decreasing the expression of both genes, only the reduction of *ANK2* leads to an AD-like phenotype, making *ANK2* the closest functional ortholog of human *ANK1* (Higham et al. 2019). The reduction of *ANK2* in Drosophila led to a shortened lifespan, memory loss and changes in neuronal excitability. These AD-relevant phenotypes were also observed after overexpression of human mutant APP and the microtubule associated protein Tau, thereby further implicating the role of *ANK1* in AD pathology (Higham et al. 2019).

> Methylome-wide association studies (MWAS) investigate differentially methylated positions (DMPs) or differentially methylated regions (DMRs).

DNAm can have various effects on gene expression. Hypermethylation in promoter regions has classically been associated with transcriptional repression (although many exceptions to this 'rule' have been reported), while intragenic DNAm can modulate alternative splicing and DNAm in the gene body has been associated with increased gene expression (Roubroeks et al. 2017; Silva et al. 2008). DNAm changes the DNA structure and, accordingly, can, e.g. affect the likelihood of transcription factors binding locally (Yin et al. 2017). Interestingly, some transcription factors preferentially bind to methylated CpGs, i.e. referred to as methyl-binding domain (MBD) proteins, which are particularly known for playing an important role in development and cell self-renewal (Mercatelli et al. 2019).

4.2.3 Non-coding RNAs

┌─ Definition ──────────────────────
│ Non-coding RNAs are a special type of
│ RNAs that are not translated into pro-
│ teins. Their sequences are known to cover
│ a considerable part of our genome and
│ have been suggested to act as master regu-
│ lators of transcription.
└───────────────────────────────────

Less than 5% of our genome consists of protein-coding genes. A considerable proportion of non-coding regions relates to non-coding RNAs (ncRNAs), which have been suggested to act as master regulators of transcription. ncRNAs can be divided into short ncRNAs (sncRNAs; < 30 nucleotides) and long ncRNAs (lncRNAs; >200 nucleotides). The former can be further divided into microRNAs (miRNAs), short interfering RNAs (siRNAs) and piwi-interacting RNAs (piRNAs). Among all ncRNAs, miRNAs

have been studied most extensively (Castanho and Lunnon 2019). Some ncRNAs are tissue-specific (e.g. marked by a relative abundance of miRNAs and lncRNAs in the brain) or species-specific (especially lncRNAs are poorly conserved) (Millan 2017). NcRNAs can regulate gene expression at both the transcriptional and posttranscriptional levels, although the exact direction of an effect is in most cases still largely unclear. For example, miRNAs are generally thought to repress translation of proteins, through partial complementarity to one or more messenger RNA (mRNA) molecules. However, if one miRNA represses the function of another miRNA (i.e. represses the repressor), then this could result in increased target protein expression. Furthermore, it is common that different miRNAs can act on the same target, while a single miRNA can have multiple targets. Interestingly, methylation and histone modification also play a role in regulating miRNAs and, vice versa, miRNAs can affect other parts of the epigenetic machinery (Van den Hove et al. 2014). For recent reviews, see Millan (2017); Van den Hove et al. (2014).

> Non-coding RNAs can regulate gene expression at the transcriptional and post-transcriptional levels by interfering with mRNA function.

As stated before, many AD-associated risk variants discovered in GWAS reside in non-coding regions, which points to the involvement of ncRNAs in the pathogenesis of AD. As such, to date, many studies have associated disrupted levels of lncRNAs and miRNAs with AD (Millan 2017). MiRNAs are largely expressed in neurons and were shown to play a role in the regulation of genes involved in Aβ function, tau metabolism, splicing and phosphorylation, as well as regulating lipid metabolism and neuroinflammation in relation to AD (Maoz et al. 2017). In the brain, lncRNAs are being found in astrocytes, oligodendrocytes, glia and neurons, where they regulate the cell cycle and synaptic plasticity, and often interact with miRNAs, indicative of a multi-level, reciprocal regula-

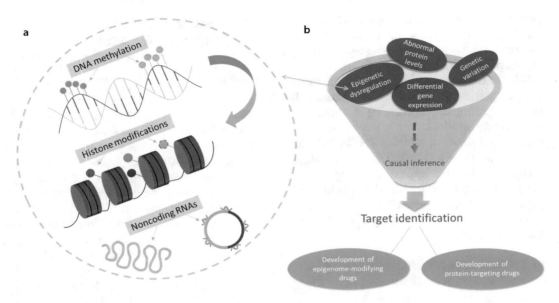

Fig. 4.1 Epigenetic-based drug discovery. **a** Epigenetic modifications. **b** Investigating epigenetic dysregulations in a disease parallel to or in combination with proteomic, genomic and transcriptomic data analyses (i.e. multi-omics approaches) can facilitate the understanding of the causal mechanisms and thus lead to the identification of more robust drug targets. Depending on the drug target identified, there is a potential to develop epigenome-modifying or protein-targeting drugs

tory circuit (Millan 2017). Although far less research has focused on the role of lncRNAs in AD, available evidence shows that several lncRNAs targeting AD-associated genes are upregulated in AD as compared to controls, amongst others effecting the regulation of Aβ (Castanho and Lunnon 2019) (■ Fig. 4.1).

4.3 Epigenetic-Based Drug Discovery for AD

How can drug development in diseases such as AD benefit from the emerging knowledge about implicated epigenetic dysregulation? First, although epigenetic mechanisms are influenced by environmental and lifestyle factors, science is still far away from pinpointing how exactly environmental insults contribute to the pathogenesis of AD. It could well be that we will never completely disentangle the complex interactions between genetic predisposition and both genotype-dependent (i.e. methylation-quantitative trait loci [mQTL]; sites at which DNAm is associated with

genetic variation) and -independent epigenetic modifications, let alone how the environment comes into play. Second, different genetic and environmental/lifestyle factors could induce a similar epigenotype, and these processes might be fine-tuned differently between individuals. Finally, epigenetic dysregulation associated with a disease may both represent a cause or consequence of its pathophysiology, or even its treatment. In fact, we are only at the very beginning of understanding the direct and indirect consequences of dysregulated epigenetic mechanisms and how they impact each other. Notwithstanding these challenges, research has provided some important insights that may translate into the identification of valid disease drug targets and the development of novel drug discovery approaches in AD.

> Different genetic and environmental/lifestyle factors could induce a similar epigenotype, and these processes might be fine-tuned differently between individuals and may represent a cause or a consequence of the disease.

Technically, histone and DNA modifications can be targeted directly in several ways. For this purpose, various enzymes responsible for transferring chemical groups to or from the DNA or histone proteins present potential drug targets as we can alter their activity. In fact, drugs targeting epigenetic modifications are already available for cancer treatment, including HDAC inhibitors (HDACis) and DNMT inhibitors (DNMTis). Interestingly, HDACis have also appeared as top hits in drug repositioning research of AD (a research paradigm discussed elsewhere in this book) (Siavelis et al. 2015; Chatterjee et al. 2018; Vargas et al. 2018).

> Enzymes involved in histone and DNA modifications present potential drug targets as we can alter their activity.

Potential treatment strategies involving histone and DNA modifications are discussed below. Virtually all enzymes involved in transferring chemical groups to or from histone proteins or the DNA present a potential drug target, and many of them have already been explored in preclinical studies and clinical trials. However, reviewing all of them is out of scope for this chapter, and rather an example of each class of epigenetic regulatory molecules will be presented. For a recent review on epigenetic drugs in clinical trials for AD, see Teijido and Cacabelos (2018). Of note, although ncRNAs also present a potential treatment target, this area of research is still in its infancy. Currently, while particularly miRNAs have been implicated as potential biomarkers for AD, owing to their AD-associated alterations that can be noninvasively measured in body fluids, e.g. in serum and plasma (Angelucci et al. 2019), due to the *in vivo* delivery difficulties and pleiotropic effects of miRNAs, as well as other ncRNAs, more research is needed in order to present them as a potential drug targets and hence, ncRNA-based therapies will not be discussed here. For recent reviews, see Gupta et al. (2017); Silvestro et al. (2019).

4.3.1 HDAC Inhibitors as a Potential Treatment for AD

Definition

HDACs are divided into class I–III. Classes I, II and III require Zn^{2+} as a cofactor and catalyse the removal of acetyl groups from histones. Class III HDACs represents nicotinamide adenine dinucleotide (NAD^+)-dependent enzymes, which are considered a separate type owing to their different mechanism of action.

Class I HDACs, mainly localised in the nucleus, consist of ubiquitously expressed HDAC 1–3 and muscle-specific HDAC 8.

Class II HDACs are divided into IIa and IIb HDACs. Class IIa HDACs 4, 5, 7 and 9 shuttles between the nucleus and the cytoplasm, while class IIb HDACs 6 and 10 localise primarily in the cytoplasm.

Class III HDACs, also known as sirtuins (SIRT1–7), are ubiquitously expressed in the nucleus, cytoplasm and mitochondria. Sirtuins are inhibited by nicotinamide, a precursor of NAD^+, but not by compounds inhibiting other classes of HDACs (e.g. Vorinostat).

Class IIII HDACs are restricted to HDAC11, which is mainly located in the nucleus and is structurally similar to class I HDACs.

There is a vast literature available discussing the potential of HDACis for the treatment of AD (Yang et al. 2017; Xu et al. 2011; Gräff et al. 2011). The field rapidly evolved in the early 2000s after several studies on various learning paradigms in rodents showed a transient increase in histone acetylation during memory formation and the concomitant increased expression of memory-related genes (Levenson et al. 2004; Chwang et al. 2007; Bredy et al. 2007; Lubin and Sweatt 2007; Koshibu et al. 2009; Lubin et al. 2008). For example, decreased H3 and H4 histone acetylation was observed after overexpression of APP in mouse cortical neurons, which also resulted in decreased cAMP-response element binding-(CREB)-binding protein (CBP) levels (Rouaux et al. 2003). CBP is a HAT involved in regulating the transcriptional activity of CREB, which in turn acts as a transcription factor and plays a key role in, e.g. neuroprotection (Rouaux et al. 2003; Vecsey et al. 2007). In a mouse model of AD, environmental enrichment led to the recovery of cognitive functions and administration of HDACis was able to mimic those effects (Fischer et al. 2007). Moreover, various preclinical studies show that some HDACis interact with Aβ and tau proteins (Xu et al. 2011).

Numerous studies support the notion of HDACis as a potential treatment for AD owing to the overall hypoacetylation of histones, although some of the studies found increased acetylation in AD (Narayan et al. 2015; Gräff et al. 2011). There are several possible explanations for this discrepancy, e.g. acetylation levels may vary throughout disease progression and/or between different brain cell-types or hypo- and hyperacetylation might occur at the same time and place, but involve different HDACs (see Definition – Classes of HDACs) and/or targeting distinct genes in concert with other epigenetic regulatory processes. Especially important is the notion of cell-type specificity with a recent study showing that histone modifications largely differ between neuronal nuclei and bulk brain tissue (Koshi-Mano et al. 2020).

HDACis have also been proposed to be used as a part of a dual multi-target inhibitor together with phosphodiesterase inhibitors (Cuadrado-Tejedor et al. 2019). As opposed to the classical drug discovery paradigm, one drug – one target, a novel approach of identifying multi-target ligands is gaining interest in the community (Ramsay et al. 2018; Prati et al. 2016; Raghavendra et al. 2018). Especially HDAC6 inhibitors were shown to have favourable properties in this respect,

most likely owing to their distinct structure and function (Ramsay et al. 2018; Zhang et al. 2013).

HDAC2 and HDAC6 show most consistent findings across *in vitro*, *in vivo* and human post-mortem brain tissue research, where both enzymes are increased compared to controls (Lardenoije et al. 2015, 2018; Chuang et al. 2009). Although HDAC1 and HDAC2 are structurally nearly identical, only HDAC2 was upregulated in two mouse models of AD, as well as in the hippocampus and entorhinal cortex of AD patients (Cruz et al. 2003; Fischer et al. 2005; Oakley et al. 2006; Gräff et al. 2012). Furthermore, HDAC2 has been associated with genes involved in synaptic plasticity and memory in mice (Guan et al. 2009). In contrast, HDAC6 interacts with tau and was shown to be increased in the hippocampus and cerebral cortex of post-mortem AD brain tissue (Ding et al. 2008). Moreover, HDAC6 inhibition or reduction led to improved cognition in mouse models of AD (Govindarajan et al. 2013; Yu et al. 2013). HDAC6 mostly targets non-histone proteins, such as alpha-tubulin, heat-shock protein (HSP)-90 and β-catenin, and is implicated in transportation of misfolded proteins, as well as in cell stress response (Fischer et al. 2010). Interestingly, findings from the largest EWAS on histone H3 acetylation to date suggest HSP90 inhibitors as potential agents able to reverse tau-driven chromatin remodelling (Klein et al. 2019).

4.3.1.1 Side Effects

There are several considerations in developing and using HDACis in the clinic. One of them is their lack of specificity since all of the memory-enhancing HDACis target multiple HDACs (Nott et al. 2013). This can affect the expression of numerous genes at the same time and lead to long-term toxicity and other side effects (Hwang et al. 2017). Achieving isoform-specificity is rather difficult considering that HDACs belonging to the same class, as well as between classes, are structurally similar, e.g. HDAC1, HDAC2 and HDAC11 (Gräff and Tsai 2013a). In addition, HDACs and HATs often act on many other proteins apart from histones, with some estimations predicting more than 2500 targets (Narayan and Dragunow 2017). Nonetheless, treatment with HDACis usually does not lead to severe side effects in animal models of AD or in cancer patients (Gräff and Tsai 2013a). Apart from gene-, protein- and isoform-specificity, there is also an issue with brain region and cell-type specificity (Hwang et al. 2017). Especially cell-type specificity should be addressed when developing or delivering HDACis since histone modifications are not evenly globally disturbed, potentially leading to undesired side effects. One of the possible ways to control for brain-region and cell-type specificity is by the means of epigenetic priming (Gräff and Tsai 2013a). Epigenetic priming in a context of cognitive processes refers to stimulating the epigenome, e.g. by the administration of HDACis, at the time of learning-induced neuronal activity (Gräff and Tsai 2013b). This primes the activated cells or brain regions and reinforces the gene expression activity occurring due to learning (Gräff and Tsai 2013a). Furthermore, various in silico computational approaches can be adopted in order to achieve isoform specificity, such as molecular docking, molecular dynamics and energy-optimised pharmacophore mapping (Ganai 2018; Ganai et al. 2015).

4.3.1.2 Clinical Trials

Some of the HDACis have already entered clinical trials for the treatment of AD. The class I and IIa HDACi valproate was evaluated on 313 patients with mild to moderate AD for 24 months (Valproate in Dementia [VALID]; NCT00071721). The study found that valproate was not effective in preventing cognitive or functional decline and reported significant toxic effects of the drug (Tariot et al. 2011). An MRI substudy including 172 participants showed that the valproate-treated group exhibited larger ventricular and whole brain volume loss as well as greater hippocampal atrophy when compared to the placebo group (Fleisher et al. 2011). One possible explana-

tion why valproate's preclinical success failed to translate into the clinic is the agent's effect on microglia (Narayan and Dragunow 2017). Additionally, valproate was first developed as a broad-spectrum anticonvulsant drug with its HDACi properties being discovered more recently, suggesting that it acts on many targets apart from HDACs (Nalivaeva et al. 2009). In rodent microglia, valproate was shown to induce caspase-3-mediated apoptosis and to stimulate phagocytic activity against amyloid peptides (Narayan and Dragunow 2017). However, in human microglia, it was shown to inhibit the phagocytosis of amyloid peptides with no effect on apoptosis.

A phase Ib clinical trial investigating Vorinostat, an inhibitor of class I, II and IV HDACs is currently recruiting participants (Clinical Trial to Determine Tolerable Dosis of Vorinostat in Patients With Mild Alzheimer Disease [VostatAD01]; NCT03056495). The study's primary endpoint is to establish the safety and tolerability of Vorinostat in AD patients.

Additionally, nicotinamide, an inhibitor of class III HDACs (i.e. sirtuins), has been investigated in a phase II clinical trial for AD treatment (Safety Study of Nicotinamide to Treat Alzheimer's Disease; NCT00580931). The phase II clinical trial failed to show meaningful improvements in cognitive functions which might be attributable to the trial's short duration (24 weeks), low sample size ($n = 31$) as well as the inclusion of patients with mild cognitive impairment (MCI) and AD (Phelan et al. 2017). However, investigators found that high doses of nicotinamide are safe in the elderly AD population, in contrast to clinical trials with other HDACis reporting that high doses were not tolerated well by AD patients (Narayan and Dragunow 2017). An additional phase II clinical trial with nicotinamide is currently underway, with an increased sample size ($n = 48$), prolonged duration (48 weeks and additional monitoring for 12 months), and only including patients with MCI or mild AD (Nicotinamide as an Early Alzheimer's Disease Treatment [NEAT]; NCT03061474).

4.3.2 DNMT Inhibitors as a Potential Treatment for AD

Definition

DNMTis can be divided in two classes, one with nucleoside analogs and second including compounds of broad heterogeneous chemical families (Stresemann and Lyko 2008). Currently approved drugs by the FDA targeting DNAm, i.e. azanucleosides such as azacytidine and decitabine, are a part of the first class. Azanucleosides, analogs of naturally occurring cytidine, act as a prodrug and thus have to be metabolised to become pharmacologically active (Erdmann et al. 2016). Phosphorylated azanucleosides are incorporated into the DNA where they substitute for cytosine during cell replication and irreversibly bind DNMTs to the DNA (Stresemann and Lyko 2008; Erdmann et al. 2016). Bound DNMTs are degraded, which leads to the depletion of cellular DNMTs (Stresemann and Lyko 2008; Erdmann et al. 2016). The second class of DNMTis includes natural compounds that interact directly with DNMTs' active domain (Stresemann and Lyko 2008).

There are three enzymes acting as DNMTs in mammals, of which DNMT1 generally maintains DNAm during cell replication, while DNMT3a and DNMT3b act as de novo methyltransferases on previously unmethylated DNA (Bayraktar and Kreutz 2018). With age, there is a global decrease in DNAm and DNMT1 expression, although some genes show increased methylation possibly related to DNMT3a/b overexpression (Wang et al. 2013). Furthermore, a SNP in the DNMT3a gene was associated with greater cognitive decline over 3 years in subjects with MCI, although another study did not find this association (Chouliaras et al. 2015; Bey et al. 2016).

In AD patients, there is evidence towards global hypo- as well as hypermethylation, although interpretation of global DNAm patterns might not be suitable due to the apparent cell-type specificity of this modification (Rao et al. 2012; Coppieters and Dragunow 2011; Mastroeni et al. 2010; Coppieters et al. 2014; Chouliaras et al. 2013; Bradley-Whitman and Lovell 2013). Moreover, most DNAm studies to date did not differentiate between 5mC and 5hmC, which introduces bias to the results (Smith et al. 2019a). Some genomic regions show hypermethylation and others hypomethylation, while, e.g. APOE shows both hypomethylation at the promoter region and hypermethylation at a 3'-CpG-island (Smith et al. 2018, 2019a; Wang et al. 2008; Mano et al. 2017; Foraker et al. 2015). A similar pattern is seen in cancer cells with global hypomethylation, but local hypo- and hypermethylation (Hervouet et al. 2018). Furthermore, differential DNAm in some AD-related genes was reported to be cell-type-specific, with, e.g. hypomethylation in glial cells and hypermethylation in neurons (Gasparoni et al. 2018).

Considering AD-related phenotypes, methylation of PSEN1 decreased Aβ production, while Aβ oligomers lowered DNMT activity leading to increased PSEN1 and APP expression in rodents (Liu et al. 2016; Scarpa et al. 2003). Additional preclinical studies suggested the involvement of DMNTs in memory, synaptic plasticity and learning (Miller et al. 2010; Feng et al. 2010; Oliveira et al. 2012; Levenson et al. 2006). In many forms of synaptic plasticity DNMT1 can functionally replace DNMT3a with an exception of the episodic memory, where DNMT3a knockout mice performed worse on an associative learning task than wildtype controls (Feng et al. 2010; Morris et al. 2014). Furthermore, chronic inhibition of DNMT1 in mice lead to fear memory impairments, and expression of both DNMT3a/b was increased after fear conditioning as well as in response to novel object-location learning (Miller and Sweatt 2007; Tunc-Ozcan et al. 2018; Mitchnick et al. 2015). In contrast, acute administration of a DNMT1 inhibitor before a pattern separation task improved performance and led to increased brain-derived neurotrophic factor (*Bdnf1*) expression in mice (Argyrousi et al. 2019). This study also hints at another important notion, as there is evidence that, apart from TET proteins, DNMTs can also actively demethylate DNA (Chen et al. 2012; van der Wijst et al. 2015). Generally the role of DNMTs is not as clear as that of HDACs and further research is needed to understand their specific effects on DNAm in more detail. The controversial mechanisms of DNAm present a major obstacle in developing DNMTis due to their off-target effects and lack of knowledge about potential interactions with other epigenetic mechanisms. Nonetheless, DNMTis are suggested as a possible treatment for AD and some natural compounds showing DNMTi properties already entered clinical trials.

4.3.2.1 Side Effects

A major drawback of nucleoside analogs is that they can incorporate anywhere in the DNA and therefore produce a global effect (Stresemann and Lyko 2008). This may lead to genotoxicity and side effects such as bleeding, anaemia and joint pain, making currently approved DNMTis unsuitable for continuous treatment (Liu et al. 2018; Giri and Aittokallio 2019). As there are target-specificity considerations with HDACis, and there are 11 different subtypes of HDACs, these considerations are even more pronounced with DNMTis targeting only 3 DNMTs. Furthermore, there might be genes regulated by multiple DNMTs and the effect of specific DNMT inhibition could be compensated for by other DNMTs stepping in, similarly as shown in (Morris et al. 2014). Finally, since DNAm is cell-type specific, there is a risk of collateral damage by disturbing DNAm levels in otherwise healthy cells.

4.3.2.2 Clinical Trials

Some natural products displaying DNMTi properties, which do not incorporate into the DNA, have been tested in humans (Teijido and Cacabelos 2018). For example, epigallocatechin gallate (EGCG), a catechin present in tea, shown to partially act as a DNMTi, was recently tested in phase II and III clinical trials, although no results are currently available (Sunphenon EGCg [Epigallocatechin-Gallate] in the Early Stage of Alzheimer's Disease [SUN-AK]; NCT00951834). Another

study testing the same compound is currently underway, where a combination of lifestyle interventions with EGCG is being tested in subjects with cognitive decline (Prevention of Cognitive Decline in ApoE4 Carriers With Subjective Cognitive Decline After EGCG and a Multimodal Intervention [PENSA]; NCT03978052). Interestingly, quercetin, another DNMTi, showed no improvement in cognitive functions, although it was shown to be well-tolerated in patients with mild to moderate AD (Hirsh et al. 2016). However, the study did not report bioavailability of quercetin or exposure levels in the blood or brain, thus exact interpretation of safety is not possible.

> ❯ HDACis and DNMTis have an unfavourable risk-efficiency ratio due to, e.g. their lack of specificity, which could be improved with the development of isoform- and cell-type-specific inhibitors.

4.3.3 Challenges

One of the main issues with EWAS is statistical power. For example, a study conducted with EPIC arrays, the largest commonly used platform thus far interrogating over 850,000 methylation sites, is estimated to need a sample size of ~1000 in order to detect small differences (Mansell et al. 2019). Although studies using smaller sample sizes may, dependent on the study design, still provide valuable data, and, when assessing DMRs instead of DMPs, a smaller sample size is often sufficient, most of the studies to date are underpowered and thus findings should be interpreted with caution. Evidently, the same holds true for studies investigating histone modifications in AD, where only one study thus far included a relatively large sample size ($n > 600$) (Klein et al. 2019).

> ❯ Cautiousness is needed while interpreting EWAS results due to the generally low sample size of the studies and understudied effects of various complimentary mechanisms (e.g. the role of 5hmC in MWAS and ubiquitination in histone modifications).

As mentioned before, most MWAS studies on brain tissue are confounded by 5hmC, since the method to distinguish between 5mC and 5hmC is fairly new and therefore not implemented in most of the research published to date (Smith et al. 2019a). Furthermore, besides 5hmC, 5fC and 5caC derivatives of 5mC are also present in the brain; however, little is known about their functionality and their levels in mental health and disease. A similar scenario can be seen in studies investigating histone modifications, where most of the post-mortem studies focused on acetylation of histone proteins, a few of them explored histone methylation or phosphorylation, whereas ubiquitination and sumoylation remain largely unexplored in relation to AD (Narayan and Dragunow 2017; Esposito and Sherr 2019).

Another major challenge in translating epigenetic research to a clinical implementation is tissue and cell-type specificity (Snijders et al. 2018). Epigenetic modifications are usually brain region-specific. Furthermore, these modifications can also be cell-type specific and, considering cellular heterogeneity within the mammalian brain, this presents a major limitation in interpreting results obtained when investigating epigenetic changes in bulk tissue (e.g. brain homogenates). This is especially important when studying AD, since AD is characterised by (specific) neuronal loss, thereby altering cellular proportions within (bulk) tissue samples. If a certain AD-specific epigenetic signature is prevalent in (specific) neurons, it could go unrecognised due to over-representation of other cell types.

Taken together, these limitations present serious challenges in identifying epigenetic-related drug targets. Furthermore, although it is known that different epigenetic processes interact with each other, it is yet unknown exactly how, and clearly, many of these epigenetic processes are largely understudied, if studied at all, in humans. Research conducted on animal models presents a valuable approach in elucidating missing links, although it also possesses many limitations and drugs active in animals often fail in humans. It is also important to keep in mind that epigenetic mechanisms work in a dynamic manner, meaning that over time, there could be a substantial number of changes (e.g. with aging; or when comparing early and late AD cases) in the epigenetic machinery, which illustrates the relevance of choosing the right (preferably longitudinal) study approach. Genomic and transcriptomic changes can introduce additional variation that goes undetected when studying a single layer of molecular information, especially when including only a single measure of disease severity and focusing on a single tissue. As such, multi-omics approaches together with advanced computational frameworks (i.e. an integrative analysis of epigenomic regulations, transcriptomics, metabolomics and genomics in various combinations) bear great potential, as they present a way to simultaneously investigate multi-layered inter-regulatory mechanisms, whilst accounting for inter-individual differences. Studies with multi-omics approaches including epigenetic regulations could therefore aid in identifying more robust classical non-epigenetic drug targets, e.g. genes or proteins that are affected or dysregulated on multiple molecular levels.

> Advances in epigenetic research may lead to the identification of epigenetic as well as non-epigenetic drug targets.

Finally, all of the abovementioned points lead to the major limitation of causal inference. It is currently extremely challenging to determine which alterations are causal in the pathogenesis of AD and which of them arise as a consequence of its development or its treatment. Nonetheless, even when in its infancy, neuroepigenetic research has contributed significantly to the AD field over the last decade and, considering ever-accelerating technological developments, it bears immense potential in aiding drug discovery for AD as well as other brain-related disorders. From the studies reviewed above, it is clear that epigenetic mechanisms have a profound effect on AD pathology and play an important role in its development. Although more research is still needed in order to fully elucidate healthy as well as pathological mechanisms of epigenetic regulation, findings thus far highlight some promising implications.

Conclusion

Epigenetic regulation provides a link between one's genetic background, environment and lifestyle. Considering that genetic predisposition cannot fully explain the occurrence of many psychiatric and neurodegenerative disorders, including AD, investigating disease-relevant epigenetic mechanisms could aid in unravelling its pathophysiology. Furthermore, as opposed to genetic variation, epigenetic mechanisms are in essence reversible, and enzymes involved in establishing epigenetic modifications bear great potential for developing epigenetics-based drugs. That said, still much more research is needed to fully understand which of these mechanisms are most relevant for the development and course of disorders such as AD and which of these are suitable in terms of drug development. Larger (longitudinal) cohort studies and new methodologies inspired by technological advances will bring us closer to pinpoint and understand the intricate, multi-levelled epigenetic dysregulation in AD. Nonetheless, several epigenetics-based molecules have already entered clinical trials for AD, and some of them show favourable safety profiles, which is one of the main challenges in epigenetic drug development. However, until now, none of these molecules have provided a clear beneficial effect on AD-related symptomology. Apart from (directly) targeting the epigenetic machinery, drug discovery can furthermore benefit from a deeper understanding of interactions between epigenetic (dys)regulation and other molecular processes in order to identify more robust non-epigenetic drug targets. In addition to DNA and histone modifications, there is increasing interest in ncRNAs and their implications for the treatment of AD. As ncRNAs act as master regulators of gene regulatory networks, targeting them in principal bears great potential in complex diseases such as AD, although their use could also accompany a less favourable safety profile. Considering the number of different molecules as part of the epigenetic machinery and the continuously increasing understanding of their interactions, this area of research already is and will increasingly be of great promise in identifying novel drug targets for psychiatric and neurodegenerative disorders.

References

Angelucci F, Cechova K, Valis M, Kuca K, Zhang B, Hort J (2019) MicroRNAs in Alzheimer's disease: diagnostic markers or therapeutic agents? Front Pharmacol 10:665. https://doi.org/10.3389/fphar.2019.00665

Argyrousi EK, de Nijs L, Lagatta DC, Schlutter A, Weidner MT, Zöller J, van Goethem NP, Joca SRL, van den Hove DLA, Prickaerts J (2019) Effects of DNA methyltransferase inhibition on pattern separation performance in mice. Neurobiol Learn Mem 159:6–15. https://doi.org/10.1016/j.nlm.2019.02.003

Bartolotti N, Lazarov O (2016) Lifestyle and Alzheimer's disease: the role of environmental factors in disease development. In: Genes, environment and Alzheimer's disease. Elsevier, San Diego, pp 197–237

Bayraktar G, Kreutz MR (2018) Neuronal DNA methyltransferases: epigenetic mediators between synaptic activity and gene expression? Neuroscientist 24(2):171–185

Bertram L, Lill CM, Tanzi RE (2010) The genetics of Alzheimer disease: back to the future. Neuron 68(2):270–281

Bey K, Wolfsgruber S, Karaca I, Wagner H, Lardenoije R, Becker J, Milz E, Kornhuber J, Peters O, Frolich L, Hull M, Ruther E, Wiltfang J, Riedel-Heller S, Scherer M, Jessen F, Maier W, van den Hove DL, Rutten BP, Wagner M, Ramirez A (2016) No association of the variant rs11887120 in DNMT3A with cognitive decline in individuals with mild cognitive impairment. Epigenomics 8(5):593–598. https://doi.org/10.2217/epi-2015-0014

Bradley-Whitman M, Lovell M (2013) Epigenetic changes in the progression of Alzheimer's disease. Mech Ageing Dev 134(10):486–495

Bredy TW, Wu H, Crego C, Zellhoefer J, Sun YE, Barad M (2007) Histone modifications around individual BDNF gene promoters in prefrontal cortex are associated with extinction of conditioned fear. Learn Mem 14(4):268–276

Castanho I, Lunnon K (2019) Epigenetic processes in Alzheimer's disease. In: Chromatin signaling and neurological disorders. Elsevier, pp 153–180

Chatterjee P, Roy D, Rathi N (2018) Epigenetic drug repositioning for Alzheimer's disease based on epigenetic targets in human Interactome. J Alzheimers Dis 61(1):53–65. https://doi.org/10.3233/jad-161104

Chen C-C, Wang K-Y, Shen C-KJ (2012) The mammalian de novo DNA methyltransferases DNMT3A and DNMT3B are also DNA 5-hydroxymethylcytosine dehydroxymethylases. J Biol Chem 287(40):33116–33121

4

Chouliaras L, Mastroeni D, Delvaux E, Grover A, Kenis G, Hof PR, Steinbusch HW, Coleman PD, Rutten BP, van den Hove DL (2013) Consistent decrease in global DNA methylation and hydroxymethylation in the hippocampus of Alzheimer's disease patients. Neurobiol Aging 34(9):2091–2099

Chouliaras L, Kenis G, Visser PJ, Scheltens P, Tsolaki M, Jones RW, Kehoe PG, Graff C, Girtler NG, Wallin AK, Rikkert MO, Spiru L, Elias-Sonnenschein LS, Ramakers IH, Pishva E, van Os J, Steinbusch HW, Verhey FR, van den Hove DL, Rutten BP (2015) DNMT3A moderates cognitive decline in subjects with mild cognitive impairment: replicated evidence from two mild cognitive impairment cohorts. Epigenomics 7(4):533–537. https://doi.org/10.2217/epi.15.22

Chuang D-M, Leng Y, Marinova Z, Kim H-J, Chiu C-T (2009) Multiple roles of HDAC inhibition in neurodegenerative conditions. Trends Neurosci 32(11):591–601

Chwang WB, Arthur JS, Schumacher A, Sweatt JD (2007) The nuclear kinase mitogen- and stress-activated protein kinase 1 regulates hippocampal chromatin remodeling in memory formation. J Neurosci 27(46):12732–12742. https://doi.org/10.1523/JNEUROSCI.2522-07.2007

Coppieters N, Dragunow M (2011) Epigenetics in Alzheimer's disease: a focus on DNA modifications. Curr Pharm Des 17(31):3398–3412

Coppieters N, Dieriks BV, Lill C, Faull RL, Curtis MA, Dragunow M (2014) Global changes in DNA methylation and hydroxymethylation in Alzheimer's disease human brain. Neurobiol Aging 35(6):1334–1344

Cruz JC, Tseng H-C, Goldman JA, Shih H, Tsai L-H (2003) Aberrant Cdk5 activation by p25 triggers pathological events leading to neurodegeneration and neurofibrillary tangles. Neuron 40(3):471–483

Cuadrado-Tejedor M, Pérez-González M, García-Muñoz C, Muruzabal D, García-Barroso C, Rabal O, Segura V, Sánchez-Arias JA, Oyarzabal J, Garcia-Osta A (2019) Taking advantage of the selectivity of histone deacetylases and phosphodiesterase inhibitors to design better therapeutic strategies to treat Alzheimer's disease. Front Aging Neurosci 11:149. https://doi.org/10.3389/fnagi.2019.00149

De Jager PL, Srivastava G, Lunnon K, Burgess J, Schalkwyk LC, Yu L, Eaton ML, Keenan BT, Ernst J, McCabe C, Tang A, Raj T, Replogle J, Brodeur W, Gabriel S, Chai HS, Younkin C, Younkin SG, Zou F, Szyf M, Epstein CB, Schneider JA, Bernstein BE, Meissner A, Ertekin-Taner N, Chibnik LB, Kellis M, Mill J, Bennett DA (2014) Alzheimer's disease: early alterations in brain DNA methylation at ANK1, BIN1, RHBDF2 and other loci. Nat Neurosci 17(9):1156–1163. https://doi.org/10.1038/nn.3786

Ding H, Dolan PJ, Johnson GVW (2008) Histone deacetylase 6 interacts with the microtubule-associated protein tau. J Neurochem 106(5):2119–2130. https://doi.org/10.1111/j.1471-4159.2008.05564.x

Erdmann A, Arimondo PB, Guianvarc'h D (2016) Structure-guided optimization of DNA methyltransferase inhibitors. In: Epi-Informatics. Elsevier, pp 53–73

Esposito MM, Sherr GL (2019) Epigenetic modifications in Alzheimer's neuropathology and therapeutics. Front Neurosci 13:476

Feng J, Zhou Y, Campbell SL, Le T, Li E, Sweatt JD, Silva AJ, Fan G (2010) Dnmt1 and Dnmt3a maintain DNA methylation and regulate synaptic function in adult forebrain neurons. Nat Neurosci 13(4):423

Fischer A, Sananbenesi F, Pang PT, Lu B, Tsai L-H (2005) Opposing roles of transient and prolonged expression of p25 in synaptic plasticity and hippocampus-dependent memory. Neuron 48(5):825–838

Fischer A, Sananbenesi F, Wang X, Dobbin M, Tsai L-H (2007) Recovery of learning and memory is associated with chromatin remodelling. Nature 447(7141):178–182. https://doi.org/10.1038/nature05772

Fischer A, Sananbenesi F, Mungenast A, Tsai L-H (2010) Targeting the correct HDAC(s) to treat cognitive disorders. Trends Pharmacol Sci 31(12):605–617. https://doi.org/10.1016/j.tips.2010.09.003

Fleisher A, Truran D, Mai J, Langbaum J, Aisen P, Cummings J, Jack C, Weiner M, Thomas R, Schneider L (2011) Chronic divalproex sodium use and brain atrophy in Alzheimer disease. Neurology 77(13):1263–1271

Foidl BM, Humpel C (2020) Can mouse models mimic sporadic Alzheimer's disease? Neural Regen Res 15(3):401–406. https://doi.org/10.4103/1673-5374.266046

Foraker J, Millard SP, Leong L, Thomson Z, Chen S, Keene CD, Bekris LM, Yu CE (2015) The APOE gene is differentially methylated in Alzheimer's disease. J Alzheimers Dis 48(3):745–755. https://doi.org/10.3233/jad-143060

Förstl H, Kurz A (1999) Clinical features of Alzheimer's disease. Eur Arch Psychiatry Clin Neurosci 249(6):288–290. https://doi.org/10.1007/s004060050101

Ganai SA (2018) Designing isoform-selective inhibitors against classical HDACs for effective anticancer therapy: insight and perspectives from in silico. Curr Drug Targets 19(7):815–824

Ganai SA, Shanmugam K, Mahadevan V (2015) Energy-optimised pharmacophore approach to identify potential hotspots during inhibition of class II HDAC isoforms. J Biomol Struct Dyn 33(2):374–387

Gasparoni G, Bultmann S, Lutsik P, Kraus TF, Sordon S, Vlcek J, Dietinger V, Steinmaurer M, Haider M, Mulholland CB (2018) DNA methylation analysis on purified neurons and glia dissects age and Alzheimer's disease-specific changes in the human cortex. Epigenetics Chromatin 11(1):41

Giri AK, Aittokallio T (2019) DNMT inhibitors increase methylation in the Cancer genome. Front

Pharmacol 10:385–385. https://doi.org/10.3389/fphar.2019.00385

Govindarajan N, Rao P, Burkhardt S, Sananbenesi F, Schluter OM, Bradke F, Lu J, Fischer A (2013) Reducing HDAC6 ameliorates cognitive deficits in a mouse model for Alzheimer's disease. EMBO Mol Med 5(1):52–63. https://doi.org/10.1002/emmm.201201923

Gräff J, Tsai L-H (2013a) The potential of HDAC inhibitors as cognitive enhancers. Annu Rev Pharmacol Toxicol 53:311–330

Gräff J, Tsai L-H (2013b) Histone acetylation: molecular mnemonics on the chromatin. Nat Rev Neurosci 14(2):97–111

Gräff J, Kim D, Dobbin MM, Tsai L-H (2011) Epigenetic regulation of gene expression in physiological and pathological brain processes. Physiol Rev 91(2):603–649. https://doi.org/10.1152/physrev.00012.2010

Gräff J, Rei D, Guan J-S, Wang W-Y, Seo J, Hennig KM, Nieland TJF, Fass DM, Kao PF, Kahn M, Su SC, Samiei A, Joseph N, Haggarty SJ, Delalle I, Tsai L-H (2012) An epigenetic blockade of cognitive functions in the neurodegenerating brain. Nature 483(7388):222–226. https://doi.org/10.1038/nature10849

Gu X, Sun J, Li S, Wu X, Li L (2013) Oxidative stress induces DNA demethylation and histone acetylation in SH-SY5Y cells: potential epigenetic mechanisms in gene transcription in Aβ production. Neurobiol Aging 34(4):1069–1079

Guan J-S, Haggarty SJ, Giacometti E, Dannenberg J-H, Joseph N, Gao J, Nieland TJF, Zhou Y, Wang X, Mazitschek R, Bradner JE, DePinho RA, Jaenisch R, Tsai L-H (2009) HDAC2 negatively regulates memory formation and synaptic plasticity. Nature 459(7243):55–60. https://doi.org/10.1038/nature07925

Gupta P, Bhattacharjee S, Sharma AR, Sharma G, Lee S-S, Chakraborty C (2017) miRNAs in Alzheimer disease–a therapeutic perspective. Curr Alzheimer Res 14(11):1198–1206

Hervouet E, Peixoto P, Delage-Mourroux R, Boyer-Guittaut M, Cartron P-F (2018) Specific or not specific recruitment of DNMTs for DNA methylation, an epigenetic dilemma. Clin Epigenetics 10:17–17. https://doi.org/10.1186/s13148-018-0450-y

Higham JP, Malik BR, Buhl E, Dawson JM, Ogier AS, Lunnon K, Hodge JJL (2019) Alzheimer's disease associated genes Ankyrin and tau cause shortened lifespan and memory loss in Drosophila. Front Cell Neurosci 13:260. https://doi.org/10.3389/fncel.2019.00260

Hirsh S, Huber L, Stein R, Schmid K, Swick A, Wand P, Brody M, Strum S, Joyal SV (2016) Open label, crossover, pilot study to assess the efficacy and safety of perispinal administration of etanercept (enbrel®) in combination with nutritional supplements versus nutritional supplements alone in subjects with mild to moderate Alzheimer's disease receiving standard care. FASEB J 30(1_supplement):lb296

Hwang J-Y, Aromolaran KA, Zukin RS (2017) The emerging field of epigenetics in neurodegeneration and neuroprotection. Nat Rev Neurosci 18(6):347–361. https://doi.org/10.1038/nrn.2017.46

Iatrou A, Kenis G, Rutten BP, Lunnon K, van den Hove DL (2017) Epigenetic dysregulation of brainstem nuclei in the pathogenesis of Alzheimer's disease: looking in the correct place at the right time? Cell Mol Life Sci 74(3):509–523. https://doi.org/10.1007/s00018-016-2361-4

Klein H-U, McCabe C, Gjoneska E, Sullivan SE, Kaskow BJ, Tang A, Smith RV, Xu J, Pfenning AR, Bernstein BE (2019) Epigenome-wide study uncovers large-scale changes in histone acetylation driven by tau pathology in aging and Alzheimer's human brains. Nat Neurosci 22(1):37–46

Kohli RM, Zhang Y (2013) TET enzymes, TDG and the dynamics of DNA demethylation. Nature 502(7472):472

Koshibu K, Gräff J, Beullens M, Heitz FD, Berchtold D, Russig H, Farinelli M, Bollen M, Mansuy IM (2009) Protein phosphatase 1 regulates the histone code for long-term memory. J Neurosci 29(41):13079–13089

Koshi-Mano K, Mano T, Morishima M, Murayama S, Tamaoka A, Tsuji S, Toda T, Iwata A (2020) Neuron-specific analysis of histone modifications with post-mortem brains. Sci Rep 10(1):3767. https://doi.org/10.1038/s41598-020-60775-z

Lardenoije R, Iatrou A, Kenis G, Kompotis K, Steinbusch HW, Mastroeni D, Coleman P, Lemere CA, Hof PR, van den Hove DL (2015) The epigenetics of aging and neurodegeneration. Prog Neurobiol 131:21–64

Lardenoije R, Pishva E, Lunnon K, van den Hove DL (2018) Chapter Four - Neuroepigenetics of aging and age-related neurodegenerative disorders. In: BPF R (ed) Progress in molecular biology and translational science, vol 158. Academic Press, Cambridge, MA, pp 49–82. https://doi.org/10.1016/bs.pmbts.2018.04.008

Lawrence M, Daujat S, Schneider R (2016) Lateral thinking: how histone modifications regulate gene expression. Trends Genet 32(1):42–56

Levenson JM, O'Riordan KJ, Brown KD, Trinh MA, Molfese DL, Sweatt JD (2004) Regulation of histone acetylation during memory formation in the hippocampus. J Biol Chem 279(39):40545–40559

Levenson JM, Roth TL, Lubin FD, Miller CA, Huang I-C, Desai P, Malone LM, Sweatt JD (2006) Evidence that DNA (cytosine-5) methyltransferase regulates synaptic plasticity in the hippocampus. J Biol Chem 281(23):15763–15773

Lithner CU, Lacor PN, Zhao W-Q, Mustafiz T, Klein WL, Sweatt JD, Hernandez CM (2013) Disruption of neocortical histone H3 homeostasis by soluble Aβ: implications for Alzheimer's disease. Neurobiol Aging 34(9):2081–2090

Liu H, Li W, Zhao S, Zhang X, Zhang M, Xiao Y, Wilson JX, Huang G (2016) Folic acid attenuates the effects of amyloid β oligomers on DNA methylation in neuronal cells. Eur J Nutr 55(5):1849–1862. https://doi.org/10.1007/s00394-015-1002-2

Liu X, Jiao B, Shen L (2018) The epigenetics of Alzheimer's disease: factors and therapeutic implications. Front Genet 9:579

Lubin FD, Sweatt JD (2007) The IκB kinase regulates chromatin structure during reconsolidation of conditioned fear memories. Neuron 55(6):942–957

Lubin FD, Roth TL, Sweatt JD (2008) Epigenetic regulation of BDNF gene transcription in the consolidation of fear memory. J Neurosci 28(42):10576–10586

Lunnon K, Smith R, Hannon E, De Jager PL, Srivastava G, Volta M, Troakes C, Al-Sarraj S, Burrage J, Macdonald R, Condliffe D, Harries LW, Katsel P, Haroutunian V, Kaminsky Z, Joachim C, Powell J, Lovestone S, Bennett DA, Schalkwyk LC, Mill J (2014) Methylomic profiling implicates cortical deregulation of ANK1 in Alzheimer's disease. Nat Neurosci 17(9):1164–1170. https://doi.org/10.1038/nn.3782

Mano T, Nagata K, Nonaka T, Tarutani A, Imamura T, Hashimoto T, Bannai T, Koshi-Mano K, Tsuchida T, Ohtomo R, Takahashi-Fujigasaki J, Yamashita S, Ohyagi Y, Yamasaki R, Tsuji S, Tamaoka A, Ikeuchi T, Saido TC, Iwatsubo T, Ushijima T, Murayama S, Hasegawa M, Iwata A (2017) Neuron-specific methylome analysis reveals epigenetic regulation and tau-related dysfunction of BRCA1 in Alzheimer's disease. Proc Natl Acad Sci U S A 114(45):E9645–e9654. https://doi.org/10.1073/pnas.1707151114

Mansell G, Gorrie-Stone TJ, Bao Y, Kumari M, Schalkwyk LS, Mill J, Hannon E (2019) Guidance for DNA methylation studies: statistical insights from the Illumina EPIC array. BMC Genomics 20(1):366

Maoz R, Garfinkel BP, Soreq H (2017) Alzheimer's disease and ncRNAs. In: Neuroepigenomics in aging and disease. Springer, pp 337–361

Mastroeni D, Grover A, Delvaux E, Whiteside C, Coleman PD, Rogers J (2010) Epigenetic changes in Alzheimer's disease: decrements in DNA methylation. Neurobiol Aging 31(12):2025–2037

Mastroeni D, Sekar S, Nolz J, Delvaux E, Lunnon K, Mill J, Liang WS, Coleman PD (2017) ANK1 is upregulated in laser captured microglia in Alzheimer's brain; the importance of addressing cellular heterogeneity. PLoS One 12(7):e0177814

Mercatelli D, Scalambra L, Triboli L, Ray F, Giorgi FM (2019) Gene regulatory network inference resources: a practical overview. Biochim Biophys Acta Gene Regul Mech 1863(6):194430. https://doi.org/10.1016/j.bbagrm.2019.194430

Millan MJ (2017) Linking deregulation of non-coding RNA to the core pathophysiology of Alzheimer's disease: an integrative review. Prog Neurobiol 156:1–68

Miller CA, Sweatt JD (2007) Covalent modification of DNA regulates memory formation. Neuron 53(6):857–869

Miller CA, Gavin CF, White JA, Parrish RR, Honasoge A, Yancey CR, Rivera IM, Rubio MD, Rumbaugh

G, Sweatt JD (2010) Cortical DNA methylation maintains remote memory. Nat Neurosci 13(6):664–666. https://doi.org/10.1038/nn.2560

Mitchnick KA, Creighton S, O'Hara M, Kalisch BE, Winters BD (2015) Differential contributions of de novo and maintenance DNA methyltransferases to object memory processing in the rat hippocampus and perirhinal cortex--a double dissociation. Eur J Neurosci 41(6):773–786. https://doi.org/10.1111/ejn.12819

Morris MJ, Adachi M, Na ES, Monteggia LM (2014) Selective role for DNMT3a in learning and memory. Neurobiol Learn Mem 115:30–37

Myung N-H, Zhu X, Kruman II, Castellani RJ, Petersen RB, Siedlak SL, Perry G, Smith MA, H-g L (2008) Evidence of DNA damage in Alzheimer disease: phosphorylation of histone H2AX in astrocytes. Age 30(4):209–215

Nalivaeva NN, Belyaev ND, Turner AJ (2009) Sodium valproate: an old drug with new roles. Trends Pharmacol Sci 30(10):509–514

Narayan P, Dragunow M (2017) Alzheimer's disease and histone code alterations. In: Neuroepigenomics in aging and disease. Springer, Cham, pp 321–336

Narayan PJ, Lill C, Faull R, Curtis MA, Dragunow M (2015) Increased acetyl and total histone levels in post-mortem Alzheimer's disease brain. Neurobiol Dis 74:281–294

Nott A, Fass DM, Haggarty SJ, Tsai L-H (2013) HDAC inhibitors as novel therapeutics in aging and Alzheimer's disease. In: Epigenetic regulation in the nervous system. Elsevier, San Diego, pp 225–248

Oakley H, Cole SL, Logan S, Maus E, Shao P, Craft J, Guillozet-Bongaarts A, Ohno M, Disterhoft J, Van Eldik L (2006) Intraneuronal β-amyloid aggregates, neurodegeneration, and neuron loss in transgenic mice with five familial Alzheimer's disease mutations: potential factors in amyloid plaque formation. J Neurosci 26(40):10129–10140

Oliveira AMM, Hemstedt TJ, Bading H (2012) Rescue of aging-associated decline in Dnmt3a2 expression restores cognitive abilities. Nat Neurosci 15(8):1111–1113. https://doi.org/10.1038/nn.3151

Phelan M, Mulnard R, Gillen D, Schreiber S (2017) Phase II clinical trial of nicotinamide for the treatment of mild to moderate Alzheimer's disease. J Geriatr Med Gerontol 3:021

Phipps AJ, Vickers JC, Taberlay PC, Woodhouse A (2016) Neurofilament-labeled pyramidal neurons and astrocytes are deficient in DNA methylation marks in Alzheimer's disease. Neurobiol Aging 45:30–42

Prati F, Cavalli A, Bolognesi ML (2016) Navigating the chemical space of multitarget-directed ligands: from hybrids to fragments in Alzheimer's disease. Molecules (Basel Switzerland) 21(4):466. https://doi.org/10.3390/molecules21040466

Raghavendra NM, Pingili D, Kadasi S, Mettu A, Prasad S (2018) Dual or multi-targeting inhibitors: the next generation anticancer agents. Eur J Med

Chem 143:1277–1300. https://doi.org/10.1016/j.ejmech.2017.10.021

Ramsay RR, Popovic-Nikolic MR, Nikolic K, Uliassi E, Bolognesi ML (2018) A perspective on multi-target drug discovery and design for complex diseases. Clin Transl Med 7(1):3–3. https://doi.org/10.1186/s40169-017-0181-2

Rao J, Keleshian V, Klein S, Rapoport S (2012) Epigenetic modifications in frontal cortex from Alzheimer's disease and bipolar disorder patients. Transl Psychiatry 2(7):e132

Rouaux C, Jokic N, Mbebi C, Boutillier S, Loeffler J-P, Boutillier A-L (2003) Critical loss of CBP/p300 histone acetylase activity by caspase-6 during neurodegeneration. EMBO J 22(24):6537–6549. https://doi.org/10.1093/emboj/cdg615

Roubroeks JAY, Smith RG, van den Hove DLA, Lunnon K (2017) Epigenetics and DNA methylomic profiling in Alzheimer's disease and other neurodegenerative diseases. J Neurochem 143(2):158–170. https://doi.org/10.1111/jnc.14148

Rudenko A, Tsai L-H (2014) Epigenetic modifications in the nervous system and their impact upon cognitive impairments. Neuropharmacology 80:70–82

Scarpa S, Fuso A, D'Anselmi F, Cavallaro RA (2003) Presenilin 1 gene silencing by S-adenosylmethionine: a treatment for Alzheimer disease? FEBS Lett 541(1–3):145–148

Semick SA, Bharadwaj RA, Collado-Torres L, Tao R, Shin JH, Deep-Soboslay A, Weiss JR, Weinberger DR, Hyde TM, Kleinman JE, Jaffe AE, Mattay VS (2019) Integrated DNA methylation and gene expression profiling across multiple brain regions implicate novel genes in Alzheimer's disease. Acta Neuropathol 137(4):557–569. https://doi.org/10.1007/s00401-019-01966-5

Siavelis JC, Bourdakou MM, Athanasiadis EI, Spyrou GM, Nikita KS (2015) Bioinformatics methods in drug repurposing for Alzheimer's disease. Brief Bioinform 17(2):322–335

Silva PNO, Gigek CO, Leal MF, Bertolucci PHF, de Labio RW, Payao SLM, Smith MAC (2008) Promoter methylation analysis of SIRT3, SMARCA5, HTERT and CDH1 genes in aging and Alzheimer's disease. J Alzheimers Dis 13(2):173–176

Silvestro S, Bramanti P, Mazzon E (2019) Role of miR-NAs in Alzheimer's disease and possible fields of application. Int J Mol Sci 20(16):3979

Smith RG, Hannon E, De Jager PL, Chibnik L, Lott SJ, Condliffe D, Smith AR, Haroutunian V, Troakes C, Al-Sarraj S (2018) Elevated DNA methylation across a 48-kb region spanning the HOXA gene cluster is associated with Alzheimer's disease neuropathology. Alzheimers Dement 14(12):1580–1588

Smith AR, Smith RG, Pishva E, Hannon E, Roubroeks JAY, Burrage J, Troakes C, Al-Sarraj S, Sloan C, Mill J, van den Hove DL, Lunnon K (2019a) Parallel profiling of DNA methylation and hydroxymethylation highlights neuropathology-associated epigenetic variation in Alzheimer's disease. Clin Epigenetics 11(1):52. https://doi.org/10.1186/s13148-019-0636-y

Smith AR, Smith RG, Burrage J, Troakes C, Al-Sarraj S, Kalaria RN, Sloan C, Robinson AC, Mill J, Lunnon K (2019b) A cross-brain regions study of ANK1 DNA methylation in different neurodegenerative diseases. Neurobiol Aging 74:70–76. https://doi.org/10.1016/j.neurobiolaging.2018.09.024

Snijders C, Bassil KC, de Nijs L (2018) Methodologies of neuroepigenetic research: background, challenges and future perspectives. In: Progress in molecular biology and translational science, vol 158. Elsevier, pp 15–27

Stresemann C, Lyko F (2008) Modes of action of the DNA methyltransferase inhibitors azacytidine and decitabine. Int J Cancer 123(1):8–13. https://doi.org/10.1002/ijc.23607

Tariot PN, Schneider LS, Cummings J, Thomas RG, Raman R, Jakimovich LJ, Loy R, Bartocci B, Fleisher A, Ismail MS, Porsteinsson A, Weiner M, Jack CR Jr, Thal L, Aisen PS (2011) Chronic divalproex sodium to attenuate agitation and clinical progression of Alzheimer disease. Arch Gen Psychiatry 68(8):853–861. https://doi.org/10.1001/archgenpsychiatry.2011.72

Teijido O, Cacabelos R (2018) Pharmacoepigenomic interventions as novel potential treatments for Alzheimer's and Parkinson's diseases. Int J Mol Sci 19(10):3199

Tunc-Ozcan E, Wert SL, Lim PH, Ferreira A, Redei EE (2018) Hippocampus-dependent memory and allele-specific gene expression in adult offspring of alcohol-consuming dams after neonatal treatment with thyroxin or metformin. Mol Psychiatry 23(7):1643–1651

Van den Hove DL, Kompotis K, Lardenoije R, Kenis G, Mill J, Steinbusch HW, Lesch K-P, Fitzsimons CP, De Strooper B, Rutten BP (2014) Epigenetically regulated microRNAs in Alzheimer's disease. Neurobiol Aging 35(4):731–745

Vargas DMD, De Bastiani MA, Zimmer ER, Klamt F (2018) Alzheimer's disease master regulators analysis: search for potential molecular targets and drug repositioning candidates. Alzheimers Res Ther 10(1):59

Vecsey CG, Hawk JD, Lattal KM, Stein JM, Fabian SA, Attner MA, Cabrera SM, McDonough CB, Brindle PK, Abel T, Wood MA (2007) Histone deacetylase inhibitors enhance memory and synaptic plasticity via CREB: CBP-dependent transcriptional activation. J Neurosci 27(23):6128–6140. https://doi.org/10.1523/jneurosci.0296-07.2007

Wang SC, Oelze B, Schumacher A (2008) Age-specific epigenetic drift in late-onset Alzheimer's disease. PLoS One 3(7):e2698. https://doi.org/10.1371/journal.pone.0002698

Wang J, Yu J-T, Tan M-S, Jiang T, Tan L (2013) Epigenetic mechanisms in Alzheimer's disease: implications for pathogenesis and therapy. Ageing Res Rev 12(4):1024–1041

van der Wijst MGP, Venkiteswaran M, Chen H, Xu G-L, Plösch T, Rots MG (2015) Local chromatin microenvironment determines DNMT activity:

from DNA methyltransferase to DNA demethylase or DNA dehydroxymethylase. Epigenetics 10(8):671–676. https://doi.org/10.1080/15592294.2015.1062204

World Health Organisation (2017) Global action plan on the public health response to dementia 2017–2025

Xu K, Dai X-L, Huang H-C, Jiang Z-F (2011) Targeting HDACs: a promising therapy for Alzheimer's disease. Oxidative Med Cell Longev 2011:143269

Yang S-S, Zhang R, Wang G, Zhang Y-F (2017) The development prospection of HDAC inhibitors as a potential therapeutic direction in Alzheimer's disease. Transl Neurodegen 6(1):19

Yin Y, Morgunova E, Jolma A, Kaasinen E, Sahu B, Khund-Sayeed S, Das PK, Kivioja T, Dave K, Zhong F, Nitta KR, Taipale M, Popov A, Ginno PA, Domcke S, Yan J, Schübeler D, Vinson C, Taipale J (2017) Impact of cytosine methylation on DNA binding specificities of human transcription factors. Science 356(6337):eaaj2239. https://doi.org/10.1126/science.aaj2239

Yu C-W, Chang P-T, Hsin L-W, Chern J-W (2013) Quinazolin-4-one derivatives as selective histone deacetylase-6 inhibitors for the treatment of Alzheimer's disease. J Med Chem 56(17):6775–6791

Zhang L, Sheng S, Qin C (2013) The role of HDAC6 in Alzheimer's disease. J Alzheimers Dis 33(2):283–295

Zhang L, Chen C, Mak MS, Lu J, Wu Z, Chen Q, Han Y, Li Y, Pi R (2020) Advance of sporadic Alzheimer's disease animal models. Med Res Rev 40(1):431–458. https://doi.org/10.1002/med.21624

Zhao Y, Garcia BA (2015) Comprehensive catalog of currently documented histone modifications. Cold Spring Harb Perspect Biol 7(9):a025064

Conventional Behavioral Models and Readouts in Drug Discovery: The Importance of Improving Translation

Kris Rutten (iD)

Contents

5.1   General Introduction in Behavioral Models and
        Readouts – 78
5.1.1   Model Versus Readout and Validity – 78

5.2   Models and Readouts in Pain
        Drug Development – 81
5.2.1   Models of Neuropathic and Inflammatory Pain – 81
5.2.2   Readouts for Assessment of Pain-Like Behavior – 83
5.2.3   An Example of Failed Translation from Animal Data
        to Clinical Trials of NK-1 Antagonist as Putative Novel
        Analgesic Drugs – 83

5.3   Improving Translation in CNS Drug
        Discovery and Development – 84
5.3.1   Neuroimaging – 84
5.3.2   PK/PD-Modeling – 85

5.4   Discussion – 87

        References – 89

This chapter describes the relevance and importance of preclinical models and readouts in the drug development process. Recently, animal behavioral models have been criticized on their role in the poor translation into novel pharmacotherapies. The first section addresses the importance of validity, ethics, and model/readout selection in preclinical behavioral pharmacology. As an example, models and readouts in the preclinical development of analgesic drugs for inflammatory and neuropathic pain are described. Furthermore, evoked and non-evoked pain readouts are reviewed. In the second section, the shortcomings of conventional preclinical models are discussed, and the necessity of improving translation in CNS drug discovery and development are also discussed. Several steps are proposed that could enhance translation from animal data to clinical efficacy. In this regard, neuroimaging and PK/PD modeling strategies are posed as crucial aspects in generating more valid, robust, reliable preclinical data resulting in more effective and translatable therapies. In conclusion, applying classical pharmacology to problems of translational medicine will aid us in improving the way we think about and use animal models. Closer collaboration and cross-over between clinical, pathological, and pharmacological research are paramount in optimizing the success of preclinical translation into novel medicines for patients in need worldwide.

Learning Objectives
- The necessity of preclinical behavioral models and readouts in drug discovery
- The importance of translational medicine for preclinical research
- The value of neuroimaging approaches and pharmacokinetic-pharmacodynamic (PK/PD) modeling strategies in preclinical drug discovery and development

5.1 General Introduction in Behavioral Models and Readouts

A crucial step in the discovery of novel drugs is proof of in vivo activity and efficacy. As described in the previous chapters, the drug discovery process is long and complicated starting from target identification and validation, followed by high throughput screening, virtual screening, numerous rounds of iterative structure activity relationship (SAR) modifications until a series of molecules is identified with drug-like properties. These molecules have demonstrated potency and selectivity at the desired target, favorable physicochemical properties, efficacy in in vitro and ex vivo model systems, and acceptable pharmacokinetic profiles. Next, target engagement, efficacy, and safety/tolerability need to be confirmed in an intact living organism, to further advance these molecules toward clinical trials and putative novel therapies. For this pivotal step, the use of animal studies is fundamentally required and inevitable.

5.1.1 Model Versus Readout and Validity

Pivotal in understanding behavioral preclinical research is to differentiate between an animal *model* and a *readout* (or test). *Animal models* or animal models of disease used in research may have an existing, inbred, or induced disease or injury that is similar to a human condition. The use of animal models allows researchers to investigate disease states in ways which would be not possible in patients. Procedures can be performed on the non-human animal that imply a level of harm that would not be considered ethical to inflict on a human. The best models of disease are similar in *etiology* and *phenotype* to the human equivalent. However, complex human diseases can often be better understood in a simplified system in which individual parts of the disease process are isolated and examined. Once an appropriate animal model of the disease or underlying part/mechanism of the disease has been established and validated, the model can be used to investigate specific hypotheses. Examples of animal models are a genetic modification in mice that mimic Aβ plaque formation in the brain of human Alzheimer's disease or a spinal nerve lesion model that mimics human neuropathic pain conditions. To test specific hypotheses in an animal model, accurate and robust *readouts* (or tests) are required, for example, a Morris water maze test for assessment of spatial learning or von Frey filaments for assessment of mechanical hypersensitivity. Thus, the model represents the implied changes to the animal to represent the disease state or mechanism of interest, whereas the readout represents the method of quantifying the effects of a (e.g., genetic or pharmacological) manipulation in that model.

Definition

Animal models are representations of human disease or conditions in a non-human species, etiology is the mechanism of cause of the disease, and phenotype stands for the respective signs and symptoms of the disease.

Preclinical development of novel drugs requires both robust models and readouts to assess disease-like behavior, underlying mechanisms, and efficacy of drug treatment. To be useful in predicting efficacy, the model and readout need to demonstrate *sensitivity*, *specificity*, and *predictivity* (Rice et al. 2008; Rutten et al. 2014).

Definition

Sensitivity is the ability to detect a true positive control, specificity stands for the ability to detect a true negative, and predictivity is the ability to predict the outcome in other model/species. Face validity assures that the biology and symptoms as seen in humans are similar in the animal model, and construct validity assures that the target exerts the same biological processes in both organisms.

Two further aspects are important in the validation of animal models, namely, *face validity* and *construct validity* (Denayer et al. 2014). A crucial aspect in preclinical development is the translation of preclinical findings into clinical efficacy. As such, *translational medicine* is the area of research that aims to improve human health by determining the relevance of novel discoveries in biological sciences to human disease. Translational medicine seeks to coordinate the use of new knowledge in clinical practice and to incorporate clinical observations and questions into scientific hypotheses in the laboratory. Thus, it is a bidirectional concept, encompassing so-called *bench-to-bedside* factors, which aim to increase the efficiency by which new therapeutic strategies developed through preclinical research are tested clinically, and *bedside-to-bench* factors, which provide feedback about the applications of new treatments and how they can be improved.

For many diseases that do not involve the central nervous system (CNS), animal models can be straightforward and clear readouts can be identified to investigate drug efficacy. For example, when developing novel anti-inflammatory drugs, a rodent model of inflammation by injection of liposaccharide (LPS) is employed and readouts such as edema or swelling can be assessed accurately and objectively (e.g., by caliper or plethysmography) and biomarkers such as the release of pro-inflammatory cytokines (e.g., TNF-alpha, or interleukins) can be measured. Both readouts can be directly inhibited by anti-inflammatory drugs. Furthermore, the same readouts (reduction of swelling and cytokine release) are employed in preclinical assays as well as in clinical trials which contributes to better translation of preclinical findings. For CNS drugs, the use of animal models and readouts is in most cases far less straightforward and several issues need to be taken into account. These will be discussed in this chapter. Indeed, for many CNS disorders the underlying mechanisms are poorly understood, the pathology is difficult to model in animals, and the readouts in animals are different from those in humans.

Because entire books have been written about all the different models and readouts for CNS disorders (e.g., McArthur and Borsini 2008), we focus here on animal models and readouts used in chronic pain and analgesic drug discovery research as an example. The drug discovery rationale, challenges, shortcomings, and discussions on translation described for the indication "pain" also apply in general to other CNS disease areas. Comprehensive reviews on animal models for specific indications have been published elsewhere (e.g., see Whiteside et al. (2013) for Pain; Puzzo et al. (2015) for Alzheimer's Disease; Soderlund and Lindskog (2018) for Depression; Harro (2018) for Anxiety).

3R Principles for Animal Experimentation
The use of animals in science is inevitable and required to unravel the underlying biology and pathology of diseases, or the mechanism of action and safety of putative novel therapies. In addition, animal studies on efficacy and safety are legally required by the regulatory agencies [e.g., the US Food and Drug Administration (FDA) and the European Medicines Agency (EMA)] that approve novel drugs to market. Importantly, all efforts must be made to restrict animal use and suffering. The *3Rs* are the guiding principles for animal experimentation, and they are adopted by ethical committees and governments across the world (first described by Russell and Burch (1992)). The 3Rs stand

for Replace, Reduce, Refine and represent a responsible approach to animal testing. The goal is to replace animal experiments whenever possible. In addition, the aim is to keep the number of animal experiments as low as possible and to only use the necessary number of animals. Finally, it is vital to ensure that the distress inflicted upon the animals is as low as possible.

- *Replace*: Replacing an animal experiment to the greatest possible extent, as long as adequate alternatives are available.
- *Reduce*: The reduction of animal experiments and the number of laboratory animals to the greatest possible extent. *Statistical power calculations* must be performed in advance to the experiments to ensure that sufficient (but not more than that) animals are used to meet the criteria for obtaining a statistically significant outcome.
- *Refine*: The methods and treatment of the animals during the experiments, and with regard to the way they are kept, should ensure that the distress caused to them is minimized to the greatest possible extent and that their well-being is taken into account as far as possible.

All three pillars are of equal importance in the 3R principles. Anyone conducting research with laboratory animals is obliged to comply with the 3R principles and regulations by local ethical authorities/government. Clearly many people object to animal research on principle (Regan 2007), and these objections have been discussed in detail elsewhere (Cohen 1986; Derbyshire 2002, 2006). This chapter will focus on animal models and readouts in CNS drug discovery, the difficulties and pitfalls, and the possible ways to improve translational medicine in the drug discovery process. A scientific rationale for the use of animal research as an important mechanism in advancing drug discovery is provided. Those that object to animal research on principle will, understandably, be unmoved by the scientific advances animal research provides.

5.2 Models and Readouts in Pain Drug Development

Experiments investigating pain in human subjects have intrinsic practical difficulties; accordingly, early analgesic development relies on animal models (Mogil 2009). Modeling human pain in experimental animals is inherently challenging for several reasons. First, pain is defined as an unpleasant sensory and emotional experience (definition on website of the International Association for Studies against Pain (IASP)), and is thus subjective, existing only in the person who experiences it (first-person perspective). Importantly, humans communicate their pain experience verbally, and pain is quantifiable via numerical or visual analogue scales (Barrot 2012). The absence of verbal communication in animals is undoubtedly a challenge to the evaluation of pain, and therefore indirect behavioral readouts must be used as *surrogate markers.* Thus, preclinical pain research must rely on stimulus-evoked responses or alterations in the behavior of an animal as readout (Barrot 2012; Deuis et al. 2017). Obviously the same issue with lack of verbal communication holds true for the evaluation of other CNS disorders such as memory-loss, anxiety, and depression (◘ Fig. 5.1). Second, pain is not

merely a somatosensory experience due to a noxious stimulus, but is influenced by a number of factors including affective, attentional, and cognitive states, all of which are dynamic in nature and difficult to model. Third, the suffering aspects of clinically relevant pain cannot be fully modeled in animals, as suffering must be minimized in animal research for ethical reasons and follow the 3R policies. Furthermore, rodents are prey animals that live in social groups and will therefore try and hide any sign of weakness, including pain, as this would make them more vulnerable to predators or lose their rank in the group hierarchy (Deacon 2006). Finally, human pain cannot be modeled as a single disease, but rather as a syndrome brought on by a wide variety or conditions (Jensen 2010).

5.2.1 Models of Neuropathic and Inflammatory Pain

Neuropathic pain is defined as "pain initiated or caused by a primary lesion or dysfunction in the nervous system" (Merskey and Bogduk 1994). *Inflammatory pain* refers to increased sensitivity due to the inflammatory response associated with tissue damage. Under these sensitized conditions for both neuropathic and inflammatory pain, an innocuous stimulus can be perceived as painful—this is known as *allodynia,* and the pain evoked by a noxious stimulus is exaggerated in both amplitude and duration—this is known as *hyperalgesia* (Sandkuhler 2009).

Over the years, many animal models for inflammatory pain using irritant agents and surgical and non-surgical animal models for neuropathic pain have been developed and used for preclinical testing. The most common models are summarized in ◘ Fig. 5.2. Briefly, to model inflammatory pain, substances that result in an immune response are injected directly into the peritoneum, paw, or into the joint, respectively, #1, #2, and #3 in ◘ Fig. 5.2. The most commonly used irritant substances are phenylbenzoquinone and other acidic compounds in writhing tests (Eckhardt et al. 1958; Vander Wende and Margolin

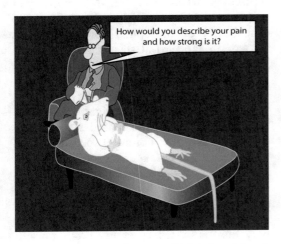

◘ **Fig. 5.1** Cartoon of a rat suffering from chronic pain. It is an impossible quest to completely mimic chronic pain, and other CNS, disorders in rodents. (Modified from: Cryan et al. (2002). Copyright license obtained from Elsevier)

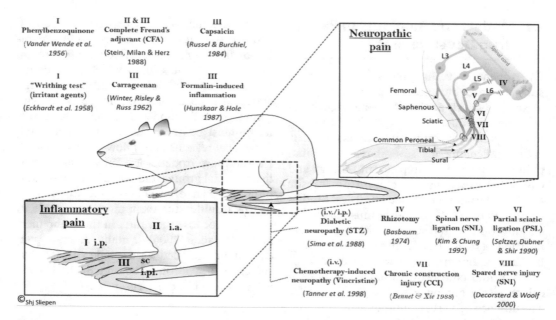

Fig. 5.2 A schematic illustration of well-known inflammatory and neuropathic pain models. i.p. intraperitoneal, i.a. intraarticular, sc subcutaneous, i.pl intraplantar, i.v. intravenously, STZ streptozotocin (From: Sliepen (2019). *Reuse permission obtained from SHJ Sliepen*)

1956). For local administration into limbs or joints heat-killed mycobacterium butyricum/tuberculosum, i.e., complete Freund's adjuvant (CFA) (Stein et al. 1988), formalin (Hunskaar et al. 1986), hot chilly pepper extract, i.e., capsaicin (Russell and Burchiel 1984), or sulphated polysaccharides from seaweed, i.e., carrageenan (Winter et al. 1962) are used. Injection of CFA into a paw, ankle, or knee joint results in local inflammation and serves as a model for human arthritic pain (Neugebauer et al. 2007). The use of genomic data is changing to a significant degree how we understand human disease. In the area of inflammation, the ability to build new genomic models in mice using information from genomic, proteomic, and metabolomic studies is growing ever more.

The majority of the neuropathic pain models are induced by unilateral surgical damage to a specific nerve (mono-neuropathic) in order to study the effects of pain-like behavior in a controlled manner. The most commonly used methods (see Fig. 5.2) are chronic constriction injury (CCI) of the

sciatic nerve (#VII; (Bennett and Xie 1988)), spared nerve injury (SNI: #VIII, (Decosterd and Woolf 2000)), and spinal nerve ligation (SNL; #V, (Kim and Chung 1992)) models as these generate the most reproducible phenotypes. In the rhyzotomy model (#IV, (Basbaum 1974)), one spinal nerve (L5) is transected, whereas in the SNL model two spinal nerves (L5 and L6) coming from the dorsal root of the spinal cord are tightly ligated. In the partial sciatic nerve ligation (PSL; #VI; (Seltzer et al. 1990)) model, a portion of the sciatic nerve is tightly ligated, whereas the CCI model involves placement of four loose chromic-gut ligatures around the sciatic nerve. In the spared nerve injury (SNI) model, the common peroneal and tibial nerves are cut, sparing the sural nerve. Alternatively, models have been designed to study poly-neuropathic pain where multiple nerves in the body are affected, such as, the streptozotocin (STZ) models for diabetic polyneuropathy (Sima et al. 1988) or vincristine models for chemotherapy-induced peripheral neuropathy (Tanner et al. 1998).

5.2.2 Readouts for Assessment of Pain-Like Behavior

Preclinical pain models may be associated with spontaneous pain-related behavior as well as allodynia and hyperalgesia, which result in enhanced responses to mechanical, heat, and/or cooling stimuli (Campbell and Meyer 2006). There are various ways to measure pain-like behavior in animals, either via a response to an applied stimulus (stimulus-evoked readout) or assessment of behavior independent of an applied stimulus (non-stimulus-evoked readout). For stimulus-evoked readouts, an external stimulus (mechanical, thermal, or chemical) is applied to a specific site on the test subjects' body to elicit a behavioral response. The nature of this response is frequently a withdrawal response to the stimulus, and the readout is either the force at which the response occurs, the latency until the response occurs, or the number of responses to the specific stimulus. Subsequently, mechanical hyperalgesia can be assessed by applying increased pressure to the paws (Randall and Selitto 1957) and mechanical allodynia via application of von Frey filaments to the plantar surface of the paw (Dixon 1980). Thermal stimuli are divided into heat stimuli, and responses are measured in a tail-flick, hot plate, or Hargreaves test, and cold stimuli, where responses are measured in a cold plate test. A main criticism of stimulus-evoked readouts is that they often rely on thresholds or latencies which do not adequately reflect clinical pain (Klinck et al. 2017). Furthermore, they are often induced by an experimenter, possibly resulting in a bias.

Thus far, translation of preclinical findings into clinical studies has been difficult and numerous examples exist where preclinically efficacious analgesic compounds did not show an effect in Phase 2 proof-of-concept clinical trials (see below). Part of this challenging translation may be due to inappropriate and unpredictable animal models and readouts. Therefore, a great effort has been made to improve alternative non-stimulus-evoked behavioral tests (Percie du Sert and Rice 2014); to standardize animal models and readouts; and to increase experimental rigidity to reduce bias in preclinical research (Knopp et al. 2015). A promising example of these alternative readouts is the burrowing test (Andrews et al. 2011; Deacon 2006), in which animals are allowed to exhibit their innate behavior of digging tunnels and burrows in a laboratory setting. Several inflammatory and neuropathic pain models result in reduced burrowing behavior, which was reversed by analgesics (Andrews et al. 2012; Huang et al. 2013; Rutten et al. 2018). A major advantage for preclinical analgesic drug development is that burrowing is less prone to generate false positives due to impaired motor skills or sedation, as opposed to traditional stimulus-evoked tests (Rutten et al. 2014).

5.2.3 An Example of Failed Translation from Animal Data to Clinical Trials of NK-1 Antagonist as Putative Novel Analgesic Drugs

Neurokin receptor 1 (NK-1) antagonists block the receptor for the neurotransmitter Substance P and boost activation of serotonin 5-HT3 receptors in order to prevent nausea and vomiting. The discovery of NK1 receptor antagonists was a turning point in the prevention of nausea and vomiting associated with cancer chemotherapy. Scientists believed that NK-1 antagonism would be a promising target for treatment of chronic pain. Unfortunately, NK-1 antagonists have become an infamous example of where preclinical efficacy did not translate into clinical efficacy in Phase 2 trials for chronic pain. Indeed, NK 1 receptor antagonists have failed to exhibit efficacy in clinical trials of a variety of clinical pain states. By contrast, there were sufficient well-conducted animal studies in which an NK1 receptor antagonist attenuated the behavioral or electrophysiological response to a noxious stimulus to justify performing clinical trials for analgesia (Hill 2000). The profile of the compounds across the behavioral tests was actually comparable to that of non-steroid anti-inflammatory drugs (NSAIDs), which are analgesic in humans. Thus, NK 1 receptor antagonists seem to be able to block behavioral responses to noxious and other stressful sensory stimuli at a level detectable in animal tests, but the translation to achieve the level of sensory blockade required to produce clinical analgesia in humans failed. The importance of supraspinal targets of analgesics has been underestimated, and most preclinical studies of analgesia are focused on the spinal dorsal horn, despite the fact that many substances elicit their analgesic effects primarily at the supraspinal level (Hill 2000). Finally, it is relevant to ask whether the failure to predict the presence or absence of analgesic properties in humans in the case of NK1 receptor antagonists has implications for the discovery and clinical evaluation of other putative analgesics. On the one hand, many examples exist of a substance exhibiting analgesia in animal models and clinical analgesia in humans. For example, ketamine is an antagonist of NMDA receptors, which are widely distributed in the CNS, and is analgesic in both animals and humans. On the other hand, the enkephalinase inhibitors, which increase the concentrations of endogenous opioid peptides, possess antinociceptive effects in animals but lack analgesic effects in humans (Villanueva 2000).

What preclinical criteria should be used to determine whether clinical trials of a new analgesic are likely to be successful? Perhaps one of the main challenges for preclinical studies of pain and other CNS diseases today is to employ holistic and integrative approaches to improve our preclinical disease understanding and to enable the building of bridges between scientists and clinicians interested in discovering novel treatment options for CNS diseases.

5.3 Improving Translation in CNS Drug Discovery and Development

Evaluating brain function by means of imaging technology in patients and respective animal models has the potential to characterize mechanisms associated with the disorder or disorder-related phenotypes and could pro-

vide a means of better *bench-to-bedside* and *bedside-to-bench* translation.

5.3.1 Neuroimaging

As discussed above, translation from behavioral models and readouts in rodents toward clinical patient–reported outcome measures is troublesome in drug discovery for pain and other CNS disorders. ☐ Figure 5.3 gives a clear representation of the difference in phenotype and CNS properties between clinical and preclinical settings in the field of pain research.

Functional magnetic resonance imaging (fMRI) is an excellent tool to study the effect of manipulations of brain function in a noninvasive and longitudinal manner. Several MRI techniques permit the assessment of functional connectivity during rest as well as brain activation triggered by sensory stimulation and/or a pharmacological challenge in both rodents and humans. Stimulation with a drug and in combination with MRI is called *pharmacological MRI (PhMRI)*, and it has a number of interesting possibilities compared to conventional fMRI. Using selective pharmacological tools, the neurotransmitter-specific brain circuitry, neurotransmitter release, and binding associated with the pharmacokinetics and/or the pharmacodynamics of drugs can be investigated (Jenkins 2012). As such, PhMRI can be characterized as a molecular imaging technique using the natural hemodynamic transduction related to neuro-receptor stimuli.

Although differences in brain size, structure, and function exist between rodents and humans, a preservation of CNS networks across species has been observed using functional brain imaging (Gozzi et al. 2006). Furthermore, using phMRI-consistent pharmacodynamic responses have been observed across species for opioids (See ☐ Fig. 5.4, (Becerra et al. 2013)) and other analgesic drugs (Borsook and Becerra 2011).

It is currently believed that neuroimaging may describe the central representation of pain or pain phenotypes and yields a basis for the development and selection of clini-

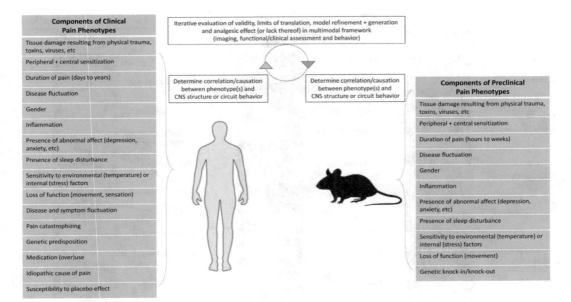

Components of Clinical Pain Phenotypes

Tissue damage resulting from physical trauma, toxins, viruses, etc

Peripheral + central sensitization

Duration of pain (days to years)

Disease fluctuation

Gender

Inflammation

Presence of abnormal affect (depression, anxiety, etc)

Presence of sleep disturbance

Sensitivity to environmental (temperature) or internal (stress) factors

Loss of function (movement, sensation)

Disease and symptom fluctuation

Pain catastrophizing

Genetic predisposition

Medication (over)use

Idiopathic cause of pain

Susceptibility to placebo effect

Iterative evaluation of validity, limits of translation, model refinement + generation and analgesic effect (or lack thereof) in multimodal framework (imaging, functional/clinical assessment and behavior)

Determine correlation/causation between phenotype(s) and CNS structure or circuit behavior

Determine correlation/causation between phenotype(s) and CNS structure or circuit behavior

Components of Preclinical Pain Phenotypes

Tissue damage resulting from physical trauma, toxins, viruses, etc

Peripheral + central sensitization

Duration of pain (hours to weeks)

Disease fluctuation

Gender

Inflammation

Presence of abnormal affect (depression, anxiety, etc)

Presence of sleep disturbance

Sensitivity to environmental (temperature) or internal (stress) factors

Loss of function (movement)

Genetic knock-in/knock-out

Fig. 5.3 Convergence of phenotypes and CNS properties in clinical and preclinical settings. A model of clinical and preclinical pain experimentation considers the use of pain-related phenotypes in conjunction with CNS function to assess and improve the overall validity of preclinical pain investigations (From: Upadhyay et al. (2018). Copyright license obtained from Elsevier)

cally relevant animal assays (For review see: Upadhyay et al. (2018)). The large numbers of molecules available, which do not require a radio-label, means that phMRI has become a very useful tool for performing drug discovery. Translational phMRI approaches may increase the probability of finding meaningful novel drugs that can help satisfy the significant unmet medical needs of patients suffering from CNS disorders.

5.3.2 PK/PD-Modeling

A crucial step in the development of novel drugs is to generate a growing understanding of the relationship between the *pharmacokinetic (PK)* profile and the *pharmacodynamic (PD)* profile. As such, *PK/PD modeling* refers to a data (PK and PD)-driven exploratory analysis, based on mathematical or statistical models. In other words, the objective of pharmacokinetic-pharmacodynamic (PK/PD) modeling is the development and application of mathematical models to describe and/or predict the time course of dose-to-concentration (PK) and concentration-to-

effect (PD) of pharmacological active agents in health and disease (Martini et al. 2011). Clinically, the rationale for measuring drug concentration is that the relationship between concentration and effect should be less variable than the relationship between dose and effect (Atkinson et al. 2007). Therefore, accurately measuring the concentration will allow for better predictions of drug effect than dose information alone.

> **Definition**
>
> The *pharmacokinetic (PK)* profile represents how the organism affects the drug by means of absorption, distribution, metabolism and excretion, and which concentrations of the drug reach the target organ. The *pharmacodynamic (PD)* profile represents how the drug affects the organism, and what dose causes which (side) effect.

This allows the observed drug effect to be related directly to the time after a given dose. Therefore, the combined PK/PD model provides a means of understanding

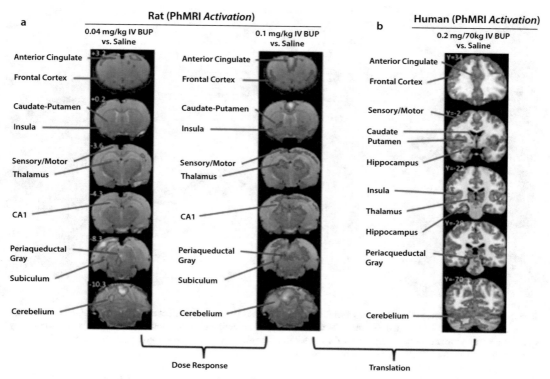

Fig. 5.4 PhMRI activation after i.v. buprenorphine administration. **a** PhMRI of 0.04 and 0.1 mg/kg i.v. buprenorphine yielded dose-dependent phMRI activation (drug. saline) in the conscious rat. **b** phMRI activation was observed in the human buprenorphine phMRI dataset with 0.2 mg/kg i.v. buprenorphine administered. The labeled brain structures highlight regions where phMRI activation was induced at the higher doses of buprenorphine tested in both species (From: Becerra et al. (2013). Copyright license obtained from ASPET Springer publishing group)

the time course of drug effect, namely, the extent, onset, and duration of drug action (Wright et al. 2011). Kinetic-dynamic reasoning should, whenever possible, be based on in vitro and in vivo concentration-time, response-time, and concentration-response relationships, with an underlying ambition to couple this to the disease state. The discipline of modeling is always data-driven, and it relies on multiple analyses of the same dataset in an iterative mode with successive and/or competing models.

PK/PD modeling and simulation can add value in all stages of the drug development process starting from the preclinical development stage up to late stage clinical development. To utilize PK/PD modeling and simulation in its optimal potential for drug development, models should be developed early in drug discovery, preferably during the preclinical phase. Such models are continuously updated and refined as more data become available. Their validation is necessary during development, and they will then provide valuable support to make important decisions, with an increased confidence level around the analyzed data.

During the preclinical phase of drug development, various in vitro and in vivo studies have been used to screen compounds for efficacy. From in vivo efficacy models, the EC_{50} *concentration* is determined, which is the average plasma concentration at which half of the subjects show a pharmacological effect of 50%. Of note, more often than not the dose-response curves in in vivo efficacy models lack dose-dependency, and EC_{50} cannot be determined (e.g., inverted U-shaped curves), in those cases the minimal effective dose/concentration (MED/MEC) will be calculated.

Further along the way, in vivo safety pharmacology assessments will be performed, often in parallel to efficacy testing, to examine side-effect profiles, and to determine the lowest dose/concentration at which the compound demonstrated no adverse effects (*NOAEL*). The efficacy EC_{50} values from the different in vivo animal models are then compared to the NOAEL levels from different safety and toxicology studies to determine *safety margins*, or *therapeutic index*.

Definition

The *EC_{50} concentration* stands for the average plasma concentration at which half of the subjects show a pharmacological effect of 50%. *NOAEL* stands for the lowest dose/concentration at which the drug demonstrated no adverse effects. The *safety margin* or *therapeutic index* describes the distance in order of magnitude between wanted effects, i.e., efficacy and unwanted effects, i.e., aversive side effects.

These values in combination with the PK/PD models are crucial in ranking compounds from a chemical series, and they are helpful techniques in understanding the complex behavior of specific drugs, especially with respect to estimation of clinical dosing protocols and assessment of therapeutic indices and safety margins based on preclinical in vitro and in vivo data. By appropriate use of PK/PD modeling the EC_{50}, MEC and safety margins are inter- or extrapolated and used to predict and determine whether a compound may proceed into further development, i.e., testing in higher species: dog, non-human primate, and eventually human clinical trials, or whether it will be stopped from further development. PK/PD modeling offers the greatest value if preclinical data can be modeled in combination with existing clinical data on related compounds (internal or competitors data) (Lesko et al. 2000).

In the later clinical stages of drug development, the PK/PD models are complemented with clinical efficacy, safety, and biomarker data in order to improve the model and enhance its predictive power. Recently, promising efforts emerge in which public domain medical knowledge about the relationship between biomarker responses and clinical outcomes for different diseases are used to build extensive PK/PD models (Pirisi 2003; Schlessinger and Eddy 2002). Large and structured databases with clinical findings are required for building such disease models and pooling patients' data from different databases that exist across the pharmaceutical industry would provide an invaluable source of information for disease modeling. If pharmaceutical companies were to collaborate on a precompetitive level to generate clinical databases and validate the disease models this would greatly benefit Phase 3 design and target population selection across the industry.

In conclusion, PK/PD modeling and simulation is an invaluable tool aiding crucial decision-making in drug development. Decisions on compound and dose selection, study design, or patient population, all of which can lead to a considerable reduction in cost of development. Thus, better implementation of PK/PD modeling throughout the drug discovery and development process could enhance translational success and result in less failed clinical trials and eventually better drugs entering the market (Gabrielsson and Weiner 2006).

5.4 Discussion

In general, preclinical CNS models are most often highly simplified representations of clinical features that are common across multiple conditions, such as tactile allodynia for both diabetic neuropathy and chemotherapy-induced pain or memory impairment for both Alzheimer's disease and schizophrenia. Of note, any combination of model and readout reflects a limited set of these clinical signs and their underlying pathophysiological mechanisms, and therefore the choice of model and readout from the battery of available assays is an important consideration (Soderlund and Lindskog 2018). A single model should not be expected to represent all aspects of the clinical conditions, but data generated in preclinical efficacy models are nevertheless useful in predicting drug efficacy when used in conjunction with other methods, ranging from drug metabolism and pharmacokinetic analysis to electrophysiology and functional imaging, biomarkers, safety margins, and PK/PD modeling.

However, recently, animal behavioral models have been increasingly scrutinized and criticized for their role in the poor translation of novel pharmacotherapies. Indeed the number of failed clinical trials and the paucity of novel market approvals for CNS disorders such as Alzheimer's disease, pain, and major depressive disorder blatantly underscore this (Bazzari et al. 2019; Mogil 2019; Soderlund and Lindskog 2018). What is important is that efforts are being made to improve the translation of preclinical findings into clinical efficacy. Recently, several proposals were made to improve translation from animal models into human clinical situations.

First, of course, better translational models are required. Employing disease models in species more relevant to humans than rodents, such as non-human primates and the implication of new technologies such as *Clustered Regularly Interspaced Short Palindromic Repeats (CRISPR)* genome DNA manipulation are progressing rapidly (See King (2018)) and may increase translational success of drug development in Alzheimer's disease and other CNS indications. Second, issues of internal validity and reproducibility of animal models must be improved. Many preclinical studies suffer from poor methodological design, lack of statistical power, and bias induced by lack of blinding and randomization (see Knopp et al. (2015)). Ideally preclinical experiments should be conducted with the same experimental rigidity and standardization as clinical studies, and strict guidelines (e.g., ARRIVE guidelines) for preclinical animal studies must always be implemented and enforced (Kilkenny et al. 2010; Rice et al. 2013).

Additionally, more efforts should be made to standardize models and readouts to allow for comparison and meta-analysis of preclinical data (See Wodarski et al. (2016)).

Third, to enhance the interaction between the clinic and neurobiology, the National Institute of Health has proposed to use Research Domain Criteria (RDoC) as a novel approach to categorizing psychiatric conditions (see ▶ http://www.nimh.nih.gov/research-priorities/rdoc/constructs/rdoc-matrix.shtml, ▶ Chap. 14) as opposed to classic diagnostic classification systems such as DSM or ICD. As such, future diagnostic systems cannot reflect ongoing advances in genetics, neuroscience, and cognitive science until a literature organized around these disciplines is available. The goal of the RDoC project is to provide a framework for research to transform the approach to the nosology of mental disorders (Cuthbert and Insel 2013). Thus, a system based on well-defined neurobiological constructs that will facilitate better communication between research and clinic should be created (Soderlund and Lindskog 2018). This could be useful not only for mental disorders but also other CNS disorders, such as pain.

Fourth, the industry has typically worked on a target and then tried to fit it to a patient population (often the prescribed regulatory patient groups). As such, conventional drug therapy typically considers large patient populations to be relatively homogeneous (the one-drug-fits-all approach). Only recently genetically based differences in response to a single-drug or multiple-drug treatment have been adopted and accepted (Vogenberg et al. 2010). *Personalized medicine* approaches stipulate that any given drug can be therapeutic in some individuals but ineffective in others, and some individuals experience adverse drug effects whereas others are unaffected. These findings should be back-translated into preclinical responder and non-responder analysis that could be helpful in better understanding efficacy.

Finally, an important step toward better translation is to create networks to learn from each other and collaborate on a non-competitive level. To be more successful in drug discovery, pharmaceutical industry, academic institutions, and healthcare practitioners need to accept failures and learn from them to find new solutions for the many patients suffering from diseases of the CNS. Initiatives such as the innovative medicines initiative (IMI) Europain (▶ www.imieuropain.org) and IMI Paincare (▶ www.imi-paincare.eu) connect scientists from clinic and preclinic as well as from academia and industry to jointly improve their research and strive for better translation and analgesic drug development. Similar initiatives exist for other disease indications of the CNS (see ▶ www.imi.europa.eu).

❯ Conclusion

In conclusion, this chapter has focused on conventional behavioral (animal) models and their usefulness and shortcomings in the drug discovery process. The need for greater understanding of the fundamental physiology underlying CNS diseases will persist at least as long as treatment of patients suffering from these diseases remains suboptimal. *From a scientific perspective, there are no short-to-medium term solutions that would lead to true advances in drug discovery, which would render animal studies obsolete. Nevertheless, the combination of human phMRI imaging (and other human) studies along with appropriate PK/PD modeling and more valid, robust, reliable animal studies will lead to far more effective and translatable science and ultimately novel drugs than has been the case thus far.* Furthermore, what used to be termed pharmacology is increasingly being labeled translational medicine and there are hopeful signs that some universities and medical schools are beginning to rethink how biomedical scientists ought to be trained (Webb 2014). Applying the sound principles of classical pharmacology to problems of translational medicine will aid us all in improving the way we think about and use animal models based on the careful cross fertilization from clinical, pathological, and pharmacological research.

References

Andrews N, Harper S, Issop Y, Rice AS (2011) Novel, nonreflex tests detect analgesic action in rodents at clinically relevant concentrations. Ann N Y Acad Sci 1245:11–13

Andrews N, Legg E, Lisak D, Issop Y, Richardson D, Harper S et al (2012) Spontaneous burrowing behaviour in the rat is reduced by peripheral nerve injury or inflammation associated pain. Eur J Pain 16(4):485–495

Atkinson AJ, Huang S, Lertora JJL, Markey SP (2007) Principles of Clinical Pharmacology. Academic Press, Amsterdam, p 652

Barrot M (2012) Tests and models of nociception and pain in rodents. Neuroscience 211:39–50

Basbaum AI (1974) Effects of central lesions on disorders produced by multiple dorsal rhizotomy in rats. Exp Neurol 42(3):490–501

Bazzari FH, Abdallah DM, El-Abhar HS (2019) Pharmacological interventions to attenuate Alzheimer's disease progression: the story so far. Curr Alzheimer Res 16(3):261–277

Becerra L, Upadhyay J, Chang PC, Bishop J, Anderson J, Baumgartner R et al (2013) Parallel buprenorphine phMRI responses in conscious rodents and healthy human subjects. J Pharmacol Exp Ther 345(1):41–51

Bennett GJ, Xie YK (1988) A peripheral mononeuropathy in rat that produces disorders of pain sensation like those seen in man. Pain 33(1):87–107

Borsook D, Becerra L (2011) CNS animal fMRI in pain and analgesia. Neurosci Biobehav Rev 35(5):1125–1143

Campbell JN, Meyer RA (2006) Mechanisms of neuropathic pain. Neuron 52(1):77–92

Cohen C (1986) The case for the use of animals in biomedical research. N Engl J Med 315(14):865–870

Cryan JF, Markou A, Lucki I (2002) Assessing antidepressant activity in rodents: recent developments and future needs. Trends Pharmacol Sci 23(5):238–245

Cuthbert BN, Insel TR (2013) Toward the future of psychiatric diagnosis: the seven pillars of RDoC. BMC Med 11:126

Deacon RM (2006) Burrowing in rodents: a sensitive method for detecting behavioral dysfunction. Nat Protoc 1(1):118–121

Decosterd I, Woolf CJ (2000) Spared nerve injury: an animal model of persistent peripheral neuropathic pain. Pain 87(2):149–158

Denayer T, Stohr T, Van Roy M (2014) Animal models in translational medicine: validation and prediction. New Horiz Transl Med 2(1):5–11

Derbyshire SWG (2002) Why animal rights are wrong. In: Lee E (ed) Animal experiments: good or bad? Hodder & Stoughton, London

Derbyshire SWG (2006) Time to abandon the three Rs. The Scientist Magazine, p 20–3

Deuis JR, Dvorakova LS, Vetter I (2017) Methods used to evaluate pain behaviors in rodents. Front Mol Neurosci 10:284

Dixon WJ (1980) Efficient analysis of experimental observations. Annu Rev Pharmacol Toxicol 20:441–462

Eckhardt ET, Cheplovitz F, Lipo M, Govier WM (1958) Etiology of chemically induced writhing in mouse and rat. Proc Soc Exp Biol Med 98(1):186–188

Gabrielsson J, Weiner D (2006) Pharmacokinetic-Pharmacodynamic data analysis: concepts and applications. Swedish Pharmaceutical Press, Stockholm

Gozzi A, Schwarz A, Reese T, Bertani S, Crestan V, Bifone A (2006) Region-specific effects of nicotine on brain activity: a pharmacological MRI study in the drug-naive rat. Neuropsychopharmacology 31(8):1690–1703

Harro J (2018) Animals, anxiety, and anxiety disorders: how to measure anxiety in rodents and why. Behav Brain Res 352:81–93

Hill R (2000) NK1 (substance P) receptor antagonists-why are they not analgesic in humans? Trends Pharmacol Sci 21(7):244–246

Huang W, Calvo M, Karu K, Olausen HR, Bathgate G, Okuse K et al (2013) A clinically relevant rodent model of the HIV antiretroviral drug stavudine induced painful peripheral neuropathy. Pain 154(4):560–575

Hunskaar S, Berge OG, Hole K (1986) Dissociation between antinociceptive and anti-inflammatory effects of acetylsalicylic acid and indomethacin in the formalin test. Pain 25(1):125–132

Jenkins BG (2012) Pharmacologic magnetic resonance imaging (phMRI): imaging drug action in the brain. NeuroImage 62(2):1072–1085

Jensen MP (2010) A neuropsychological model of pain: research and clinical implications. J Pain 11(1):2–12

Kilkenny C, Browne WJ, Cuthill IC, Emerson M, Altman DG (2010) Improving bioscience research reporting: the ARRIVE guidelines for reporting animal research. PLoS Biol 8(6):e1000412

Kim SH, Chung JM (1992) An experimental model for peripheral neuropathy produced by segmental spinal nerve ligation in the rat. Pain 50(3):355–363

King A (2018) The search for better animal models of Alzheimer's disease. Nature 559(7715):S13–SS5

Klinck MP, Mogil JS, Moreau M, Lascelles BDX, Flecknell PA, Poitte T et al (2017) Translational pain assessment: could natural animal models be the missing link? Pain 158(9):1633–1646

Knopp KL, Stenfors C, Baastrup C, Bannon AW, Calvo M, Caspani O et al (2015) Experimental design and reporting standards for improving the internal validity of pre-clinical studies in the field of pain: consensus of the IMI-Europain consortium. Scand J Pain 7(1):58–70

Lesko LJ, Rowland M, Peck CC, Blaschke TF (2000) Optimizing the science of drug development: opportunities for better candidate selection and accelerated evaluation in humans. Pharm Res 17(11):1335–1344

Martini C, Olofsen E, Yassen A, Aarts L, Dahan A (2011) Pharmacokinetic-pharmacodynamic modeling in acute and chronic pain: an overview of the recent literature. Expert Rev Clin Pharmacol 4(6):719–728

McArthur RA, Borsini F (2008) In: McArthur RA, Borsini F (eds) Psychiatric disorders. Academic Press, Burlington

Merskey H, Bogduk N (1994) Classification of chronic pain: descriptions of chronic pain syndromes and definitions of pain terms. 2 ed. IASP Press, Seattle

Mogil JS (2009) Animal models of pain: progress and challenges. Nat Rev Neurosci 10(4):283–294

Mogil JS (2019) The translatability of pain across species. Philos Trans R Soc Lond B Biol Sci 374(1785):20190286

Neugebauer V, Han JS, Adwanikar H, Fu Y, Ji G (2007) Techniques for assessing knee joint pain in arthritis. Mol Pain 3:8

Percie du Sert N, Rice AS (2014) Improving the translation of analgesic drugs to the clinic: animal models of neuropathic pain. Br J Pharmacol 171(12):2951–2963

Pirisi A (2003) Can a supercomputer help doctors manage patients? American Diabetes Association hopes online computer consultant will improve diabetes care. Lancet 362(9382):496

Puzzo D, Gulisano W, Palmeri A, Arancio O (2015) Rodent models for Alzheimer's disease drug discovery. Expert Opin Drug Discov 10(7):703–711

Randall LO, Selitto JJ (1957) A method for measurement of analgesic activity on inflamed tissue. Arch Int Pharmacodyn Ther 111(4):409–419

Regan T (2007) Defending animal rights. University of Illinois Press, Champaign, IL

Rice AS, Cimino-Brown D, Eisenach JC, Kontinen VK, Lacroix-Fralish ML, Machin I et al (2008) Animal models and the prediction of efficacy in clinical trials of analgesic drugs: a critical appraisal and call for uniform reporting standards. Pain 139(2):243–247

Rice ASC, Morland R, Huang W, Currie GL, Sena ES, Macleod MR (2013) Transparency in the reporting of in vivo pre-clinical pain research: the relevance and implications of the ARRIVE (Animal Research: Reporting In Vivo Experiments) guidelines. Scand J Pain 4(2):58–62

Russell LC, Burchiel KJ (1984) Neurophysiological effects of capsaicin. Brain Res 320(2–3):165–176

Russell WMS, Burch RL (1992) The principles of humane experimental technique (special ed.). Universities Federation for Animal Welfare, South Mimms, Potters Bar, Herts, UK

Rutten K, Robens A, Read SJ, Christoph T (2014) Pharmacological validation of a refined burrowing paradigm for prediction of analgesic efficacy in a rat model of sub-chronic knee joint inflammation. Eur J Pain 18(2):213–222

Rutten K, Gould SA, Bryden L, Doods H, Christoph T, Pekcec A (2018) Standard analgesics reverse burrowing deficits in a rat CCI model of neuropathic pain, but not in models of type 1 and type 2 diabetes-induced neuropathic pain. Behav Brain Res 350:129–138

Sandkuhler J (2009) Models and mechanisms of hyperalgesia and allodynia. Physiol Rev 89(2):707–758

Schlessinger L, Eddy DM (2002) Archimedes: a new model for simulating health care systems--the mathematical formulation. J Biomed Inform 35(1):37–50

Seltzer Z, Dubner R, Shir Y (1990) A novel behavioral model of neuropathic pain disorders produced in rats by partial sciatic nerve injury. Pain 43(2):205–218

Sima AA, Zhang WX, Tze WJ, Tai J, Nathaniel V (1988) Diabetic neuropathy in STZ-induced diabetic rat and effect of allogeneic islet cell transplantation. Morphometric analysis. Diabetes 37(8):1129–1136

Sliepen SHJ (2019) The role of the nociceptin/orphanin FQ system in an animal model of bone cancer pain (PhD Thesis). University of Copenhagen, Copenhagen

Soderlund J, Lindskog M (2018) Relevance of rodent models of depression in clinical practice: can we overcome the obstacles in translational neuropsychiatry? Int J Neuropsychopharmacol 21(7):668–676

Stein C, Millan MJ, Herz A (1988) Unilateral inflammation of the hindpaw in rats as a model of prolonged noxious stimulation: alterations in behavior and nociceptive thresholds. Pharmacol Biochem Behav 31(2):445–451

Tanner KD, Reichling DB, Levine JD (1998) Nociceptor hyper-responsiveness during vincristine-induced painful peripheral neuropathy in the rat. J Neurosci 18(16):6480–6491

Upadhyay J, Geber C, Hargreaves R, Birklein F, Borsook D (2018) A critical evaluation of validity and utility of translational imaging in pain and analgesia: Utilizing functional imaging to enhance the process. Neurosci Biobehav Rev 84:407–423

Vander Wende C, Margolin S (1956) Analgesic tests based upon experimentally induced acute abdominal pain in rats. Fed Proc 15:494

Villanueva L (2000) Is there a gap between preclinical and clinical studies of analgesia? Trends Pharmacol Sci 21(12):461–462; author reply 5

Vogenberg FR, Isaacson Barash C, Pursel M (2010) Personalized medicine: part 1: evolution and development into theranostics. P T 35(10):560–576

Webb DR (2014) Animal models of human disease: inflammation. Biochem Pharmacol 87(1):121–130

Whiteside GT, Pomonis JD, Kennedy JD (2013) An industry perspective on the role and utility of animal models of pain in drug discovery. Neurosci Lett 557 Pt A:65–72

Winter CA, Risley EA, Nuss GW (1962) Carrageenin-induced edema in hind paw of the rat as an assay for antiinflammatory drugs. Proc Soc Exp Biol Med 111:544–547

Wodarski R, Delaney A, Ultenius C, Morland R, Andrews N, Baastrup C et al (2016) Cross-centre replication of suppressed burrowing behaviour as an ethologically relevant pain outcome measure in the rat: a prospective multicentre study. Pain 157(10):2350–2365

Wright DF, Winter HR, Duffull SB (2011) Understanding the time course of pharmacological effect: a PKPD approach. Br J Clin Pharmacol 71(6):815–823

Safety and Drug Metabolism: Toward NCE and First in Human

Jacco J. Briedé

Contents

6.1 Introduction: Human Drug Safety of Psychoactive Compounds: The Challenges – 94

6.2 Drug Design and Safety Regulations – 95

6.3 Testing for Human Neurotoxicity: From Animals to In Vitro Models – 95

6.4 Pharmacokinetics: Relevant Dose Selection for In Vitro Tests – 97

6.5 Metabolism of Neuroactive Compounds – 98

6.6 From Toxicogenomics to Systems Biology – 99

References – 100

The attrition rate of new chemical entities (NCEs) for CNS drugs is relatively high, making the development of new CNS drugs a challenge. A significant cause of this is related to safety issues arising either during animal toxicity testing or in the human clinical test phases. To reduce these risks, and increase approval rates by regulatory agencies, by implementing sophisticated human-cell based experimental models, e.g. organoids or stem-cells based models, in safety testing strategy of potential candidate drugs will result in an improved understanding and prevent safety issues in humans. Especially when these approaches are combined with omics tools in order to better understand the biological mechanisms involved, discovery of additional adverse health effects of existing and newly developed CNS drugs in the clinical test phase will be lowered.

Learning Objectives
- The high attrition rate of new chemical entities (NCEs) in preclinical and clinical phases is due to insufficient efficacy, bioavailability, safety, toxicological issues, and economic reasons. A significant cause of attrition is due to safety issues arising either during animal toxicity testing or in the human clinical test phases.
- For CNS drugs, other more sophisticated approaches are necessary to assess the physicochemical properties of lead molecules for new candidate drugs.
- Although regulatory toxicology still relies on observational toxicology tests in animals, incorporation or complete replacement by in vitro and other non-animal tests that enable understanding of human toxicity mechanisms is expected.
- Applying omics approaches on sophisticated in vitro models will help to better understand and prevent the adverse effects of existing and newly developed CNS drugs.

6.1 Introduction: Human Drug Safety of Psychoactive Compounds: The Challenges

The discovery of new drugs that target the central nervous system (CNS) is extremely challenging. This was clearly illustrated by numbers indicating that the success rates for CNS drugs, defined as final marketing approval by the US Food and Drug Administration (FDA), were less than half of the approval rates for non-CNS drugs for the period 1995 until 2007 (6.2% vs. 13.3%, respectively) (Gribkoff and Kaczmarek 2017). Also for CNS drugs the time to approval following submission of an application for marketing approval was about 30% longer than for non-CNS drugs (19.3 months vs. 14.7 months, respectively). Relative to non-CNS drugs, the mean development time of CNS drugs was longer, and the number of CNS drugs given priority review by the FDA was significantly lower.

The attrition rate also depends on the discovery stage and the therapeutic area. In the areas of CNS and oncology, it seems that compounds tend to fail more than in other therapeutic areas. Drug discovery and development are inherently risky, with figures indicating that less than 11% of new pharmaceutical agents entering clinical development reach the marketplace across all therapeutic areas. But the success rates vary considerably between the different therapeutic areas: for instance, cardiovascular disorders have a ~20% success rate, whereas oncology and central nervous system (CNS) disorders have ~5% and ~8% success rates, respectively (Kola and Landis 2004). For disease-modifying treatments of chronic neurodegenerative disorders, the failure rate has been 100%, with the exception of multiple sclerosis (MS) (Gribkoff and Kaczmarek 2017). The high attrition rate of new chemical entities (NCEs) in preclinical and clinical phases can be due to many factors (Kola and Landis 2004). NCEs fail mainly due to insufficient efficacy, bioavailability, safety, toxicological issues, and economic reasons. A significant cause of attrition is due to safety issues arising either during animal toxicity testing or in the human clinical test phases (Walker 2004).

6.2 Drug Design and Safety Regulations

Different regulatory agencies per continent, like the FDA in the USA, and the European Medicines Agency (EMA) in the EU, are involved in advising and reviewing the process for ensuring the safety and efficacy of new drugs. Therefore, pharmaceutical companies, research institutions, and other organizations that are responsible for developing a new drug must show the results of preclinical testing in toxicity testing and what they propose to do for human testing in order to receive approval to continue to phase I clinical testing in humans.

In 1997, "the rule of five" (RO5) was created to help medicinal chemists to design drugs with improved physicochemical properties (Lipinski et al. 2001). It is based on a database of clinical candidates that had reached phase II trials or further. It defined end points for a set of four physicochemical properties that described 90% of orally active drugs that achieved phase II clinical status: (i) molecular weight, MW <500 Da; (ii) lipophilicity, log P or the calculate of 1-octanol–water partition coefficient, ClogP <5; (iii) number of hydrogen-bond donors, OH plus NH count, <5; and (iv) number of hydrogen-bond acceptors, O plus N atoms, <10. These four physicochemical parameters and their criteria describe the fundamental attributes that are associated with acceptable aqueous solubility and intestinal permeability, key factors for the first step of oral bioavailability. However, specifically for CNS drug, these parameters only describe oral availability in a very simple way, and other more sophisticated approaches (Wager et al. 2010) are necessary to assess the physicochemical properties of lead molecules for new candidate drugs.

6.3 Testing for Human Neurotoxicity: From Animals to In Vitro Models

In general, when a new drug is taken, this will be dealt with just like a potential toxicant that enters the body. So this will limit absorption, prevent distribution and quickly starts to metabolize this into a form that can be secreted in urine and/or feces. In order to reduce attrition rates due to human toxicity, getting a complete picture of the human drug toxicity profile in the pre-clinical phase of drug development is vital. Strategies to assess the complete pharmacokinetic profile which includes understanding the absorption, distribution, metabolism, and elimination (ADME) of the compound and its toxic metabolites (ADME-Tox or ADMET) in humans are very important.

> **Definition**
>
> Toxicokinetics describes how the human body handles the drugs and their metabolites.

In order to test parameters that potentially predict toxic effects in humans, multiple model systems are available, ranging from specific isolated proteins and receptors to isolated cell organelles, monoculture cell lines, complex multicellular cellular systems (organoids), organs-on-a-chip up to different animal models. New drugs are routinely screened for toxicity by first applying a battery of in vitro models to assess, for example, target binding and metabolization, followed by in vivo confirmation using an appropriate animal model. Although regulatory toxicology still relies on observational toxicology tests in animals, increased attention for in vitro and other non-animal tests enable a better understanding of toxicity mechanisms (Malloy et al. 2017; Scholz et al. 2013). Multiple additional reasons support this transition, such as the following facts: i) the translation of results from animal testing into prediction of human safety is poor, ii) more realistic than the standard high dose applied in animals can be used, iii) high throughput testing in animals for a low costs is complex, and iv) there are many ethical reasons to reduce and replace animal experiments.

If we focus on model systems for performing predictive neurotoxicity of CNS drugs in humans, there are several features that are specifically related to the nervous system only if compared to other organs: (i) the blood–brain barrier (BBB) forms a very selective outer layer, (ii) the brain and nerves are lipid dense, (iii) it uses high energy and is a metabolically active system, (iv) it involves in highly specific intracellular signaling via neurotransmitters, (v) nerve cells have a very specific morphology and biochemistry, and (vi) dysfunctionality of only a small number of cells a has a major impact.

Once more information about relevant doses are available, toxicity of compounds of interest can be tested for (neuro)toxicity in different model systems in order to obtain information about the potential hazardous effect of a compound in the human body. Initially, information about a potential new neurodrug will be gathered based on in vitro tests. These can, for example, be explored in different cellular tests using in vitro cell models.

Examples of different cellular test systems are as follows:

1. Non-human cell lines and primary neuronal cells

Neuronal-like cell lines prevent that cells have to be taken from animals every time. Cell lines derived from, for example, rat or murine origins are available. For example, immortalized mouse neuronal cells (M4b) were derived from the spinal cord of a fetal mouse (Cardenas et al. 2002). M4b is useful in high-throughput neuronal cytotoxicity studies. Primary cultures of cortical neurons and of cerebellar granule cells are also used in neurotoxicity studies. Primary cultures of cortical neurons are constituted by around 40% of GABAergic neurons, whereas primary cultures of cerebellar granule cells are mainly constituted by glutamatergic neurons (Sunol et al. 2008).

2. Human cell lines and primary neural cells

The most popular applied neuronal cell lines in toxicity tests are the SH-SY5Y human neuroblastoma cell lines. Originally these are derived from a metastatic bone tumor biopsy. SH-SY5Y cells can be differentiated into a more mature neuron-like phenotype characterized by neuronal markers. Several methods exist to differentiate SH-SY5Y cells of which retinoic acid is the most commonly used (Kovalevich and Langford 2013). Access to human primary cell lines is limited, since these cells are difficult to acquire from aborted fetuses or brain surgery materials. From these sources, specific cells like neurons, microglia, oligodendrocytes, and astrocytes can be isolated (Jana et al. 2007).

3. Non-human embryonic stem cells

Embryonic stem cells (ESCs) are stem cells derived from the undifferentiated inner mass cells of an embryo. Embryonic stem cells are pluripotent, meaning they are able to grow and differentiate into different cells of the body or animal. These ESCs can be differentiated into each of the more than 220 cell types in the adult body and therefore represent a valuable platform for neurotoxicity screening of drugs. Embryonic stem cells are distinguished by two distinctive properties: their pluripotency and their ability to replicate indefinitely. Pluripotency distinguishes embryonic stem cells from adult stem cells found in adults – while embryonic stem cells can generate all cell types in the body, adult stem cells are multipotent and can produce only a limited number of cell types. Murine ESCs (mESCs) have been widely applied in developmental neurotoxicity testing (Kuegler et al. 2010). Human ESCs (hESCs) can be differentiated into a range of different types of neurons, e.g., glutaminergic and dopaminergic. Because of their human characteristics, these hESCs can be used for new drug toxicity screening, e.g., the screening of novel anesthetics (Bosnjak 2012).

4. Human-induced pluripotent stem cells

Induced pluripotent stem cell which share the advantage that in contrast to ESCs, these can be directly generated from adult cells. In addition, these can also be made in a patient-matched manner, so that each individual could be represented by their own pluripotent stem cell line.

Also the in vitro culturing methods went through constant developments towards models that better mimic the in vivo situation. Cell culture lines that are routinely used for toxicity screening are conventional 2D cell monolayers. These simplistic cell cultures are lacking the complex multicellular type of interactions including the complex physiological architecture and functionalities. 2D culture models also miss the signaling cues (mechanical, autocrine, paracrine, and endocrine) via microvessel structures, as occurs in native tissues, which direct the cultured cells to form multilayer (3D) tissue phenotype. So 2D cellular cell models do not display functional maturity, exhibit a short lifespan, and respond to pharmacological and toxicological challenges differently from their native counterparts and, therefore, data from 2D cell studies on drug candidates may not provide adequate human safety data. Therefore, development of improved cell models focuses on different 3D models, which include the development of spheroids, organoids, mini-organs, and organs-on-a-microfluidic chip (Breslin 2013; Cavero et al. 2019).

> For a better understanding of neurotoxicity of NCEs and existing compounds, challenges lie in improving stem cell-derived models and organoid-like cell models for the brain and CNS, so that these can at least partly replace animal models with the advantage that the prediction of human safety will be improved, costs will be reduced, results will be rapidly available, and the number of animal experiments will be reduced.

6.4 Pharmacokinetics: Relevant Dose Selection for In Vitro Tests

Generally the drug concentration parameters used are based on blood plasma concentrations because this is the routinely sampled body fluid in clinical practice. However, these plasma concentrations do not reflect the tissue levels in different organs, so the different tissue levels of these drugs and their metabolites might significantly differ in different organs compared to blood levels, which is due to differences in the physicochemistry and biological properties. One possibility to better model actual tissue concentrations is by making use of the so-called physiologically based pharmacokinetics (PBPK) models (Kuepfer et al. 2018) to account for the role of different organs in drug-related ADME.

> **Definition**
>
> In PBPK models, the fate of drug concentrations is estimated in physiologically realistic compartmental structure, e.g. organs, by using sets of equations based on blood flow, metabolism, etc.

Specifically for the brain these models have been built taking also into account specific-brain-related parameters, such as blood brain barrier (BBB) permeability, plasma protein binding, and brain tissue binding for assessing the time to reach brain equilibrium for different model compounds like caffeine, CP-141938 [methoxy-3-[(2-phenyl-piperadinyl-3-amino)-methyl]-phenyl-N-methyl-methane-sulfonamide], fluoxetine, NFPS [N[3-(4-fluorophenyl)-3-(4-phenylphenoxy)propyl]sarcosine], propranolol, theobromine, and theophylline (Liu et al. 2005).

In general the brain is hard to access by drugs due to its protection by an extensive network of proteins that forms collectively the BBB. This layer provides selective passage of nutrients, hormones, and waste products, while restricting passage of toxins, pathogens, and xenobiotics. Disruption or dysfunction of the BBB has been linked to numerous neurological disorders, including brain tumors, epilepsy, ischemic stroke, Alzheimer's disease, and multiple sclerosis. Before reaching a brain cell to exert its effect, a compound first has to pass through endothelial cells forming the wall of capillaries, the basement membrane, and the brain extracellular fluid. In this kinetic transport process, many interactions with cellular receptors and intracellular organelles can be involved. Only unbound compounds can exert their pharmacodynamics actions. Of the antipsychotic drugs for example, chlorpromazine binds 99.8% to brain tissue, while only 0.2% binds to brain receptors in order to induce its pharmacological effect (Watson et al. 2009). Overall, it is irrelevant to consider total concentrations without taking these kinds of effects into account.

Different in vitro models are used for gaining knowledge about compound-induced changes in the physiology of the blood–brain barrier. The most commonly used in vitro model is the transwell model, in which one or more cell types are cultured on semipermeable microporous inserts, allowing for separate culture compartments within the same well. In these co-culture models, usually endothelial cells are cultured on the upper side to form luminal layer. Other cells, such as astrocytes and/or pericytes, are cultured on the lower compartment. But transwell models do not include the complex architecture of the blood–brain barrier because these are simplified co-culture systems and also lack fluid flow. In recent years, models have been developed that lack this caveat. These include the so-called fiber-based dynamic in vitro blood–brain barrier models and microfluidic chips, especially the latter has become increasingly popular. Fiber-based dynamic in vitro blood–brain barrier models consist of co-cultured endothelial cells and other cells on the inner and outer walls of hollow fibers. Culture medium is pulsated through the tube, generating shear stress on the surface of the endothelial cells, causing an increased expression of tight junction proteins. Microfluidic chips consist of channels, chambers, and valves on the sub-micrometer scale on a platform. The channels in microfluidic chips are of comparable size to biological systems, such as cells and transport vessels. On these chips, (3D) cell co-culture devices and organs-on-a-chip can be designed. Also this can be combined with new technologies as bio-printing which can be used to create 3D models layer by layer on these chips. Compared with conventional microfluidic chip fabrication methods, bio-printing can create more complex, uniform, and reproducible model structures. Many different designs are available for modeling the blood–brain barrier on microfluidic chips (Jiang et al. 2019). These models have been used to study the function of blood–brain barrier, screen the safety of novel drug candidates, and predict the clearance of pharmaceuticals.

> When considering relevant concentrations of NCEs, it must be taken into account that the brain is protected by a functional BBB which is normally difficult to pass through by drugs, although this barrier can become less effective in neurological disorders.

6.5 Metabolism of Neuroactive Compounds

Although the brain has a large metabolic capacity, the liver is the main organ involved in the elimination and detoxification in the body. An important class of enzymes involved in the metabolism of drugs and xenobiotics are cytochromes P450 (CYPs). CYP-mediated metabolism can lead to the detoxification of drugs rendering them pharmacologically inactive, or, on the other hand, bioactivation can take place wherein a drug can be metabolized to pharmacologically active metabolite(s) that may have longer or shorter half-life compared to parent drug. CYP enzymes are present and active in the brain, despite that the total brain CYP expression being a fraction of hepatic CYP expression (McMillan and Tyndale 2018). But unique human CYPs are present in the brain and these can metabolize xenobiotics including drugs to active/inactive metabolites through biotransformation pathways that are different from the well-characterized ones in the liver (Ravindranath et al. 2006). For example, alprazolam, an extensively used anti-anxiety drug, is activated by different CYPs into pharmacologically more active metabolites both in the liver and brain (Pai et al. 2002). Another example is psilocybin, a psychedelic tryptamine, which following oral or parenteral administration is rapidly dephosphorylated to its active metabolite psilocin (Brown et al. 2017) in the gut, which can be further metabolized into another metabolite, psilocin-glucuronide.

6.6 From Toxicogenomics to Systems Biology

An important contribution for improving and obtaining better toxicity profiles and responses is the rapid development of the new, so-called omics technologies, in combination with bioinformatics approaches for data analysis.

> **Definition**
>
> Omics technologies measure some characteristics of a large family of cellular molecules, such as genes, proteins, or small metabolites, to explore the roles, relationships, and actions of various types of molecules that make up the cells of an organism.

These omics tools can assess changes in the cellular biomolecules at various levels, e.g., transcriptomics for changes in RNA, proteomics for proteins, lipidomics for lipids, and metabolomics for metabolites. These tools will enable to establish responses at different molecular levels with higher sensitivity than most classical effect markers, providing information on the involved molecular mechanisms of action. As such, toxicogenomics research combines toxicology with omics approaches in order to obtain more accurate understanding of toxicological processes. The future challenges lie within systems biology approaches that combine the data from different omics technologies (Zhang et al. 2018) (see ◘ Fig. 6.1) and the translation of this data into mechanisms that will support the improved understanding of the complex biological and toxicological processes. Insight into this can also be improved by applying single-cell omics technologies (Qi et al. 2018), for example, in brain cells for gaining insight into cellular processes that lead to toxicity, dysfunction, or neurodegenerative responses.

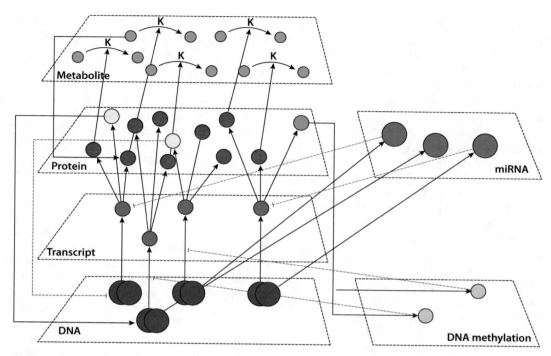

🔲 **Fig. 6.1** *Integration of multiple omics level.* Simplified overview of the integration of omics data at the DNA, transcript, and protein levels including miRNA and DNA methylation levels (Unger 2014). According to the "central dogma of molecular biology," information is transferred from DNA (genes, blue) to the RNA level (transcripts, green) and then to the protein level (proteins, red) in a linear manner. Proteins, i.e., enzymes, then catalyze biochemical reactions in which metabolites are processed. The metabolites are indicated by gray circles while the κ sign symbolizes that this process follows certain kinetics. The concentration of metabolites is well measured by receptor proteins; therefore, there is a strong communication between the protein and metabolite level. However, both transcription and translation and the lifetime of transcripts and proteins are regulated by other levels such as DNA methylation (cyan) and miRNAs (pink). Mediated by transcription factors (yellow), there is also feedback from the protein level back to the DNA level. Another molecular switch are proteins (DNA methyltransferases and DNA demethylases, light-blue) that control transcription by changing the levels of methylation of histones; therefore, there is a feedback from the protein level to the DNA methylation

❯ Conclusions

New CNS drugs seem to fail more than non-CNS drugs. Especially drugs intended to treat chronic neurodegenerative disorders fail in preclinical and clinical phases during animal toxicity testing or in the human clinical test phases. Regulatory toxicology still relies on observational toxicology tests in animals, while designing and application of novel in vitro models enabling a better understanding of toxicity mechanisms are necessary. Also design of in vitro test systems can be improved based on the information of relevant doses, e.g. based on pharmacokinetics, and a better understanding of compound metabolism.

Applying omics approaches in sophisticated in vitro models will help to better understand the reasons for high attrition rates caused by toxic responses in humans and will prevent adverse effects of existing and newly developed CNS drugs in the clinical test phase.

References

Bosnjak ZJ (2012) Developmental neurotoxicity screening using human embryonic stem cells. Exp Neurol 237:207–210

Breslin S, O'Driscoll L (2013) Three-dimensional cell culture: the missing link in drug discovery. Drug Discov Today 18:240–249

Brown RT, Nicholas CR, Cozzi NV, Gassman MC, Cooper KM, Muller D, Thomas CD, Hetzel SJ, Henriquez KM, Ribaudo AS, Hutson PR (2017) Pharmacokinetics of escalating doses of Oral psilocybin in healthy adults. Clin Pharmacokinet 56:1543–1554

Cardenas AM, Allen DD, Arriagada C, Olivares A, Bennett LB, Caviedes R, Dagnino-Subiabre A, Mendoza IE, Segura-Aguilar J, Rapoport SI, Caviedes P (2002) Establishment and characterization of immortalized neuronal cell lines derived from the spinal cord of normal and trisomy 16 fetal mice, an animal model of down syndrome. J Neurosci Res 68:46–58

Cavero I, Guillon JM, Holzgrefe HH (2019) Human organotypic bioconstructs from organ-on-chip devices for human-predictive biological insights on drug candidates. Expert Opin Drug Saf 18:651–677

Gribkoff VK, Kaczmarek LK (2017) The need for new approaches in CNS drug discovery: why drugs have failed, and what can be done to improve outcomes. Neuropharmacology 120:11–19

Jana M, Jana A, Pal U, Pahan K (2007) A simplified method for isolating highly purified neurons, oligodendrocytes, astrocytes, and microglia from the same human fetal brain tissue. Neurochem Res 32:2015–2022

Jiang L, Li S, Zheng J, Li Y, Huang H (2019) Recent Progress in microfluidic models of the blood-brain barrier. Micromachines (Basel) 10(6):375

Kola I, Landis J (2004) Can the pharmaceutical industry reduce attrition rates? Nat Rev Drug Discov 3:711–715

Kovalevich J, Langford D (2013) Considerations for the use of SH-SY5Y neuroblastoma cells in neurobiology. Methods Mol Biol 1078:9–21

Kuegler PB, Zimmer B, Waldmann T, Baudis B, Ilmjarv S, Hescheler J, Gaughwin P, Brundin P, Mundy W, Bal-Price AK, Schrattenholz A, Krause KH, van Thriel C, Rao MS, Kadereit S, Leist M (2010) Markers of murine embryonic and neural stem cells, neurons and astrocytes: reference points for developmental neurotoxicity testing. ALTEX 27:17–42

Kuepfer L, Clayton O, Thiel C, Cordes H, Nudischer R, Blank LM, Baier V, Heymans S, Caiment F, Roth A, Fluri DA, Kelm JM, Castell J, Selevsek N, Schlapbach R, Keun H, Hynes J, Sarkans U, Gmuender H, Herwig R, Niederer S, Schuchhardt J, Segall M, Kleinjans J (2018) A model-based assay design to reproduce in vivo patterns of acute drug-induced toxicity. Arch Toxicol 92:553–555

Lipinski CA, Lombardo F, Dominy BW, Feeney PJ (2001) Experimental and computational approaches to estimate solubility and permeability in drug discovery and development settings. Adv Drug Deliv Rev 46:3–26

Liu X, Smith BJ, Chen C, Callegari E, Becker SL, Chen X, Cianfrogna J, Doran AC, Doran SD, Gibbs JP, Hosea N, Liu J, Nelson FR, Szewc MA, Van Deusen J (2005) Use of a physiologically based pharmacokinetic model to study the time to reach brain equilibrium: an experimental analysis of the role of blood-brain barrier permeability, plasma protein binding, and brain tissue binding. J Pharmacol Exp Ther 313:1254–1262

Malloy T, Zaunbrecher V, Beryt E, Judson R, Tice R, Allard P, Blake A, Cote I, Godwin H, Heine L, Kerzic P, Kostal J, Marchant G, McPartland J, Moran K, Nel A, Ogunseitan O, Rossi M, Thayer K, Tickner J, Whittaker M, Zarker K (2017) Advancing alternatives analysis: the role of predictive toxicology in selecting safer chemical products and processes. Integr Environ Assess Manag 13:915–925

McMillan DM, Tyndale RF (2018) CYP-mediated drug metabolism in the brain impacts drug response. Pharmacol Ther 184:189–200

Pai HV, Upadhya SC, Chinta SJ, Hegde SN, Ravindranath V (2002) Differential metabolism of alprazolam by liver and brain cytochrome (P4503A) to pharmacologically active metabolite. Pharmacogenomics J 2:243–258

Qi M, Philip MC, Yang N, Sweedler JV (2018) Single cell neurometabolomics. ACS Chem Neurosci 9:40–50

Ravindranath V, Kommaddi RP, Pai HV (2006) Unique cytochromes P450 in human brain: implication in disease pathogenesis. J Neural Transm Suppl (70):167–171

Scholz S, Sela E, Blaha L, Braunbeck T, Galay-Burgos M, Garcia-Franco M, Guinea J, Kluver N, Schirmer K, Tanneberger K, Tobor-Kaplon M, Witters H, Belanger S, Benfenati E, Creton S, Cronin MT, Eggen RI, Embry M, Ekman D, Gourmelon A, Halder M, Hardy B, Hartung T, Hubesch B, Jungmann D, Lampi MA, Lee L, Leonard M, Kuster E, Lillicrap A, Luckenbach T, Murk AJ, Navas JM, Peijnenburg W, Repetto G, Salinas E, Schuurmann G, Spielmann H, Tollefsen KE, Walter-Rohde S, Whale G, Wheeler JR, Winter MJ (2013) A European perspective on alternatives to animal testing for environmental hazard identification and risk assessment. Regul Toxicol Pharmacol 67:506–530

Sunol C, Babot Z, Fonfria E, Galofre M, Garcia D, Herrera N, Iraola S, Vendrell I (2008) Studies with neuronal cells: from basic studies of mechanisms of neurotoxicity to the prediction of chemical toxicity. Toxicol In Vitro 22:1350–1355

Unger K (2014) Integrative radiation systems biology. Radiat Oncol 9:21

Wager TT, Chandrasekaran RY, Hou X, Troutman MD, Verhoest PR, Villalobos A, Will Y (2010) Defining desirable central nervous system drug space through the alignment of molecular properties, in vitro

ADME, and safety attributes. ACS Chem Neurosci 1:420–434

Walker DK (2004) The use of pharmacokinetic and pharmacodynamic data in the assessment of drug safety in early drug development. Br J Clin Pharmacol 58:601–608

Watson J, Wright S, Lucas A, Clarke KL, Viggers J, Cheetham S, Jeffrey P, Porter R, Read KD (2009)

Receptor occupancy and brain free fraction. Drug Metab Dispos 37:753–760

Zhang Y, Yuan S, Pu J, Yang L, Zhou X, Liu L, Jiang X, Zhang H, Teng T, Tian L, Xie P (2018) Integrated metabolomics and proteomics analysis of Hippocampus in a rat model of depression. Neuroscience 371:207–220

6

The Various Forms of Non-invasive Brain Stimulation and Their Clinical Relevance

Tom A. de Graaf, Alix Thomson, Felix Duecker, and Alexander T. Sack

Contents

7.1   Introduction – 104
7.1.1   NIBS Techniques – 105
7.1.2   The Biological Mechanisms
        in the CNS Targeted by NIBS – 108
7.1.3   Clinical Applications of NIBS – 110

        References – 111

© The Author(s), under exclusive license to Springer Nature Switzerland AG 2021
R. Schreiber (ed.), *Modern CNS Drug Discovery*, https://doi.org/10.1007/978-3-030-62351-7_7

As one of the promising upcoming alternatives to traditional drug therapy in central nervous system disorders and psychiatric disease, medical devices for neuromodulation have received a lot of attention. In addition to the invasive neural implant technologies used for deep-brain stimulation, a range of non-invasive brain stimulation (NIBS) techniques have recently been at the center of interest in research and therapy. Here, we provide an overview of the ever-growing family of NIBS methodologies, their clinical applications, and mechanisms of action involved. We suggest that NIBS technologies can be classified based on (1) the underlying technique (magnetic: transcranial magnetic stimulation, or electrical: transcranial electrical stimulation), (2) the targeted neurobiological process (ongoing processing, excitability/plasticity, oscillatory entrainment), or (3) the clinical domain of application (treatment, diagnosis, or prognosis). The current overview should prove valuable in understanding along which dimensions NIBS can be compared with traditional or alternative upcoming CNS modulation technologies.

Learning Objectives

- Gain an overview of the most commonly used NIBS techniques and protocols, including transcranial magnetic stimulation (TMS) and transcranial electrical stimulation (TES).
- Understand the physical and physiological mechanisms of action of TMS and TES.
- Understand NIBS through three different classification schemes: (1) mechanisms of action, (2) targeted brain process, and (3) clinical application.

7.1 Introduction

We are now two decades from the "decade of the brain" (1990–2000), but progress in neuroscience has continued unabated. With no end in sight as of yet, the numbers of meaningful breakthroughs and neuroscience publications have only grown. In the wake of these developments, the last decade has seen an increase in more applied research as well as commercialization of brain-based treatment and products. For instance, treatment of depression with commercially available brain stimulation devices is now established, and the question is no longer whether such treatment is effective, but rather whether it should be considered as a first-line or even first-choice treatment. Such rapid progress was perhaps no surprise, as better understanding of the central nervous system (CNS) naturally led to developments of a range of techniques and tools that allow CNS interventions/modulation. Developments aiming both to facilitate ever more sophisticated research, and to measure (diagnosis, prognosis) and directly modulate (treatment) CNS activity in clinical applications.

Besides the invasive neural implant technologies used for deep-brain-stimulation, "non-invasive brain stimulation" (NIBS) has been referred to as one of the most promising families of devices/techniques. These tools have been around for a few decades now, but some of their recent incarnations and applications have exploded onto the clinical and research landscape only recently. Some of the newer applications are outlined in the companion chapter 8, where we also critically discuss limitations and unknowns. But roughly speaking, NIBS has in recent years demonstrated equivalent or even superior effects relative to alternative (e.g., drug) treatments of certain brain-based disorders, such as major depression disorder, with only minimal side effects. So, it seems useful to provide an overview of the NIBS toolkit in this chapter and then discuss the applications and limitations in the next chapter.

NIBS differs from other neuromodulation techniques in several key aspects. Firstly, it is *non-invasive*, in the sense that it does not penetrate (the skin remains intact) or introduce external substances into (e.g., drugs, neural implants, or electrodes) the body.

This sets it apart from otherwise conceptually overlapping techniques such as deep brain stimulation (Aum and Tierney 2018), electroconvulsive therapy (ECT) (Lisanby 2007), optogenetics (Kim et al. 2017; Henderson et al. 2009), or chemical neuromodulation (Robbins 2000). Secondly, NIBS is considered a local (*brain*) neuromodulation approach, for instance able to target a cortical site of around a centimeter squared (Deng et al. 2013), as opposed to neuromodulation with chemicals that flood the system or system-level approaches such as neurofeedback (Sitaram et al. 2017). Thirdly, NIBS involves direct neuromodulation (*stimulation*), in the sense that it actively induces action potentials or modulates membrane potentials (Romero et al. 2019; Jackson et al. 2016). As such, NIBS directly affects activity in the building blocks of the CNS: neuronal firing.

Yet, even if NIBS is a more restrictive term than "neuromodulation," it still encompasses a wide, and rapidly growing, family of techniques and approaches. In fact, the range of NIBS techniques and applications has grown to the point where it is not trivial to decide what sort of taxonomy (classification scheme) makes most sense. At the same time, it is crucial to maintain a meaningful overview and to allow evaluations and comparisons of NIBS techniques relative to other options for both research and clinical application within the dynamic landscape of CNS medical devices and approaches. Depending on one's question, it might be useful to classify NIBS techniques according to the underlying physical mechanisms. This is a good starting point and perhaps the classical approach to categorizing the various techniques. Alternatively, it might actually be more valuable to classify NIBS techniques according to the biological mechanism they target/modulate. Indeed, it

turns out that very different NIBS approaches might be interchangeable for certain interventions, at least conceptually (Dunlop et al. 2017; Blumberger et al. 2013). Lastly, it makes sense to evaluate the contributions of NIBS in different clinical settings, discussing NIBS approaches in the context of diagnosis, prognosis, or treatment. All three of these taxonomy schemes have merit, they are largely orthogonal, and together they offer a complete picture of the physical mechanisms underlying the modulation, the biological mechanisms modulated, and the range of resulting applications of NIBS in the lab and the clinic.

7.1.1 NIBS Techniques (◻ Table 7.1)

From the perspective of mechanisms of action, NIBS can be divided into magnetic and electrical techniques. *Transcranial magnetic stimulation (TMS)* allows magnetic stimulation (Barker et al. 1985).

TMS hardware includes a generally non-portable stimulation device positioned on a table or trolley, with a port to which different TMS coils can be connected. Inside the device, a large capacitor can charge up to high voltage, leading to a strong electrical current if a TMS coil is connected and an internal switch is flipped to close the circuit. The current flows through the TMS coil, which consists of one (circular coil) or two neighboring (butterfly or figure-8 coil) windings, housed in a synthetic protective casing which is placed on the skull of the patient (or over peripheral nervous system). Due to well-established physical principles of electromagnetic induction, the following sequence of events occurs: (1) the electric cur-

◻ Table 7.1   NIBS techniques (simplified)

Technique	Protocol	Stimulation parameters	Effects	Applications (conceptual)
Repetitive transcranial magnetic stimulation (rTMS)	1 Hz	1 pulse per second	Inhibitory (LTD-like)	"Virtual lesion"; research clinical treatment
	10 Hz	10 pulses per second	Excitatory (LTP-like)	"Neuro-enhancement"; research clinical treatment
	cTBS	3 pulses at 50 Hz, repeated at 5 Hz. Continuous for 600 pulses	Inhibitory (LTD-like)	Same as 1 Hz rTMS
	iTBS	3 pulses at 50 Hz, repeated at 5 Hz for 2 seconds on, 8 seconds off 600 pulses	Excitatory (LTP-like)	Same as 10 Hz rTMS
Transcranial magnetic stimulation (TMS)	Single pulse	1 pulse, delivered manually every 5–7 seconds	Excitation	Diagnosis/prognosis
	Paired pulse	2 pulses, different inter-pulse intervals	CNS interactions	Diagnosis/prognosis
Transcranial electrical stimulation (TES)	Anodal tDCS	Continuous, direct current stimulation	Depolarize membrane potential (also, LTP-like)	Increase regional excitability, similar as 10 Hz rTMS
	Cathodal tDCS	Continuous, direct current stimulation	Hyperpolarize membrane potential (also, LTD-like)	Decrease regional excitability, similar as 1 Hz rTMS
	tACS	User-defined frequency. Polarity alternates between electrodes	Sinusoidally affects membrane potentials	Align/amplify/impose neuronal oscillations
	tRNS	Random high-frequency, polarity alternates between electrodes	May increase neuronal excitability	May increase regional excitability

rent through the TMS coil gives rise to a perpendicular magnetic field. Since the electric current is rapidly changing, the magnetic field is rapidly changing in proportion, becoming a magnetic "pulse." (2) The magnetic pulse crosses the scalp and skull unhindered, noninvasively. (3) In the conductive neuronal tissue reached by the magnetic field, electrical activity is again induced (Polson et al. 1982; Walsh and Rushworth 1999; Jalinous 1991; Hallett 2007; Rossini et al. 2015; Kammer et al. 2001).

The strength of the induced electric field and electric currents is proportional to the rate of change of the magnetic field, rather than the strength of the magnetic field directly (which, incidentally, is why an MRI scan does not stimulate the neurons) (Jalinous 1991; Barker 1991). This is also why different "waveforms" of the electric currents through the TMS coil can have different effects on the affected neurons (Kammer et al. 2001; Groppa et al. 2012). Irrespectively, the mechanism of TMS is that the magnetic pulse and associated induced electric field are sufficient to depolarize (most likely the axons of cortico-cortical inter-) neurons to achieve action potentials with each TMS pulse (Romero et al. 2019; Pashut et al. 2011).

> Single suprathreshold TMS pulse and associated electric field are sufficient to depolarize neurons.

As outlined below, even such single TMS pulses can be used both in research and clinically. However, much of the excitement surrounding TMS as a research and treatment tool is based on repetitive application of single pulses. Repetitive TMS (rTMS) involves multiple pulses applied rhythmically in patterns of single or even multiple frequencies. Classical rTMS protocols such as 1 Hz rTMS and 10 Hz rTMS were soon found to affect targeted brain regions even beyond the period of stimulation (Pascual-Leone et al. 1994; Muellbacher et al. 2000). Recent patterned protocols, such as the 40-second continuous theta burst stimulation (cTBS) and the 3-minute intermittent theta burst stimulation (iTBS) protocols, were shown to have much longer after-effects on cortical excitability than the protocol duration; for cTBS up to an hour after the end of the stimulation (Huang et al. 2005). The cTBS protocol involves triplets of single pulses at 50 Hz, which themselves are presented in a 5 Hz rhythm, until 600 pulses are delivered. The iTBS protocol adds another pattern, that is, 2 seconds of such theta burst stimulation followed by 8 seconds of rest, until again 600 pulses are administered in total. Importantly, effects on excitability are (on average) inhibitory (1 Hz rTMS, cTBS) or excitatory (10 Hz rTMS, iTBS), depending on the precise parameters of the rTMS protocols. It has been suggested that both classical and patterned rTMS protocols engage synaptic plasticity mechanisms, such as long-term potentiation (LTP) and long-term depression (LTD), to achieve these impressive modulations of neuronal activity (Huang et al. 2007; Teo et al. 2007; Cirillo et al. 2017). However, the precise mechanisms involved remain unclear and may differ between rTMS protocols.

> rTMS protocols are capable of affecting targeted brain regions beyond the period of stimulation. Depending on stimulation parameters, effects on excitability can be (on average) inhibitory (1 Hz rTMS, cTBS) or excitatory (10 Hz rTMS, iTBS).

The second family of NIBS applications involves transcranial electrical stimulation (TES).

Definition

TES: Transcranial electrical stimulation. Also referred to as low-intensity transcranial current stimulation, among other labels. Low-intensity current is administered to a brain region, not sufficiently strong to cause pain or to induce action potentials, yet sufficiently strong to modulate membrane potentials to change excitability. It can also have after-effects on excitability.

TES requires a, usually portable, stimulation device that primarily includes a battery and contact points for two connected electrodes. The latter come in different shapes and sizes, which directly affect the spatial configuration and intensity of stimulation in the brain (Woods et al. 2016). Low-intensity electric current flows between both electrodes, from the "anodal" electrode to the "cathodal" electrode (Paulus 2011). If continuous, these are referred to as anodal or cathodal transcranial direct current stimulation (tDCS) (Nitsche et al. 2008). Electric current can also be directed back and forth between both electrodes, which rhythmically switch polarity at a user-defined frequency, in transcranial alternating current stimulation (tACS) (Paulus 2011). A different setting on these machines causes current to switch direction much more often and at differing frequencies, which is called "transcranial random noise stimulation" (tRNS) (Terney et al. 2008). No matter which of these protocols are applied, in contrast to TMS (and also to electroconvulsive therapy (ECT), for example), TES primarily does not actually excite neurons to the point of action potentials. Instead, it achieves neuromodulation by changing the resting membrane potentials of affected neurons (Jackson et al. 2016; Radman et al. 2009). For instance, anodal tDCS depolarizes neurons slightly, bringing them "closer to threshold," which means fewer excitatory inputs will be required to induce action potentials (Liebetanz et al. 2002). Cathodal tDCS instead hyperpolarizes, achieving the opposite effect on excitability. tRNS may increase excitability, but by less straightforward mechanisms (Antal and Herrmann 2016). tACS sinusoidally changes membrane potentials, mimicking naturally occurring neuronal oscillations (Krause et al. 2019). Interestingly, all of these modulations of cortical excitability seem to last beyond the period of stimulation, just as in rTMS (Bindman et al. 1964).

> TES primarily does not actually excite neurons to the point of action potentials. Instead, it achieves neuromodulation by changing the resting membrane potentials of affected neurons.

There are ongoing developments in both these families of NIBS methodologies. For instance, TES is becoming more sophisticated by the use of more focal electrode montages (e.g., a small center electrode surrounded by a ring electrode or several small surrounding electrodes) (Sehm et al. 2013) and research in computational modeling to better understand the achieved distributions of induced electric fields for different electrode montages (Saturnino et al. 2018; Thielscher et al. 2015). In TMS, research to better understand the precise biological mechanisms affected by different protocols; how to tailor rTMS protocols to individual physiology; and how best to design coils and waveforms to achieve reliable or deeper (subcortical) modulation is underway (Deng et al. 2013; Romero et al. 2019; Cirillo et al. 2017; Cuypers et al. 2014; Banerjee et al. 2017; Medaglia et al. 2019). Fundamentally different tools that do not involve TMS or TES are arising as well. Still under investigation is "static magnetic stimulation" which involves a strong local static magnet (Oliviero et al. 2011), and recently "focused ultrasound" stimulation (Legon et al. 2014) has been receiving attention as a new alternative to TMS/TES with similar potential applications and a new set of pros and cons. Here, ultrasound at particular frequencies is directed toward particular brain regions, possibly mechanically causing neurites to "vibrate" and depolarize through entirely different mechanisms as compared to TMS (Krasovitski et al. 2011). While not yet as mainstream as TMS or TES, this ultrasound approach is an interesting avenue to follow going forward. We will restrict our further discussion to TMS and TES technologies.

7.1.2 The Biological Mechanisms in the CNS Targeted by NIBS

We provided an overview of NIBS approaches, mainly TMS and TES with various specifications, based on how these techniques work: the underlying physics. A very different perspective, and indeed different classification scheme, arises when we focus on their effects in the brain. In other words, we can also classify NIBS technologies according to the biological mechanism affected. We find it useful to delineate three targeted biological mechanisms, or three neuromodulation targets, to capture most NIBS applications.

Non-repetitive TMS is often referred to as "single-pulse TMS" or "event-related TMS," but this seems a bit restrictive. The point is that NIBS is used to excite neurons briefly. Every such administration of NIBS is momentary and delivers a datapoint. In research, event-related TMS can be used for "chronometric studies" for example, where a particular event (e.g., the presentation of a visual image on screen) is time-locked to TMS pulses (e.g., a single pulse 100 milliseconds after image presentation) to evaluate the causal role of the targeted cortical region (e.g., occipital cortex) for a particular function (e.g., image discrimination: an occipital pulse around 100 ms will impair discrimination or even make the image invisible) (de Graaf et al. 2011). Clinically, single TMS pulses can be applied to motor cortex to elicit motor-evoked potentials (MEPs) that can be measured with electromyography (EMG) (Rothwell et al. 1999). The amplitude of the MEP, which is simply a quantification of a TMS-induced motor twitch, has clinical implications. As we discuss below, this can be in diagnosis and prognosis, in isolation or coupled with neuroimaging techniques. Also under this category, one might consider paired-pulse TMS, which involves two TMS pulses administered in quick succession (Valls-Sole et al. 1992). When applied to the motor cortex, a subthreshold TMS pulse preceding a suprathreshold TMS pulse by around 2–7 milliseconds will reduce the MEP elicited by the latter pulse (Valls-Sole et al. 1992; Kujirai et al. 1993). This reduction is called short-interval cortical inhibition (SICI) and has clinical implications, as do long-interval cortical inhibition (LICI) and intracortical facilitation (ICF), all of which involve paired TMS pulses applied to motor cortex in different configurations of TMS intensity and inter-pulse intervals (Berardelli et al. 2008; McClintock et al. 2011; Rossini et al. 1994). By using two TMS coils, one over motor cortex and the other over anatomically/functionally connected regions, it is possible to quantify similar modulations of MEP by prior excitation of other motor network nodes (Hampson and Hoffman 2010). Interesting work has taken similar approaches beyond the motor system to understand cortical information flows (Pascual-Leone and Walsh 2001). Ultimately, these are all instances of momentary NIBS excitation of neurons to achieve different goals.

Plasticity targeted NIBS refers to the collection of NIBS tools and protocols that likely engage either LTD or LTP to transiently decrease or increase cortical excitability, respectively. To decrease cortical excitability, there are cathodal tDCS, classical inhibitory rTMS (1 Hz), and cTBS. To increase excitability, there are anodal tDCS, tRNS, classical excitatory rTMS (10 Hz), and iTBS (Rossini et al. 2015). It is not unequivocally established that all these protocols indeed similarly engage LTD and LTP, or that they rely on the same mechanisms of action in the brain, but that is not what is relevant here. What is relevant is that they all can serve the same

functional purpose. If one wishes to increase excitability beyond the period of stimulation, or decrease it, there are these various – in many ways very different – options. One can weigh the pros and cons according to the use case, or the patient. In research, temporarily changing cortical excitability in a local brain region allows assessment of the causal contribution that region makes to various tasks (e.g., decreasing excitability might induce a task impairment). In clinical applications, treatment with NIBS builds on this foundation, though an additional mechanism of plasticity is somehow involved which remains imperfectly understood. After all, the effects of these NIBS protocols are in the range of minutes to hours, not weeks to years (Cirillo et al. 2017). Yet, the clinical efficacy of repeating such protocols over weeks has indeed been reported to last for such extended periods of time (Dunlop et al. 2017; Sonmez et al. 2019; Blumberger et al. 2018).

Entrainment is the final NIBS application to mention here. The human brain operates in large part by means of naturally occurring oscillations. The power and phase of these oscillations, in different frequency bands, have been related to various sorts of motor, cognitive, and perceptual functions (Ward 2003). In turn, neuromodulation of these oscillations is possible. Especially for research, short bursts of TMS pulses can briefly increase the power of oscillations in a particular frequency band in a particular region to evaluate the causal contribution of these oscillations to a task of interest (Thut et al. 2011). tACS can achieve the same for a longer period of time (Polania et al. 2012; Pogosyan et al. 2009). In fact, alpha power was increased even beyond the period of alpha-frequency tACS (Helfrich et al. 2014). Both methods allow the investigation and modulation of both oscillatory power and phase. Since many psychiatric disorders have been related to oscillatory/connectivity dysfunction, the direct NIBS-targeting of oscillations has the potential to make unique neuromodulatory contributions not only to the lab but also to the clinic (Hong et al. 2010; Michelini et al. 2018; Schnitzler and Gross 2005).

7.1.3 Clinical Applications of NIBS

In our companion Chapter 8, we place NIBS in the context of traditional clinical tools. Here, we would like to delineate the three core clinical applications of NIBS, providing a classification scheme along a third, more applied, dimension. These applications are diagnosis, prognosis, and treatment.

Diagnosis and *prognosis* currently primarily involve non-repetitive TMS, that is, either single-pulse TMS or paired-pulse paradigms as outlined above. In all cases, the principal idea is to assess responses to TMS pulses to obtain information about the current brain state (diagnosis) which may also have predictive value regarding the further development of a disorder/disease (prognosis). To illustrate, single-pulse TMS applied over primary motor cortex can typically elicit muscle twitches (motor-evoked potentials; MEP). However, damage to the corticospinal tract can cause subtle changes of MEP amplitudes and latencies, or even a complete absence of motor responses (Kobayashi and Pascual-Leone 2003). TMS thus allows probing the integrity of the motor system, which is a critical diagnostic step in the acute phase after stroke, a useful tool for monitoring changes during stroke rehabilitation, and even of prognostic value as the presence/absence of MEPs is indicative of the potential for long-term functional recovery (Di Pino et al. 2014).

Similarly, the minimum intensity required to observe a TMS-induced motor response (motor threshold) is an established measure of cortical excitability. While inter-individual variability of cortical excitability currently poses some limits in terms of specificity, there are promising applications of TMS as a diagnostic marker in epilepsy (Kimiskidis et al. 2014), and monitoring excitability changes over time can help in determining which particular antiepileptic drug effectively decreases cortical excitability without relying on the occurrence of seizures as a marker of treatment success (Badawy et al. 2012).

Lastly, the vast majority of studies have focused on diagnostic and prognostic applications of TMS in the motor system because of the simplicity of MEP recordings. However, the potential of TMS dramatically increases when combined with neuroimaging. In recent years, EEG has become very popular to assess brain responses to TMS pulses outside the motor system. In a pioneering research line, the simultaneous combination of TMS and EEG has been used to reveal how TMS-induced activity spreads throughout the brain in various disorders of consciousness. Strikingly, the complexity of the brain network response was sufficient to allow researchers to accurately classify individual patients as being in unresponsive wakefulness syndrome, a minimally conscious state, or locked-in syndrome (Sarasso et al. 2014). Admittedly, the technical complexities of such multimodal approaches currently constrain their application in clinical practice, but fully integrated systems to record TMS-induced changes in EEG activity are already emerging on the market.

Not only can NIBS be used to help establish a diagnosis, or inform a prognosis, it is perhaps most well-known for its application as actual brain-based *treatment*, in neurorehabilitation but especially also for psychiatric disorders. This application relies on the lasting effects on plasticity described above. In its currently most widespread clinical application, NIBS is applied to either increase cortical excitability in left frontal cortex or decrease cortical excitability in right frontal cortex to treat depression (Santre et al. 1995; O'Reardon et al. 2007; Hoppner et al. 2003). The evidence for efficacy is strongest for high-frequency left frontal rTMS (O'Reardon et al. 2007). As discussed above, anodal tDCS may have similar effects, since it should achieve the same thing: an increase in excitability (Brennan et al. 2017). This particular example at the same time exemplifies that things are never as straightforward as they seem: left frontal rTMS works well for treatment-resistant depression patients, while left frontal anodal tDCS actually receives more evidence for efficacy in non-treatment-resistant patients (Blumberger et al. 2013). This, as well as NIBS treatment efficacy in a range of other disorders, is under intense investigation. It seems NIBS might be helpful in the treatment of not only mood disorders but also neuropathic pain, motor disorders, anxiety disorders, and a range of other brain-based malfunctions (Chen et al. 2017; Chalah and Ayache 2019). Excellent, and at the same time exhaustive, reviews on the precise level of evidence for both rTMS and tDCS can be found in overview articles by Lefaucheur et al. (Lefaucheur et al. 2017; Lefaucheur et al. 2014) for rTMS and by O'Reardon et al. (2007) for tDCS. In fact, as elaborated in the companion chapter, an updated overview was recently published by an overlapping group of experts (Lefaucheur et al. 2020).

> **Conclusion**

Non-invasive brain stimulation (NIBS) is an umbrella term for a wide and growing range of techniques and applications. There are so many techniques, and in fact we have presented three different classification schemes to create an overview of NIBS. NIBS applications were classified according to their physical principles, the biological mechanisms they targeted, and clinical applications. The value of NIBS in research is established. Its value in clinical applications is becoming increasingly clear and has received sufficient empirical support that implementation is widespread. But such rapid growth and acknowledgment come with a risk. NIBS to "improve" the healthy human brain (neuroenhancement) and as a sort of mental panacea (to cure all brain problems) has captured the public imagination, while at least some clinicians remain more wary. NIBS is a technique with such wide applications, at low cost, with minimal side effects, with minimal risks involved – does it not sound too good to be true? We address this question in the companion Chapter 8.

References

Antal A, Herrmann CS (2016) Transcranial alternating current and random noise stimulation: possible mechanisms. Neural Plast 2016:3616807

Aum DJ, Tierney TS (2018) Deep brain stimulation: foundations and future trends. Front Biosci (Landmark Ed) 23(1093–4715 (Electronic)):162–182

Badawy RAB et al (2012) Cortical excitability and refractory epilepsy: a three-year longitudinal transcranial magnetic stimulation study. Int J Neural Syst 23(01):1250030

Banerjee J et al (2017) Immediate effects of repetitive magnetic stimulation on single cortical pyramidal neurons. PLoS One 12(1):e0170528

Barker AT (1991) An introduction to the basic principles of magnetic nerve stimulation. J Clin Neurophysiol 8(1):26–37

Barker AT, Jalinous R, Freeston IL (1985) Non-invasive magnetic stimulation of human motor cortex. Lancet 325(0140–6736 (Print)):1106–1107

Berardelli A et al (2008) Consensus paper on short-interval intracortical inhibition and other transcranial magnetic stimulation intracortical paradigms in movement disorders. Brain Stimul 1(3):183–191

Bindman LJ, Lippold OC, Redfearn JW (1964) The action of brief polarizing currents on the cerebral cortex of the rat (1) during current flow and (2) in the production of long-lasting after-effects. J Physiol 172:369–382

Blumberger DM, Mulsant BH, Daskalakis ZJ (2013) What is the role of brain stimulation therapies in the treatment of depression? Curr Psychiatry Rep 15(7):368

Blumberger DM et al (2018) Effectiveness of theta burst versus high-frequency repetitive transcranial magnetic stimulation in patients with depression (THREE-D): a randomised non-inferiority trial. Lancet 391(10131):1683–1692

Brennan S et al (2017) Anodal transcranial direct current stimulation of the left dorsolateral prefrontal cortex enhances emotion recognition in depressed patients and controls. J Clin Exp Neuropsychol 39(4):384–395

Chalah MA, Ayache SS (2019) Noninvasive brain stimulation and psychotherapy in anxiety and depressive disorders: a viewpoint. Brain Sci 9(4):82

Chen ML et al (2017) Non-invasive brain stimulation interventions for management of chronic central neuropathic pain: a scoping review protocol. BMJ Open 7(10):e016002

Cirillo G et al (2017) Neurobiological after-effects of non-invasive brain stimulation. Brain Stimul 10(1):1–18

Cuypers K, Thijs H, Meesen RL (2014) Optimization of the transcranial magnetic stimulation protocol by defining a reliable estimate for corticospinal excitability. PLoS One 9(1):e86380

Deng Z-D, Lisanby SH, Peterchev AV (2013) Electric field depth–focality tradeoff in transcranial magnetic stimulation: simulation comparison of 50 coil designs. Brain Stimul 6(1):1–13

Di Pino G et al (2014) Modulation of brain plasticity in stroke: a novel model for neurorehabilitation. Nat Rev Neurol 10(10):597–608

Dunlop K, Hanlon CA, Downar J (2017) Noninvasive brain stimulation treatments for addiction and major depression. Ann N Y Acad Sci 1394(1):31–54

de Graaf TA, Herring J, Sack AT (2011) A chronometric exploration of high-resolution 'sensitive TMS masking' effects on subjective and objective measures of vision. Exp Brain Res 209(1):19–27

Groppa S et al (2012) A practical guide to diagnostic transcranial magnetic stimulation: report of an IFCN committee. Clin Neurophysiol 123(5):858–882

Hallett M (2007) Transcranial magnetic stimulation: a primer. Neuron 55(2):187–199

Hampson M, Hoffman RE (2010) Transcranial magnetic stimulation and connectivity mapping: tools for studying the neural bases of brain disorders. Front Syst Neurosci 4:40

Helfrich RF et al (2014) Entrainment of brain oscillations by transcranial alternating current stimulation. Curr Biol 24(3):333–339

Henderson JM, Federici T, Boulis N (2009) Optogenetic neuromodulation. Neurosurgery 64(5):796–804

Hong LE et al (2010) Gamma and delta neural oscillations and association with clinical symptoms under subanesthetic ketamine. Neuropsychopharmacology 35(3):632–640

Hoppner J et al (2003) Antidepressant efficacy of two different rTMS procedures. High frequency over left versus low frequency over right prefrontal cortex compared with sham stimulation. Eur Arch Psychiatry Clin Neurosci 253(2):103–109

Huang YZ et al (2005) Theta burst stimulation of the human motor cortex. Neuron 45(2):201–206

Huang Y-Z et al (2007) The after-effect of human theta burst stimulation is NMDA receptor dependent. Clin Neurophysiol 118(5):1028–1032

Jackson MP et al (2016) Animal models of transcranial direct current stimulation: methods and mechanisms. Clin Neurophysiol 127(11):3425–3454

Jalinous R (1991) Technical and practical aspects of magnetic nerve stimulation. J Clin Neurophysiol 8(1):10–25

Kammer T et al (2001) Motor thresholds in humans: a transcranial magnetic stimulation study comparing different pulse waveforms, current directions and stimulator types. Clin Neurophysiol 112(2):250–258

Kim CK, Adhikari A, Deisseroth K (2017) Integration of optogenetics with complementary methodologies in systems neuroscience. Nat Rev Neurosci 18(4):222–235

Kimiskidis VK, Valentin A, Kalviainen R (2014) Transcranial magnetic stimulation for the diagnosis and treatment of epilepsy. Curr Opin Neurol 27(2):236–241

Kobayashi M, Pascual-Leone A (2003) Transcranial magnetic stimulation in neurology. Lancet Neurol 2(3):145–156

Krasovitski B et al (2011) Intramembrane cavitation as a unifying mechanism for ultrasound-induced bioeffects. Proc Natl Acad Sci U S A 108(8):3258–3263

Krause MR et al (2019) Transcranial alternating current stimulation entrains single-neuron activity in the primate brain. Proc Natl Acad Sci U S A 116(12):5747–5755

Kujirai T et al (1993) Corticocortical inhibition in human motor cortex. J Physiol 471:501–519

Lefaucheur JP et al (2014) Evidence-based guidelines on the therapeutic use of repetitive transcranial magnetic stimulation (rTMS). Clin Neurophysiol 125(11):2150–2206

Lefaucheur JP et al (2017) Evidence-based guidelines on the therapeutic use of transcranial direct current stimulation (tDCS). Clin Neurophysiol 128(1):56–92

Lefaucheur J-P et al (2020) Evidence-based guidelines on the therapeutic use of repetitive transcranial magnetic stimulation (rTMS): an update (2014–2018). Clin Neurophysiol 131(2):474–528

Legon W et al (2014) Transcranial focused ultrasound modulates the activity of primary somatosensory cortex in humans. Nat Neurosci 17(2):322–329

Liebetanz D et al (2002) Pharmacological approach to the mechanisms of transcranial DC-stimulation-induced after-effects of human motor cortex excitability. Brain 125(Pt 10):2238–2247

Lisanby SH (2007) Electroconvulsive therapy for depression. N Engl J Med 357(19):1939–1945

McClintock SM et al (2011) Transcranial magnetic stimulation: a neuroscientific probe of cortical function in schizophrenia. Biol Psychiatry 70(1):19–27

Medaglia JD et al (2019) Personalizing neuromodulation. Int J Psychophysiol 154:101–110

Michelini G et al (2018) Shared and disorder-specific event-related brain oscillatory markers of attentional dysfunction in ADHD and bipolar disorder. Brain Topogr 31(4):672–689

Muellbacher W et al (2000) Effects of low-frequency transcranial magnetic stimulation on motor excitability and basic motor behavior. Clin Neurophysiol 111(1388–2457 (Print)):1002–1007

Nitsche MA et al (2008) Transcranial direct current stimulation: state of the art 2008. Brain Stimul 1(3):206–223

O'Reardon JP et al (2007) Efficacy and safety of transcranial magnetic stimulation in the acute treatment of major depression: a multisite randomized controlled trial. Biol Psychiatry 62(11):1208–1216

7

Oliviero A et al (2011) Transcranial static magnetic field stimulation of the human motor cortex. J Physiol 589(Pt 20):4949–4958

Pascual-Leone A, Walsh V (2001) Fast backprojections from the motion to the primary visual area necessary for visual awareness. Science 292(5516): 510–512

Pascual-Leone A et al (1994) Responses to rapid-rate transcranial magnetic stimulation of the human motor cortex. Brain 117(Pt 4):847–858

Pashut T et al (2011) Mechanisms of magnetic stimulation of central nervous system neurons. PLoS Comput Biol 7(3):e1002022

Paulus W (2011) Transcranial electrical stimulation (tES - tDCS; tRNS, tACS) methods. Neuropsychol Rehabil 21(5):602–617

Pogosyan A et al (2009) Boosting cortical activity at Beta-band frequencies slows movement in humans. Curr Biol 19(19):1637–1641

Polania R et al (2012) The importance of timing in segregated theta phase-coupling for cognitive performance. Curr Biol 22(14):1314–1318

Polson MJ, Barker AT, Freeston IL (1982) Stimulation of nerve trunks with time-varying magnetic fields. Med Biol Eng Comput 20(2):243–244

Radman T et al (2009) Role of cortical cell type and morphology in subthreshold and suprathreshold uniform electric field stimulation in vitro. Brain Stimul 2(4):215–228, 228 e1-3

Robbins TW (2000) Chemical neuromodulation of frontal-executive functions in humans and other animals. Exp Brain Res 133(1):130–138

Romero MC et al (2019) Neural effects of transcranial magnetic stimulation at the single-cell level. Nat Commun 10(1):2642

Rossini PM et al (1994) Non-invasive electrical and magnetic stimulation of the brain, spinal cord and roots: basic principles and procedures for routine clinical application. Report of an IFCN committee. Electroencephalogr Clin Neurophysiol 91: 79–92

Rossini PM et al (2015) Non-invasive electrical and magnetic stimulation of the brain, spinal cord, roots and peripheral nerves: basic principles and procedures for routine clinical and research application. An updated report from an I.F.C.N. committee. Clin Neurophysiol 126(6):1071–1107

Rothwell JC et al (1999) Magnetic stimulation: motor evoked potentials. The International Federation of Clinical Neurophysiology. Electroencephalogr Clin Neurophysiol Suppl 52:97–103

Santre C et al (1995) Amikacin levels in bronchial secretions of 10 pneumonia patients with respiratory support treated once daily versus twice daily. Antimicrob Agents Chemother 39(1):264–267

Sarasso S et al (2014) Quantifying cortical EEG responses to TMS in (un)consciousness. Clin EEG Neurosci 45(1):40–49

Saturnino GB et al (2018) SimNIBS 2.1: a comprehensive pipeline for individualized electric field modelling for transcranial brain stimulation. bioRxiv:500314

Schnitzler A, Gross J (2005) Normal and pathological oscillatory communication in the brain. Nat Rev Neurosci 6(4):285–296

Sehm B et al (2013) A novel ring electrode setup for the recording of somatosensory evoked potentials during transcranial direct current stimulation (tDCS). J Neurosci Methods 212(2):234–236

Sitaram R et al (2017) Closed-loop brain training: the science of neurofeedback. Nat Rev Neurosci 18(2):86–100

Sonmez AI et al (2019) Accelerated TMS for depression: a systematic review and meta-analysis. Psychiatry Res 273:770–781

Teo JT, Swayne OB, Rothwell JC (2007) Further evidence for NMDA-dependence of the after-effects of human theta burst stimulation. Clin Neurophysiol 118(7):1649–1651

Terney D et al (2008) Increasing human brain excitability by transcranial high-frequency random noise stimulation. J Neurosci 28(52):14147–14155

Thielscher A, Antunes A, Saturnino GB (2015) Field modeling for transcranial magnetic stimulation: a useful tool to understand the physiological effects of TMS? In: 2015 37th Annual International Conference of the IEEE Engineering in Medicine and Biology Society (EMBC)

Thut G et al (2011) Rhythmic TMS causes local entrainment of natural oscillatory signatures. Curr Biol 21(14):1176–1185

Valls-Sole J et al (1992) Human motor evoked responses to paired transcranial magnetic stimuli. Electroencephalogr Clin Neurophysiol 85(6): 355–364

Walsh V, Rushworth M (1999) A primer of magnetic stimulation as a tool for neuropsychology. Neuropsychologia 37(2):125–135

Ward LM (2003) Synchronous neural oscillations and cognitive processes. Trends Cogn Sci 7(12):553–559

Woods AJ et al (2016) A technical guide to tDCS, and related non-invasive brain stimulation tools. Clin Neurophysiol 127(2):1031–1048

Is Non-invasive Brain Stimulation the Low-Hanging Fruit?

Tom A. de Graaf, Shanice E. W. Janssens, and Alexander T. Sack

Contents

8.1   Introduction – 116

8.2   Focus: Major Depression – 117
8.2.1   rTMS Depression Treatment – 117
8.2.2   Conventional Depression Treatments – 121
8.2.3   The (Difficult) Comparison – 121
8.2.4   Room for Growth, in the Example of Depression
        Treatment – 122
8.2.5   rTMS for Treatment of Other Conditions:
        Rapid Developments – 123

        References – 125

© The Author(s), under exclusive license to Springer Nature Switzerland AG 2021
R. Schreiber (ed.), *Modern CNS Drug Discovery*, https://doi.org/10.1007/978-3-030-62351-7_8

Mental health is a growing concern, with increases in diagnosed psychopathologies and associated costs. Traditionally, psychopathology has been treated with psychotherapy, drug therapy, or combinations of both. But recent years have seen the development of new approaches to directly target the biological basis for mental illness. For a range of disorders, neuromodulation using non-invasive brain stimulation (NIBS) constitutes a promising avenue to direct, comfortable, and focused CNS-targeted treatment (central nervous system). At the same time, neuromodulation using direct brain stimulation is sometimes touted as a tool with almost unlimited future and range of applications. We here discuss to what extent NIBS can indeed contribute to brain-based medicine, particularly clinical treatment. Is NIBS a passing trend? Or is NIBS the low-hanging fruit?

🔘 Learning Objectives
- Gain an overview of the promise and value of repetitive transcranial magnetic stimulation (rTMS) as a clinical treatment option.
- Know the method and efficacy of rTMS treatment of depression.
- Contrast along several dimensions rTMS treatment of depression with other depression treatments.
- Gain an impression of the rate of progress in clinical rTMS research and applications.

8.1 Introduction

Non-invasive brain stimulation (NIBS) is a family of neuromodulation tools including primarily transcranial magnetic stimulation (TMS) and transcranial electric stimulation (TES). In a preceding companion chapter, we provided no less than three classification schemes to capture the increasingly wide and sophisticated range of tools and applications in the NIBS arsenal. Depending on the stimulation parameters, NIBS can momentarily activate, disrupt, or for a longer period of time inhibit or facilitate local brain regions by decreasing or increasing cortical excitability,

respectively. One classification scheme evaluated NIBS in terms of their physical and physiological mechanisms, such as magnetic versus electrical stimulation tools. A second taxonomy focused on the modulated neuronal/functional process, such as instantaneous neuronal excitation versus lasting excitability changes. The third discussed how NIBS can be clinically applied, for diagnosis, prognosis, and treatment. Especially with respect to the latter, treatment of brain-based disorders with NIBS, this chapter will go in some more depth.

With such diverse applications, offering distinct possibilities and opportunities, it is no wonder that excitement about NIBS runs high not only among researchers and clinicians but even among the general public. In recent years, a new neurotechnology market has boomed, leading to a wealth of direct-to-consumer offerings including wearable brain stimulation devices whose efficacy and marketing approaches have been called into question (Coates McCall et al. 2019). Noting such trends, we previously addressed questions, including ethical considerations, about commercial and even home-made brain stimulation devices (Duecker et al. 2014). We pointed out that one should separately assess the promise and efficacy of NIBS in a research setting (definitely valuable), in a clinical setting (definitely valuable for some applications, not yet clear for others), and in a commercial, consumer, public setting (see (Duecker et al. 2014) and (Coates McCall et al. 2019) for discussion). But for each of these domains, it is important to ask what the short- and long-term value of NIBS is. It is unlikely to be a panacea, but sometimes new technologies really can have a transformative impact. So, is NIBS the low-hanging fruit? How justified are the sometimes amazing promises, publicly imagined wide range of applications, and exciting painted pictures of a neuromodulated future?

This is a broad question, and one that could and should be asked about the wider range of neuromodulation tools and approaches, as well as other central nervous system (CNS) medical devices including invasive options. In this chapter, we ask it specifically for NIBS, and primarily for clinical applications of NIBS. To be explicit, this refers only to professionally supervised NIBS treatment of diagnosed (brain-based) disorders. Later in this chapter, we will shortly review evidence and developments of TES and TMS as clinical treatment options for a wider range of disorders. But as a case study, we first focus on the currently most established NIBS treatment approach; repetitive TMS (rTMS) treatment of (major, unipolar) depression. Conventional depression treatments include psychotherapeutic and pharmacological intervention, as well as electro-convulsive therapy (ECT). What is the state-of-the-art understanding and evidence for rTMS as an alternative treatment option? Is it a viable alternative? Could it become a superior, possibly preferred, alternative?

> **Definition**
>
> TMS: transcranial magnetic stimulation. The non-invasive delivery of magnetic pulses to a brain region, inducing electric field/current that can depolarize neurons and induce action potentials.

> **Definition**
>
> rTMS: repetitive transcranial magnetic stimulation. This technique involves the rhythmic delivery of magnetic pulses, often at either 10 Hertz or 1 Hertz, to respectively enhance or suppress the excitability of the targeted region. Repeated rTMS can have longer-lasting and clinically meaningful effects on the targeted, as well as connected, regions in the brain.

8.2 Focus: Major Depression

NIBS *is* effective in the treatment of depression. As a matter of fact, NIBS, and particularly rTMS, has in the last three decades slowly but surely worked its way into the established range of treatment options for depression, with TMS clinics sprouting globally and TMS machines finding their way into the offices of experienced psychologists and psychiatrists. Mostly, clinical research to establish its efficacy has focused on unipolar

major depression, initially primarily testing efficacy in treatment-resistant patients. Below, we briefly summarize the primary NIBS treatment protocol and procedure. For this example of its application, NIBS could only be considered low-hanging fruit if it compares favorably to established/alternative treatment options on a few relevant dimensions, including (1) *efficacy* (e.g., response, remission, relapse rates), (2) *tolerability* (e.g., comfort, risk, side effects), and (3) *room for growth*. To allow comparison, we will evaluate established depression treatment options along these same dimensions. As mentioned briefly below, there are certainly additional considerations, including cost, availability, and ease of use, which are not focused on in this chapter.

8.2.1 rTMS Depression Treatment

Depression has been linked to a variety of biological underpinnings, but several of these point to frontal cortex, its connections to the limbic system, and an (im)balance in baseline activity between left and right frontal cortex. Currently, the primary NIBS approach to treat depression, which received Food and Drug Administration (FDA) approval in the United States in 2008, and health insurance coverage in increasingly many countries worldwide, is daily high-frequency repetitive TMS (HF-rTMS) applied to left dorsolateral prefrontal cortex (DLPFC) for several weeks (Lefaucheur et al. 2020).

The traditional protocol and procedure are as follows: (1) localize the scalp position at which a TMS pulse elicits a measurable/observable response in the first dorsal interosseous (FDI) muscle of the right hand, and determine the TMS intensity required to elicit such a response on half of the trials (motor threshold), (2) move the TMS coil forward 5 cm, (3) and apply 4 seconds of 10-Hertz rTMS at 120% of the motor threshold, followed by 26 seconds of rest, repeated until 3000 pulses have been administered (~37 minutes). An rTMS depression treatment administers such a session 5 days per week (weekdays), for 4–6 weeks. ◻ Figure 8.1 visualizes this conventional depression treatment protocol.

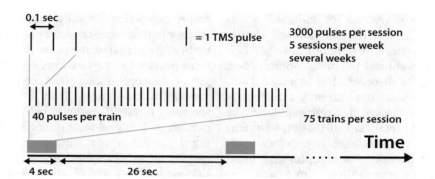

0.1 sec

= 1 TMS pulse

3000 pulses per session
5 sessions per week
several weeks

40 pulses per train

75 trains per session

Time

4 sec   26 sec

Fig. 8.1 Conventional rTMS protocol for depression treatment. TMS is applied to left dorsolateral prefrontal cortex, often (but not exclusively) located relative to the left "motor hotspot" ("5 centimeter rule"). 10-Hertz rTMS is administered in trains of 4 seconds, separated by 26 seconds; 75 trains per session, and 5 sessions per week for often (but not exclusively) 4–6 weeks

There have been many excellent studies evaluating the efficacy of left frontal rTMS for depression treatment, and we refer to Lefaucheur et al. (2014, 2020) for exhaustive overviews. These same authors presented a classification scheme, identifying studies as Class I, II, III, or IV, with decreasing value of evidence based on a list of criteria including randomization concealment, sample size, and other relevant considerations broadly considered to contribute to scientific rigor. Since Class I studies in this scheme should provide the most informative results, ◻ Table 8.1 presents details of 10 Class I studies of excitatory left frontal TMS in depression. Note that, even in this small selection of studies, there are quite some methodological variations in the administered protocol.

This protocol should "stimulate" the left frontal cortex, but another promising approach is the inverse on right frontal cortex, using low-frequency repetitive TMS on this contralateral hemisphere. One rTMS session increases cortical excitability (see companion chapter) in left frontal cortex (or decreases excitability in the right frontal cortex), but efficacious modulation of depression may have multiple potential (not mutually exclusive) underlying mechanisms of action. After all, depression has been linked to a hypometabolic left frontal cortex, and/or hypermetabolic right frontal cortex (Grimm et al. 2008). But the DLPFC is also connected to limbic system regions (Ferrarelli et al. 2004; Li et al. 2004; Tik et al. 2017). And empirical evidence

suggests that the left frontal rTMS protocol moreover affects multiple neurotransmitter systems, neurotrophic factors, blood flow, etc. (see (Lefaucheur et al. 2014), p. 29). Future research will undoubtedly shed more light on the precise mechanisms of action, for different TMS protocols and different patients.

There have been several variations and developments of the left frontal excitatory rTMS protocol, as is also evident from ◻ Table 8.1. A recent version of this treatment protocol simply shortens the 26-second break between 4-second rTMS trains to 11 seconds, essentially halving the duration of each session without changing the relevant stimulation parameters. A form of "patterned rTMS," called intermittent theta burst stimulation (iTBS) (Huang et al. 2005), presents 600 pulses as 50-Hertz triplets in a 5-Hertz rhythm for 2 seconds, followed by an 8-second break, which means a single session takes only just over 3 minutes. A non-inferiority study revealed similar efficacy as compared to "classical" HF-rTMS (Blumberger et al. 2018), and iTBS has been FDA approved since 2018 (Lefaucheur et al. 2020).

In terms of where to place the coil, there have been reports that the 5-centimeter landmark approach to locating left DLPFC from the cortical FDI-representation is not always optimal (Herwig et al. 2001; Herbsman et al. 2009; Nauczyciel et al. 2011). Alternative approaches include simply changing this rule to a 6- or even 7-centimeter rule (Fitzgerald et al. 2009b; Ahdab et al. 2010), using elec-

(continued)

Table 8.1 Class I studies – 10 Hz rTMS to left DLPFC to treat depression

Study	Method	Design	Sample size	Response variable	Results	Conclusion
Avery et al. (2006)	5 cm landmark 110% SCD-corrected rMT	Randomized, sham-controlled	68 (35 active, 33 sham)	HRSD	Significantly larger reduction in HRSD scores for active vs. sham TMS ($p = 0.002$) Significantly larger response rate (i.e., $\geq 50\%$ reduction in HRSD score) for active vs. sham TMS ($p = 0.008$) Significantly larger remission rate (i.e., HRSD score <8) for active vs. sham TMS ($p = 0.033$)	Positive
Blumberger et al. (2016)	Neuronavigation 120% SCD-corrected rMT	Randomized, sham-controlled	81 (40 active, 41 sham)	HRSD	HRSD score reduction did not significantly differ between active vs. sham TMS Response rates did not significantly differ between active vs. sham TMS Remission rates did not significantly differ between active vs. sham TMS	No evidence
Herbsman et al. (2009)	5 cm landmark 120% SCD-corrected rMT	Randomized, sham-controlled	54 (35 active, 33 sham)	HRSD	Active vs. sham TMS significantly reduced HDRS scores ($p = 0.017$) Remission rates were significantly higher in the active vs. sham group ($p = 0.004$) Patients with more anterio-lateral coil positioning were more likely to respond to TMS treatment	Positive
Herwig et al. (2007)	F3, 10–20 system 110% rMT	Randomized, double-blind, sham-controlled, multi-center	127 (62 active, 65 sham)	HRSD, BDI, MADRS	No significant difference in rating scores between active vs. sham TMS No significant difference in remission (i.e., score ≤ 10 on all three scales) between active vs. sham TMS No significant difference in responders (i.e., $\geq 50\%$ score improvement on at least two out of three rating scales) between active vs. sham TMS	No evidence

Table 8.1 (continued)

Study	Method	Design	Sample size	Response variable	Results	Conclusion
O'reardon et al. (2007)	5 cm landmark 120% rMT	Randomized, double-blind, sham-controlled, multi-center	301 (155 active, 146 sham)	MADRS, HRSD, CGI-S	Significant difference in MADRS score reduction between active vs. sham TMS ($p = 0.038$) Significant difference in HRSD score reduction between active vs. sham TMS ($p = 0.006$) Significantly greater improvement in CGI-S in the active vs. sham group ($p = 0.012$) Significantly larger response rates in the active vs. sham group ($p < 0.05$)	Positive
Jorge et al. (2008)	MRI localization 110% rMT	Randomized, sham-controlled	92 (48 active, 44 sham)	HRSD	Significantly larger decrease in HRSD scores in the active vs. sham group ($p < 0.001$) Significantly higher response rates in the active vs. sham group ($p = 0.003$) Significantly higher remission rates in the active vs. sham group ($p = 0.01$)	Positive
Lisanby et al. (2009)	5 cm landmark 120% rMT	Randomized, double-blind sham-controlled, multi-center	301 (155 active, 146 sham)	MADRS	Significant decrease in MADRS scores in the active vs. sham group ($p = 0.006$) in patients that previously received one antidepressant treatment in the current depressive episode	Positive
George et al. (2010)	120% rMT	Randomized, double-blind, sham-controlled, multi-center	190 (92 active, 98 sham)	HRSD	Significantly larger remission rates in the active vs. sham group ($p = 0.02$)	Positive

Table based in part on Lefaucheur et al. (2014) with permission from the publisher

Abbreviations: *SCD* scalp-to-cortex distance, *rMT* resting motor threshold, *HRSD* Hamilton Rating Scale for Depression, *BDI* Beck Depression Inventory, *MADRS* Montgomery-Asberg Depression Rating Scale, *CGI-S* clinician-rated global illness severity

trode location "F3" in the international 10–20 coordinate system originally developed for EEG (Herwig et al. 2003), the related "Beam F3" method (Beam et al. 2009; Mir-Moghtadaei et al. 2015), and even individual stereotactic navigation ('neuronavigation') to anatomically defined regions based on an MRI scan (Fitzgerald et al. 2009a; Peleman et al. 2010; Rusjan et al. 2010; Wall et al. 2016). Finally, other cortical targets and protocols have been explored, including the aforementioned low-frequency rTMS on the right frontal cortex, bilateral rTMS, rTMS to parietal cortex (Schutter and van Honk 2005; Schutter et al. 2009), as well as TES protocols (Fregni et al. 2006). Some of these are even implemented routinely. Yet, for this chapter we continue to focus on the most established, left frontal HF-rTMS, depression treatment protocol. It remains unclear exactly how and to what extent the mechanisms of action and treatment effects differ between some of these protocols. As discussed below, research is ongoing on an increasingly large scale, but many questions remain.

8.2.2 Conventional Depression Treatments

The standard conventional depression treatments, against which rTMS should be evaluated, primarily include psychotherapy (an umbrella term to refer to cognitive-behavioral therapy, psychoanalysis therapy, etc.), pharmacology (antidepressant drugs), and ECT.

Pharmacological treatment offers many options. Cipriani et al. (2018) compared 21 drugs versus placebo in a meta-analysis, assessing response rates (a minimally 50% reduction in symptoms on a depression measure) and acceptability (drop-out rates). A useful measure is the odds ratio, which contrasts an observed outcome measure (e.g., response rate) for both the intervention (i.e., drug) and placebo directly, as a relative measure. For response rate, for example, a positive odds ratio would indicate that response rate was higher for the active intervention, while a negative odds ratio would indicate that response rate was higher for the placebo. Odds

ratios (95% confidence interval) for the 21 assessed drugs ranged from 1.37 (1.16–1.63) to 2.13 (1.89–2.41) in favor of the active intervention, and this response was statistically significant for all tested drugs, indicating that they all "worked," on the group level. Unfortunately, only two had a significantly positive odds ratio on the measure of acceptability, as captured by drop-out rates. Side effects for antidepressants are diverse, depending on the drug in question as well as the patient, including among others nausea, weight gain, fatigue, insomnia, constipation, anxiety, etc. (Bet et al. 2013). An antidepressant cannot, at least not in conventional treatment, be taken locally. This means that the active chemical ingredient affects relevant, but also irrelevant brain regions and systems and even physical systems outside the brain.

Psychotherapy again includes multiple forms, such as cognitive-behavioral therapy (CBT), psychodynamic therapy, supportive counseling, etc. This makes it difficult to assign single "scores" to the efficacy of psychotherapy. Nevertheless, Cuijpers et al. (2014) performed a meta-analysis of psychotherapy, and Barth et al. (2013) a network meta-analysis. In these analyses, efficacy did not seem to differ much, if at all, between the main psychotherapy approaches. All main approaches outperform waitlist, or care-as-usual, control groups. But in these analyses, both control groups and intervention groups show rather impressive improvements. To provide one example, staying with the outcome measure of odds ratio of response rate between intervention and control, a recent meta-analysis on CBT by Santoft et al. (2019) reported an odds ratio of 2.47 (95% CI: 1.60–3.80) in favor of CBT. This seems not too far removed from the highest odds ratio of 2.13 that Cipriani et al. (2018) reported for their list of reviewed antidepressants.

8.2.3 The (Difficult) Comparison

So how does rTMS perform, in comparison? Although it was acknowledged as an established conventional treatment, we here do not focus much on ECT. This is not because ECT is a less viable option, in fact it may be the most efficacious. Indeed, one direct comparison of rTMS versus ECT efficacy confirmed that ECT was more effective, especially in psychotic depression (Ren et al. 2014). Yet, rTMS is more often compared to psychotherapy and especially antidepressants, perhaps because these are all considered good candidates for first-line, or at least early, treatment. Still, also the comparison between rTMS and psycho- or pharmacological therapy is not easy, since only a limited number of direct clinical comparison studies have been performed (e.g., as in Bares (2009), who compared right frontal rTMS to venlafaxine and found no difference). The odds ratio of response rate for the standard left frontal rTMS protocol was 3.3 (95% CI: 2.35–4.64), in a meta-analysis of 29 randomized sham-controlled trial (RCT) studies by Berlim et al. (2014). On this measure, rTMS comes out favorably as compared to the previously mentioned 2.47 for psychotherapy and 2.13 for antidepressants. But this is just one measure (odds ratio of response rate), which we here selected randomly from many outcome measures that might be compared (e.g., remission rates, relapse rates, drop-out rates, absolute response/remission rates rather than odds ratios). What is the best way to compare rTMS efficacy with that of the conventional treatments? Which is the most relevant measure? Different choices here might provide different insights. A second problem with any of these comparisons is that rTMS is still the newcomer. Understandably, early clinical trials, and indeed still most clinical applications, included "treatment-resistant" patients, who had not responded to conventional treatment such as common antidepressants. To what extent are the patient samples in current clinical trial literature, for rTMS and for conventional treatment, comparable groups of patients? If the patient samples are not comparable, absolute efficacy rates from separate clinical trials are difficult to meaningfully compare also.

While we should keep in mind these limitations, rTMS seems to fare rather well in comparisons. Response rates and remission rates vary between clinical trials, but for instance in the meta-analysis previously referred to, Berlim et al. (2014) reported 29.3% (versus 10.4% for placebo) response rate and 18.6% (versus 5% for placebo) remission rate. Again keeping in mind the difficulties, Baeken et al. (2019) selected "the largest studies and datasets available" to attempt a direct comparison of response and remission rates in (1) psychotherapy monotherapy, (2) psychotherapy + antidepressants, (3) antidepressants, as reported in the large STAR*D trial (Rush et al. 2006), and rTMS in (4) monotherapy, or (5) combined with psychotherapy. For details, please refer Figure 1 of Baeken et al. (2019), but their overview suggests that rTMS efficacy is not clearly inferior to conventional treatments (leaving out ECT here), and possibly superior to antidepressants particularly after failed pharmacological treatment. Generally speaking, rTMS appears to compare very favorably to at least some antidepressants for at least some patients, in the relative mildness of side effects. The most common reported side effect is headaches, which are alleviated with common painkillers, and the most serious adverse effect often mentioned is the potential seizure. However, a recent estimate of the rate of occurrence of such rTMS-induced seizures when operating within published safety guidelines was about 1 in 60.000 TMS sessions (Lerner et al. 2019), which we would consider acceptable in nearly any cost-benefit analysis.

8.2.4 Room for Growth, in the Example of Depression Treatment

In our example of depression, we have so far concluded that rTMS treatment compares relatively well to other available treatments, in terms of efficacy and tolerability. But it is at least possible, if not certain, that there is still room for growth.

- Many clinical trials used coil targeting approaches that were reported to be suboptimal (e.g., 5 cm rule versus 6 or 7 cm, or versus F3, F3 Beam, or neuronavigation approaches (Herwig et al. 2001; Fitzgerald et al. 2009b; Herbsman et al. 2009).
- We have only scratched the surface of all possible rTMS protocols. Future options include "accelerated" rTMS protocols

(Holtzheimer et al. 2010; Baeken et al. 2014; McGirr et al. 2015; Duprat et al. 2016; Baeken 2018; Fitzgerald et al. 2018), real-time EEG-driven rTMS (Zrenner et al. 2019, 2020), or multi-region rTMS (Prasser et al. 2015; Lefaucheur et al. 2020), primed protocols (Fitzgerald et al. 2008, 2013), and many more possible tweaks or reconfigurations of the current treatment.

- The hot topic at the moment is personalized medicine, and it applies to rTMS treatment as well. Research is ongoing to classify patients into sub-groups, find biomarkers, or adapt protocols based on measured individual brain activity. Recent reports even suggest that treatment response might be predicted based on quantifiable TMS-induced heart rate modulations (Iseger et al. 2017, 2019, 2020).

- There are few published accounts of rTMS combined with psychotherapy. But at least one naturalistic study to which we contributed reported that, in their patient sample, rTMS combined with psychotherapy (psychotherapy during rTMS administration) led to an impressive 66% response rate and 56% remission rate, with 60% sustained remission at follow-up (Donse et al. 2018).

These are just examples of where current research focuses its efforts. Added to these should be the exciting possibility of rTMS as a first-line treatment, which might result in still higher response and remissions rates than previously found in large clinical trials.

8.2.5 rTMS for Treatment of Other Conditions: Rapid Developments

In 2014, a group of European experts on clinical TMS presented an exhaustive overview of rTMS efficacy as a treatment approach for nearly all explored avenues of application (Lefaucheur et al. 2014). As mentioned above, based on systematic literature search on the Pubmed database, with each result screened and reviewed by multiple authors, clinical trials were classified as Class I, II, III, IV, reflect-

ing the strength of evidence. But in a second step, per rTMS application (a specific rTMS protocol to a specific cortical target site for a specific brain-based disorder), the number and combination of Class I, II, III studies with positive or negative results was summarized as constituting "Level A, B, or C evidence" for efficacy. Level A classification means "definitively effective," or the inverse "definitely ineffective." Level B means "probably (in)effective," and Level C "possibly (in) effective."

> **Definition**
>
> Levels of Evidence: Lefaucheur et al. (2014, 2020) introduced Level A, B, C classifications to particular rTMS treatments/protocols, to indicate the (at the time) strength of empirical evidence that a particular rTMS treatment is effective or ineffective. Level A: definitely (in)effective. Level B: probably (in)effective. Level C: possibly (in)effective.

For the exhaustive list of these classifications and details, we refer to the original publication. But to present some of the highlights: based on all clinical rTMS studies fulfilling inclusion criteria published until March 2014, Level A evidence was awarded to rTMS treatment of treatment-resistant major depression (high-frequency TMS, left DLPFC). But Level A evidence was also found for efficacy of high-frequency rTMS treatment of neuropathic pain, targeting primary motor cortex of the hemisphere contralateral to the pain. Furthermore, there was Level B evidence (probably effective) for right-hemisphere low-frequency rTMS treatment efficacy of depression, and a longer list of Level C evidence for treatment of disorders including motor stroke, tinnitus, obsessive-compulsive disorder, schizophrenia, addiction, and craving.

The point here is not to list all possible applications. The point is that largely the same group of experts recently published an update, including new evidence from 2014 to 2018 (Lefaucheur et al. 2020). In just those few years, the additional body of published clinical trials changed the level of evidence for several of these rTMS treatments considerably. ◘ Figure 8.2 shows side-by-side bar graphs of the number of rTMS protocols receiving Level A, B, or C evidence for protocol (in)efficacy, as suggested in the 2014 review and the 2020 review by Lefaucheur and colleagues. Note that an "rTMS protocol" here includes the combination of stimulation parameters, neural target site, and treatable disorder, meaning that rTMS treatment for depression can be included more than once if different rTMS treatment approaches received a Level A, B, C classification (e.g., high-frequency rTMS left and low-frequency rTMS right). This figure is mostly intended to convey the impressive accumulation of new data in recent years, since the 2014 overview included all the evidence from all previous years, and the 2020 update reflects insights collected in just the few years since. Note also that this figure does not include many additional rTMS treatments/protocols studied and receiving positive or negative reports, because they simply did not (yet) receive minimally Level C classification from Lefaucheur et al. (2014, 2020).

Again, it is simply not possible in this chapter to exhaustively present all the information, for which we refer to the cited publications. But we can share a few subjective highlights. As of 2020, there is also Level A evidence for neurorehabilitation in post-acute motor stroke, as well as deep high-frequency rTMS over left frontal cortex in depression. As of 2020, there is now Level B evidence for rTMS treatment efficacy for quality-of-life improvement in fibromyalgia and analgesic effects in fibromyalgia, motor symptoms in Parkinson's disease, post-stroke aphasia recovery, lower limb plasticity in multiple sclerosis, and post-traumatic stress disorder. Furthermore, several new treatment applications have reached Level C status. This suggests that rTMS for depression treatment really was just an example. It is not possible to do justice to the state of the art in this field in just these few paragraphs. But clearly, some of the early promises of NIBS are coming to fruition. Non-invasive neuromodulation with real, meaningful clinical efficacy, seems here to stay. And moreover, development in this field has proceeded at an impressive pace. TMS in its current form was

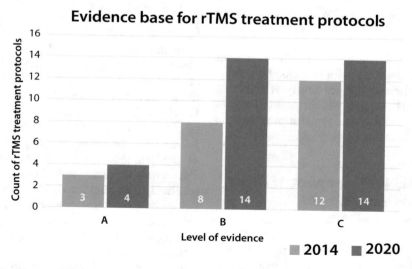

◘ Fig. 8.2 Rapid pace of clinical research. Lefaucheur and colleagues presented an exhaustive overview of rTMS treatment efficacy, across a wide range of brain-based disorders, rTMS stimulation parameters, and rTMS target sites. Depending on the results and the scientific rigor of underlying clinical trials, rTMS treatments/protocols could receive a "level of evidence" rating of A (definitely (in)effective), B (probably (in)

effective), C (possibly (in)effective). In 2020, an update was published including new evidence, along the same criteria. This figure shows bar graphs of the numbers of rTMS treatments/protocols receiving Level A, B, C ratings for (in)efficacy, in 2014 and in the 2020 update. Judging by these developments, in just a few years our state of knowledge of rTMS efficacy across the range of clinical applications was much enhanced

introduced 35 years ago. The list of evidence-based applications has expanded impressively in just the last several years.

⟩ Conclusion

NIBS fills a clinical niche, with its combination of localized, targeted modulation of intrinsic brain mechanisms, wide range of potential applications, coupled with only minimal adverse/side effects and generally high tolerability, and of course its inherently non-invasive nature. It does have its limitations, which we did not focus on in this chapter. TMS machines are not cheap, and in contrast to antidepressants, they cannot (yet) be taken home. More generally, it remains important to be careful. Developments of such pace can pique excitement to unreasonable levels, and extrapolation of development in previous years offers no guarantees for future success. There is "only" Level A evidence for a, as of yet, small range of rTMS applications. These, surely, should be widely implemented if it makes ethical, medical, and financial sense in comparison to or in concert with conventional pharmacological and psychotherapeutic treatment options. The Level B classifications are promising, but "promising" does not mean established. Moreover, this all concerns rTMS. A second large branch of NIBS is transcranial electric brain stimulation (TES), and here the current maximal classification by a partially overlapping group of European experts is Level B evidence, again for treatment of depression (Lefaucheur et al. 2017). Interestingly, there is actually Level B evidence *against* tDCS efficacy for *treatment-resistant* depression, showing at once that tDCS and rTMS may be complementary tools in the mental medicine repertoire, and that we still have much to learn in understanding the mechanisms of action underlying clinical NIBS effects.

All in all, from the overview in this chapter, one might conclude that, perhaps, clinical NIBS is indeed low-hanging fruit.

For some applications, Level A evidence for efficacy suggests the fruit is ripe for the plucking. The pace of development suggests that it hangs low, still. How good it really tastes, time will tell.

References

Ahdab R, Ayache SS, Brugières P, Goujon C, Lefaucheur JP (2010) Comparison of "standard" and "navigated" procedures of TMS coil positioning over motor, premotor and prefrontal targets in patients with chronic pain and depression. Clin Neurophysiol 40:27–36

Avery D. H, Holtzheimer III P. E, Fawaz W, Russo J, Neumaier J, Dunner D. L, Roy-Byrne P (2006) A controlled study of repetitive transcranial magnetic stimulation in medication-resistant major depression. Biological psychiatry, 59(2):187–194

Baeken C (2018) Accelerated rTMS: a potential treatment to alleviate refractory depression. Front Psychol 9:2017

Baeken C, Marinazzo D, Wu G-R, Van Schuerbeek P, De Mey J, Marchetti I, Vanderhasselt M-A, Remue J, Luypaert R, De Raedt R (2014) Accelerated HF-rTMS in treatment-resistant unipolar depression: insights from subgenual anterior cingulate functional connectivity. World J Biol Psychiatry 15:286–297

Baeken C, Brem AK, Arns M, Brunoni AR, Filipčić I, Ganho-Ávila A, Langguth B, Padberg F, Poulet E, Rachid F, Sack AT, Vanderhasselt MA, Bennabi D (2019) Repetitive transcranial magnetic stimulation treatment for depressive disorders: current knowledge and future directions. Curr Opin Psychiatry 32:409–415

Barth J, Munder T, Gerger H, Nüesch E, Trelle S, Znoj H, Jüni P, Cuijpers P (2013) Comparative efficacy of seven psychotherapeutic interventions for patients with depression: a network meta-analysis. PLoS Med e1001454:10

Beam W, Borckardt JJ, Reeves ST, George MS (2009) An efficient and accurate new method for locating the F3 position for prefrontal TMS applications. Brain Stimul 2:50–54

Berlim MT, Van Den Eynde F, Tovar-Perdomo S, Daskalakis ZJ (2014) Response, remission and drop-out rates following high-frequency repetitive transcranial magnetic stimulation (rTMS) for treating major depression: a systematic review and meta-analysis of randomized, double-blind and sham-controlled trials. Psychol Med 44:225–239

Bet PM, Hugtenburg JG, Penninx BWJH, Hoogendijk WJG (2013) Side effects of antidepressants during long-term use in a naturalistic setting. Eur Neuropsychopharmacol 23:1443–1451

Blumberger D. M, Maller J. J, Thomson L, Mulsant B. H, Rajji T. K, Maher M, Daskalakis Z. J (2016). Unilateral and bilateral MRI-targeted repetitive transcranial magnetic stimulation for treatment-resistant depression: a randomized controlled study. Journal of psychiatry & neuroscience: JPN, 41(4):E58

Blumberger DM, Vila-Rodriguez F, Thorpe KE, Feffer K, Noda Y, Giacobbe P, Knyahnytska Y, Kennedy SH, Lam RW, Daskalakis ZJ, Downar J (2018) Effectiveness of theta burst versus high-frequency repetitive transcranial magnetic stimulation in patients with depression (THREE-D): a randomised non-inferiority trial. Lancet 391:1683–1692

Cipriani A, Furukawa TA, Salanti G, Chaimani A, Atkinson LZ, Ogawa Y, Leucht S, Ruhe HG, Turner EH, Higgins JPT, Egger M, Takeshima N, Hayasaka Y, Imai H, Shinohara K, Tajika A, Ioannidis JPA, Geddes JR (2018) Comparative efficacy and acceptability of 21 antidepressant drugs for the acute treatment of adults with major depressive disorder: a systematic review and network meta-analysis. Focus (Madison) 16:420–429

Coates McCall I, Lau C, Minielly N, Illes J (2019) Owning ethical innovation: claims about commercial wearable brain technologies. Neuron 102:728–731

Cuijpers P, Karyotaki E, Weitz E, Andersson G, Hollon SD, Van Straten A (2014) The effects of psychotherapies for major depression in adults on remission, recovery and improvement: a meta-analysis. J Affect Disord 159:118–126

Donse L, Padberg F, Sack AT, Rush AJ, Arns M (2018) Simultaneous rTMS and psychotherapy in major depressive disorder: clinical outcomes and predictors from a large naturalistic study. Brain Stimul 11:337–345

Duecker F, de Graaf TA, Sack AT (2014) Thinking caps for everyone? The role of neuro-enhancement by non-invasive brain stimulation in neuroscience and beyond. Front Syst Neurosci 8:71

Duprat R, Desmyter S, Rudi DR, Van Heeringen K, Van Den Abbeele D, Tandt H, Bakic J, Pourtois G, Dedoncker J, Vervaet M, Van Autreve S, Lemmens GMD, Baeken C (2016) Accelerated intermittent theta burst stimulation treatment in medication-resistant major depression: a fast road to remission. J Affect Disord 200:6–14

Ferrarelli F, Haraldsson HM, Barnhart TE, Roberts AD, Oakes TR, Massimini M, Stone CK, Kalin NH, Tononi G (2004) A [17F]-fluoromethane PET/TMS study of effective connectivity. Brain Res Bull 64:103–113

Fitzgerald PB, Hoy K, McQueen S, Herring S, Segrave R, Been G, Kulkarni J, Daskalakis ZJ (2008) Priming stimulation enhances the effectiveness of low-frequency right prefrontal cortex transcranial magnetic stimulation in major depression. J Clin Psychopharmacol 28:52–58

Fitzgerald PB, Hoy K, McQueen S, Maller JJ, Herring S, Segrave R, Bailey M, Been G, Kulkarni J, Daskalakis ZJ (2009a) A randomized trial of rTMS targeted with MRI based neuro-navigation in treatment-resistant depression. Neuropsychopharmacology 34:1255–1262

Fitzgerald PB, Maller JJ, Hoy KE, Thomson R, Daskalakis ZJ (2009b) Exploring the optimal site for the localization of dorsolateral prefrontal cortex in brain stimulation experiments. Brain Stimul 2:234–237

Fitzgerald PB, Hoy KE, Singh A, Gunewardene R, Slack C, Ibrahim S, Hall PJ, Daskalakis ZJ (2013) Equivalent beneficial effects of unilateral and bilateral prefrontal cortex transcranial magnetic stimulation in a large randomized trial in treatment-resistant major depression. Int J Neuropsychopharmacol 16:1975–1984

Fitzgerald PB, Hoy KE, Elliot D, Susan McQueen RN, Wambeek LE, Daskalakis ZJ (2018) Accelerated repetitive transcranial magnetic stimulation in the treatment of depression. Neuropsychopharmacology 43:1565–1572

Fregni F, Boggio PS, Nitsche MA, Marcolin MA, Rigonatti SP, Pascual-Leone A (2006) Treatment of major depression with transcranial direct current stimulation. Bipolar Disord 8:203–204

George M. S, Lisanby S. H, Avery D, McDonald W. M, Durkalski V, Pavlicova M, Sackeim H. A (2010). Daily left prefrontal transcranial magnetic stimulation therapy for major depressive disorder: a sham-controlled randomized trial. Archives of general psychiatry, 67(5):507–516

Grimm S, Beck J, Schuepbach D, Hell D, Boesiger P, Bermpohl F, Niehaus L, Boeker H, Northoff G (2008) Imbalance between left and right dorsolateral prefrontal cortex in major depression is linked to negative emotional judgment: an fMRI study in severe major depressive disorder. Biol Psychiatry 63:369–376

Herbsman T, Avery D, Ramsey D, Holtzheimer P, Wadjik C, Hardaway F, Haynor D, George MS, Nahas Z (2009) More lateral and anterior prefrontal coil location is associated with better repetitive transcranial magnetic stimulation antidepressant response. Biol Psychiatry 66:509–515

Herwig U, Padberg F, Unger J, Spitzer M, Schönfeldt-Lecuona C (2001) Transcranial magnetic stimulation in therapy studies: examination of the reliability of "standard" coil positioning by neuronavigation. Biol Psychiatry 50(1):58–61

Herwig U, Satrapi P, Schönfeldt-Lecuona C (2003) Using the international 10-20 EEG system for positioning of transcranial magnetic stimulation. Brain Topogr 16:95–99

Herwig U, Fallgatter A. J, Höppner J, Eschweiler G. W, Kron M, Hajak G, Schönfeldt-Lecuona, C. (2007) Antidepressant effects of augmentative transcranial magnetic stimulation: randomised multicentre trial. The British Journal of Psychiatry, 191(5):441–448

Holtzheimer PE, McDonald WM, Mufti M, Kelley ME, Quinn S, Corso G, Epstein CM (2010) Accelerated repetitive transcranial magnetic stimulation for treatment-resistant depression. Depress Anxiety 27:960–963

Huang YZ, Edwards MJ, Rounis E, Bhatia KP, Rothwell JC (2005) Theta burst stimulation of the human motor cortex. Neuron 45:201–206

Iseger TA, Padberg F, Kenemans JL, Gevirtz R, Arns M (2017) Neuro-cardiac-guided TMS (NCG-TMS): probing DLPFC-sgACC-vagus nerve connectivity using heart rate – first results. Brain Stimul 10(5):1006–1008

Iseger T, Vila-Rodriguez F, Padberg F, Downar J, Daskalakis Z, Blumberger D, Kenemans L, Arns M (2019) The heart-brain pathway in depression: optimizing TMS treatment for depression using cardiac response (Neuro-cardiac-guided-TMS). Brain Stimul 12:491–492

Iseger TA, van Bueren NER, Kenemans JL, Gevirtz R, Arns M (2020) A frontal-vagal network theory for major depressive disorder: implications for optimizing neuromodulation techniques. Brain Stimul 13(1):1–9

Jorge R. E, Moser D. J, Acion L, & Robinson R. G. (2008) Treatment of vascular depression using repetitive transcranial magnetic stimulation. Archives of General Psychiatry, 65(3):268–276

Lefaucheur JP, André-Obadia N, Antal A, Ayache SS, Baeken C, Benninger DH, Cantello RM, Cincotta M, de Carvalho M, De Ridder D, Devanne H, Di Lazzaro V, Filipović SR, Hummel FC, Jääskeläinen SK, Kimiskidis VK, Koch G, Langguth B, Nyffeler T, Oliviero A, Padberg F, Poulet E, Rossi S, Rossini PM, Rothwell JC, Schönfeldt-Lecuona C, Siebner HR, Slotema CW, Stagg CJ, Valls-Sole J, Ziemann U, Paulus W, Garcia-Larrea L (2014) Evidence-based guidelines on the therapeutic use of repetitive transcranial magnetic stimulation (rTMS). Clin Neurophysiol 125(11):2150–2206

Lefaucheur JP, Antal A, Ayache SS, Benninger DH, Brunelin J, Cogiamanian F, Cotelli M, De Ridder D, Ferrucci R, Langguth B, Marangolo P, Mylius V, Nitsche MA, Padberg F, Palm U, Poulet E, Priori A, Rossi S, Schecklmann M, Vanneste S, Ziemann U, Garcia-Larrea L, Paulus W (2017) Evidence-based guidelines on the therapeutic use of transcranial direct current stimulation (tDCS). Clin Neurophysiol 128(1):56–92

Lefaucheur JP, Aleman A, Baeken C, Benninger DH, Brunelin J, Di Lazzaro V, Filipović SR, Grefkes C, Hasan A, Hummel FC, Jääskeläinen SK, Langguth B, Leocani L, Londero A, Nardone R, Nguyen JP, Nyffeler T, Oliveira-Maia AJ, Oliviero A, Padberg F, Palm U, Paulus W, Poulet E, Quartarone A, Rachid F, Rektorová I, Rossi S, Sahlsten H, Schecklmann M, Szekely D, Ziemann U (2020) Evidence-based guidelines on the therapeutic use of repetitive transcranial magnetic stimulation (rTMS): an update (2014–2018). Clin Neurophysiol 131(2):474–528

Lisanby S. H, Husain M. M, Rosenquist P. B, Maixner D, Gutierrez R, Krystal A, George M. S (2009). Daily left prefrontal repetitive transcranial magnetic stimulation in the acute treatment of major depression: clinical predictors of outcome in a multisite, randomized controlled clinical trial. Neuropsychopharmacology, 34(2):522–534

Li X, Nahas Z, Kozel FA, Anderson B, Bohning DE, George MS (2004) Acute left prefrontal transcranial magnetic stimulation in depressed patients is associated with immediately increased activity in prefrontal cortical as well as subcortical regions. *Biol. Psychiatry* 55:882–890

McGirr A, Van Den Eynde F, Tovar-Perdomo S, Fleck MPA, Berlim MT (2015) Effectiveness and acceptability of accelerated repetitive transcranial magnetic stimulation (rTMS) for treatment-resistant major depressive disorder: an open label trial. J Affect Disord 173:216–220

Mir-Moghtadaei A, Caballero R, Fried P, Fox MD, Lee K, Giacobbe P, Daskalakis ZJ, Blumberger DM, Downar J (2015) Concordance between BeamF3 and MRI-neuronavigated target sites for repetitive transcranial magnetic stimulation of the left dorsolateral prefrontal cortex. Brain Stimul 8:965–973

Nauczyciel C, Hellier P, Morandi X, Blestel S, Drapier D, Ferre JC, Barillot C, Millet B (2011) Assessment of standard coil positioning in transcranial magnetic stimulation in depression. Psychiatry Res 186:232–238

O'Reardon, J. P, Solvason H. B, Janicak P. G, Sampson S, Isenberg K. E, Nahas Z, Sackeim H. A (2007) Efficacy and safety of transcranial magnetic stimulation in the acute treatment of major depression: a multisite randomized controlled trial. Biological psychiatry, 62(11):1208–1216

Peleman K, Van Schuerbeek P, Luypaert R, Stadnik T, De Raedt R, De Mey J, Bossuyt A, Baeken C (2010) Using 3D-MRI to localize the dorsolateral prefrontal cortex in TMS research. World J Biol Psychiatry 11:425–430

Prasser J, Schecklmann M, Poeppl TB, Frank E, Kreuzer PM, Hajak G, Rupprecht R, Landgrebe M, Langguth B (2015) Bilateral prefrontal rTMS and theta burst TMS as an add-on treatment for depression: a randomized placebo controlled trial. World J Biol Psychiatry 16:57–65

Ren J, Li H, Palaniyappan L, Liu H, Wang J, Li C, Rossini PM (2014) Repetitive transcranial magnetic stimulation versus electroconvulsive therapy for major depression: a systematic review and meta-analysis. Prog Neuro-Psychopharmacol Biol Psychiatry 51:181–189

Rush AJ, Trivedi MH, Wisniewski SR, Nierenberg AA, Stewart JW, Warden D, Niederehe G, Thase ME, Lavori PW, Lebowitz BD, McGrath PJ, Rosenbaum JF, Sackeim HA, Kupfer DJ, Luther J, Fava M (2006) Acute and longer-term outcomes in depressed outpatients requiring one or several treatment steps: a STAR*D report. Am J Psychiatry 163:1905–1917

Rusjan PM, Barr MS, Farzan F, Arenovich T, Maller JJ, Fitzgerald PB, Daskalakis ZJ (2010) Optimal transcranial magnetic stimulation coil placement for targeting the dorsolateral prefrontal cortex using novel magnetic resonance image-guided neuronavigation. Hum Brain Mapp 31:1643–1652

Santoft F, Axelsson E, Öst L-G, Hedman-Lagerlöf M, Fust J, Hedman-Lagerlöf E (2019) Psychological medicine cognitive behaviour therapy for depression in primary care: systematic review and meta-analysis. Psychol Med 49:1266–1274

Schutter DJLG, van Honk J (2005) A framework for targeting alternative brain regions with repetitive transcranial magnetic stimulation in the treatment of depression. J Psychiatry Neurosci 30:91–97

Schutter DJLG, Martin Laman D, van Honk J, Vergouwen AC, Frank Koerselman G (2009) Partial clinical response to 2 weeks of 2 Hz repetitive transcranial magnetic stimulation to the right parietal cortex in depression. Int J Neuropsychopharmacol 12:643

Tik M, Hoffmann A, Sladky R, Tomova L, Hummer A, Navarro de Lara L, Bukowski H, Pripfl J, Biswal B, Lamm C, Windischberger C (2017) Towards understanding rTMS mechanism of action: stimulation of the DLPFC causes network-specific increase in functional connectivity. NeuroImage 162:289–296

Wall CA, Croarkin PE, Maroney-Smith MJ, Haugen LM, Baruth JM, Frye MA, Sampson SM, Port JD (2016) Magnetic resonance imaging-guided, open-label, high-frequency repetitive transcranial magnetic stimulation for adolescents with major depressive disorder. J Child Adolesc Psychopharmacol 26:582–589

Zrenner B, Gordon P, Kempf A, Belardinelli P, McDermott E, Soekadar S, Fallgatter A, Zrenner C, Ziemann U, Dahlhaus FM (2019) Alpha-synchronized stimulation of the left DLPFC in depression using real-time EEG-triggered TMS. Brain Stimul 12:532

Zrenner B, Zrenner C, Gordon PC, Belardinelli P, McDermott EJ, Soekadar SR, Fallgatter AJ, Ziemann U, Müller-Dahlhaus F (2020) Brain oscillation-synchronized stimulation of the left dorsolateral prefrontal cortex in depression using real-time EEG-triggered TMS. Brain Stimul 13:197–205

8

Translational Medicine and Technology

Contents

Chapter 9   Electrophysiology: From Molecule to
              Cognition, from Animal to Human – 131
              Anke Sambeth

Chapter 10   MRI in CNS Drug Development – 149
              Mitul A. Mehta

Chapter 11   Positron Emission Tomography in Drug
              Development – 165
              Frans van den Berg and A. (Ilan) Rabiner

Chapter 12   Application of Cognitive Test Outcomes for
              Clinical Drug Development – 183
              Chris J. Edgar

Chapter 13   Exciting Research in the Field of Sexual
              Psychopharmacology: Treating Patients with
              Inside Information – 199
              Paddy Janssen

Chapter 14   A Paradigm Shift from DSM-5 to Research
              Domain Criteria: Application to Translational
              CNS Drug Development – 211
              William Potter and Bruce Cuthbert

Electrophysiology: From Molecule to Cognition, from Animal to Human

Anke Sambeth

Contents

9.1   Introduction – 132

9.2   Electrophysiology: From Molecule to Network
        Activity – 133
9.2.1   Patch-Clamp Techniques – 133
9.2.2   Single-Cell Recordings – 135
9.2.3   Multi-Unit Recordings and Local Field Potentials – 136
9.2.4   Electroencephalography – 138

9.3   Electrophysiology in Humans and Animals – 141

        References – 145

© The Author(s), under exclusive license to Springer Nature Switzerland AG 2021
R. Schreiber (ed.), *Modern CNS Drug Discovery*, https://doi.org/10.1007/978-3-030-62351-7_9

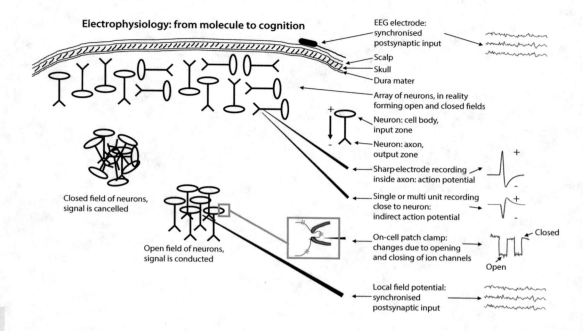

Electrophysiology: from molecule to cognition

EEG electrode: synchronised postsynaptic input

Scalp
Skull
Dura mater

Array of neurons, in reality forming open and closed fields

Neuron: cell body, input zone

Neuron: axon, output zone

Sharp-electrode recording inside axon: action potential

Single or multi unit recording close to neuron: indirect action potential

On-cell patch clamp: changes due to opening and closing of ion channels

Closed

Open

Local field potential: synchronised postsynaptic input

Closed field of neurons, signal is cancelled

Open field of neurons, signal is conducted

9

Electrophysiology is the field of study that examines physiological signals of the brain. It ranges from assessment of molecular activity using patch clamp techniques to population measurements such as electroencephalography (EEG). This chapter deals with those methods and their role in drug discovery. The on-cell patch is the most widely used patch clamp technique, which measures the activity of one ion in a community of intracellular ions. Single-cell recordings are done to assess changes in action potentials, the output of a system. Intracranial electrodes can also be used to detect activity from multiple or populations of neurons. Depending on how the signal is filtered, indirect action potentials (i.e., output from a region) or postsynaptic signals (i.e., input to a region) are detected. A final type of measurement described in this chapter is EEG. The measured signal, obtained from the scalp, is similar to that obtained from local field potentials that are recorded intracranially. EEG measured in humans has frequently been compared to local field potentials recorded from living animals, showing a number of similarities between species. However, due to genetic or anatomical differences, in combination with differing electrode locations, it is still unclear how comparable the

signals are. Future research should try to optimize animal methodology to increase the translational potential.

Learning Objectives
- To understand the various electrophysiological techniques available in humans and animals
- To analyze the translational potential of electrophysiology
- To use knowledge of electrophysiology in developing a (pre)clinical trial that uses an electrophysiological technique

9.1 Introduction

Neurons communicate with electrical and chemical signals. Ion channels inside and outside the neuron are responsible for the electrical activity as they move within the intracellular and extracellular fluid and move across the membrane. This causes a variation of physiological signals. Electrophysiology is the field in which we assess those physiological signals.

Electrophysiology encompasses a large field of research, ranging from molecular activity using patch clamp techniques to population

Table 9.1 Electrophysiological techniques
and their outcome measures

Technique	Outcome measure
Patch-clamp	Ionic currents/spikes
Single-unit recording	Spikes
Multi-unit recording	Spikes of several neurons
Local field potentials	Postsynaptic activity
EEG	Various frequencies (e.g., theta and alpha) Event-related potentials Event-related (de)synchronization

Table 9.1 Electrophysiological techniques and their outcome measures

measurements using electroencephalography (EEG, for an overview, see ◘ Table 9.1). When used in animal research, electrophysiology is always invasive, with the need to measure activity *in vitro* or to implant electrodes into living animals. In humans, most electrophysiological measures are noninvasive, like the measurement of an EEG. However, situations in which highly invasive measurements are needed may occur in patients with, for example, epilepsy. In this case, it may be necessary to look into the activity of a specific population of neurons using electrocortigography (ECoG) (e.g., Motoi et al. 2019), or even single neurons. In the latter case, patients with pharmacologically intractable epilepsy often take part in more fundamental research targeted toward learning more about the role of particular neurons of specific brain regions in cognition or behavior. A particularly interesting example is that of Quian Quiroga et al. (2005), who discovered a subset of medial temporal lobe neurons that are selectively activated by strikingly different pictures of well-known individuals, in this case the actors Jennifer Aniston and Halle Berry.

Electrophysiological measures offer an insight into the exact timing of neural activity, as activity is measured in milliseconds. However, the spatial resolution, in other words the ability to link a particular response to a particular brain area, is far from optimal and can much better be recorded using neuro-

imaging methods such as functional magnetic resonance imaging (fMRI). The more invasive methods like ECoG, however, measure the best of everything, as they can measure intracortical activity with a millisecond time resolution at specific target areas in the brain (Rizzolatti et al. 2018). Other opportunities to obtain fairly good temporal and spatial resolution are by combining electrophysiological recordings with noninvasive brain stimulation (NIBS) such as transcranial magnetic stimulation (TMS; see de Graaf et al. 2012). In this case, specific target sites in the brain are stimulated and their change in activity is recorded with, for instance, EEG. When aiming to define the underlying pathophysiology of neuropsychiatric disorders, this combination of methods seems particularly promising (Noda 2020).

In this chapter, a variety of measures will be introduced. In addition, for each of these methods, I will tap into the role they may play in drug development or assessment. Next, I discuss the translational potential of the various electrophysiological techniques from animal to human. Finally, schizophrenia will serve as an example to demonstrate which electrophysiological techniques can be used to characterize the disorder and how this may guide the drug discovery process.

9.2 Electrophysiology: From Molecule to Network Activity

9.2.1 Patch-Clamp Techniques

Patch-clamp PC can be used to measure the neuronal ion flow. Four different types of patch-clamp exist (see ◘ Fig. 9.1), of which the on-cell patch is one of the most often used in pharmacology. With this method, by sealing a micropipette to the membrane and applying some gentle suction, ions are isolated from the neuron while keeping the cell intact. The micropipette, being filled with a saline solution that resembles the extracellular fluid, also called electrolyte, can be injected with a drug of choice. By doing so, the external cell

On-cell patch   Inside-out patch   whole-cell patch   Outside-out patch

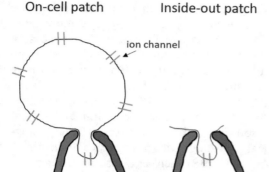
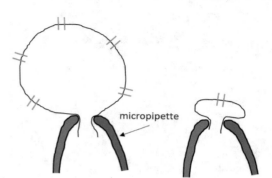

▢ Fig. 9.1 The four patch-clamp methods used in electrophysiology. An on-cell patch enables you to measure activity of one ion in relation to others in the neuron, the inside-out patch is used to manipulate an ion from the inside of the membrane, the whole-cell patch measures activity from the entire cell, and the outside-out patch offers the opportunity to measure one single ion in isolation

surface is challenged, which may change the electrical currents. These currents can be measured with an electrode that is placed in the micropipette. An advantage of this technique is that interactions of drugs can be measured from single ions that act in a community of ions. However, drug concentrations often cannot be changed in a single patch, as exact drug concentrations cannot be determined when different doses are combined due to dialyzing taking place. Therefore, several patches are needed in order to determine a dose-response curve.

> **Definition**
>
> *Patch-clamp recording.* A method used to assess ionic currents in individual isolated patches of a cell membrane in order to characterize the electrical properties of a membrane.

If the on-cell configuration has been established, and if you quickly withdraw the micropipette from the neuron, part of the membrane will be torn off and an inside-out patch is established. In this situation, one single ion can be manipulated by exposing the previously intracellular surface of the membrane (see ▢ Fig. 9.1). In other words, the intracellular (inside) surface now is on the outside (out). This measurement is relevant if the experimenter wants to learn about ions that are normally influenced by molecules inside the neuron.

When in the on-cell configuration, and when applying much suction at once, the membrane of the cell breaks and electrical changes inside the neuron can be measured. This provides us with a whole-cell patch that can be used to measure the entire cell. But we can go one step further. If the micropipette is slowly removed from the neuron, part of the membrane will be pulled away until it separates from the cell. It will automatically make a new seal, forming a mini neuron. In this configuration, what was previously outside stays outside, consequently calling this an outside-out patch. If this newly formed vesicle is small enough and only contains one ion channel, true activity of this single ion can be measured, making this the largest difference compared to the on-cell patch, where ions are measured in interaction with the other neuronal ions. Another difference is that, because of the recording electrode being on the inside, drugs can be applied on the outside. This can be controlled much better, providing opportunity to measure the same patch with different dosages. This way, an entire dose-response curve can be established.

Traditionally, patch-clamp techniques have been technically demanding and low throughput, which prevented the method from being valuable in drug discovery (Wickenden 2014).

Namely, these methods relied heavily on the skills of independent researchers. More recently, however, automated procedures were developed, such as the planar patch-clamp electrode (Liu et al. 2019). Using the newest techniques, hundreds to thousands of samples can be acquired daily as compared to a few when clamping was done manually. For example, using the IonFlux, it was shown that effects of agonists and antagonists of nicotinic acetylcholine receptors can reliably be recorded (Yehia and Wei 2020). This opens new avenues for drug discovery. For example, automated patch clamp methods now allow for higher throughput functional screening campaigns to find new drugs that target ion channels.

> To summarize, the on-cell patch is the most widely used patch clamp technique for drug discovery, which has become even more promising with the development of automated patch clamp methodology.

9.2.2 Single-Cell Recordings

Various techniques exist to measure the activity of one single neuron. First of all, sharp-electrode recordings can be done to measure the (changes in) membrane potential inside a neuron. Micropipettes similar to those of the patch clamp recordings are commonly used. In fact, the whole-cell patch described before also measures the membrane potential of the entire cell. The key difference between the two methods is that the opening of the tip is larger for the whole-cell patch as compared to the sharp-electrode recording. The advantage of the whole-cell patch, therefore, is that better electrical access to the cell is guaranteed. This occurs because the resistance is relatively low. However, the cell will be dialyzed rather quickly, as the electrolyte in the micropipette will replace the intracellular fluid due to volume conduction.

Whole-cell patch and sharp-electrode recordings may be used to measure minor changes in potentials, also called subthreshold potentials. These are potentials that are relatively small and will not lead to the generation of an action potential. Sharp-electrode recordings can also be used to measure action potentials (i.e., spikes), the signals that downstream lead to the release of neurotransmitters. If positively charged sodium ions cross the membrane (or dilute from the micropipette for that matter) and enter the neuron, the membrane potential rises. Once a certain threshold is reached at the axon hillock, an action potential occurs and propagates toward the axon terminal, where in turn neurotransmitters can be released. The amplitude and duration of an action potential may influence the neurotransmitter release. For example, when injecting dorsal root ganglion neurons with interleukin 1β, amplitude and duration of the neuronal action potentials increased (Noh et al. 2019), which affected calcium influx. Given the role of interleukin 1β, Noh and colleagues argued that this may be the process that induces pain. In other words, changes in action potentials may play a role in the development of pain symptoms.

Calcium also plays a role in other neuronal processes. Whole-cell patch clamp was used to determine the role of calcium deficiencies in the hippocampal region in symptoms seen in Alzheimer's disease (Pourbadie et al. 2017). The researchers first injected amyloid beta for six days into the entorhinal cortex of living rats. Next, hippocampal slices were prepared and whole-cell recordings were done from single neurons in a slice. Pourbadie and colleagues (2017) found that calcium currents were reduced in the dentate gyrus after injecting amyloid beta into the entorhinal cortex. This led to the conclusion that amyloid beta triggers electrophysiological alterations in cells relevant for learning and memory.

┌─ **Definition** ─────────────────────
│ *Single-unit/multi-unit recording.* A recording from an electrode tip nearby one or more neuronal membranes and which records action potentials indirectly.
└──────────────────────────────────

Another method to record activity of one single neuron is a single-unit recording, which can be obtained by implanting a microelectrode into a living brain and measuring activity from neurons close to the electrode tip. When using such a setup, indirect action potentials are measured. They are indirect, because the actual action potential takes place inside the neuron, being represented by a steep depolarization of the membrane. Simultaneously, because the positively charged ions flow from the extracellular fluid toward the intracellular fluid, a negative charge is left behind, also called a current sink. This negative charge is measured by the single-unit electrode and is somewhat smaller in amplitude as compared to the intracellular positive charge. Once the action potential has decayed, positively charged ions move out of the cell again, causing a current source near the recording site.

An advantage of a single-unit recording over the sharp-electrode recording or whole-cell patch-clamp is the fact that measurements can be done in the brain of living, although usually anesthetized, animals. As intracellular action potentials eventually lead to the release of neurotransmitters, measuring the indirect action potentials with a single unit is highly relevant for drug research. Here are some examples. It was shown that injection of the nor-adrenaline and dopamine inhibitor bupropion into the ventral tegmental area (VTA) highly decreased the firing rate (Amirabadi et al. 2014). Thus, it reduced the number of action potentials, an effect that could be related to the sides effects found when treating patients with antidepressants (Amirabadi et al. 2014). For example, the VTA is part of the brain's reward system and its manipulation could lead to drug dependence or tolerance. Firing rate was also recorded in the study by Wang and co-workers (2019), who showed that injection of orexin-A and orexin-B into the globus pallidus of parkinsonian mice significantly increased spontaneous firing rates of their neurons, which suggests that orexins may decrease Parkinson's disease-related motor deficits.

> In sum, when assessing action potentials, single-unit recordings are preferred over the sharp-electrode recording or whole-cell patch-clamp as they can be done in the brain of living animals.

9.2.3 Multi-Unit Recordings and Local Field Potentials

Multi-unit recordings are those in which activity from various neurons near the electrode tip is measured. They are not only obtained from unconscious animals but can also be measured in awake animals, providing an opportunity to assess changes in electrical signals while an animal is performing a certain activity. Awake animals can be recorded this way, because the electrode is somewhat larger as compared to that for the single-unit recordings, making it less fragile and sensitive to movement of the animal. The output of such a recording is the sum of activities of each single neuron that is close to the electrode tip. In other words, a mixture of spikes from all the neurons is shown in one graph. These spikes represent the output from the area.

In order to know which of the recorded signals comes from which neuron, a method called spike sorting is used. The idea here is that each class of neurons elicits unique action potentials. For instance, one neuron type should more or less always reach the same amplitude after action potential generation, which makes spike amplitude the most common feature in spike sorting (Heinricher 2004). Another key characteristic is the shape of a spike, which may differ from neuron to neuron (Buzsaki 2004). A third feature is the inter-spike interval, so the time between two action potentials (Leblois and Pouzat 2017). For example, your electrode tip sees three different neurons, two of class I and one of class II. It is unlikely that all those neurons are equally close to the tip. The more distant the neuron is from the electrode, the smaller the amplitude becomes. This means that a variation in amplitude can be used for the sorting. After all, the spikes of the neurons from class I should differ based on that. Variation in the shape of the spike or inter-spike interval can further be used to detect the class II neuron.

Neuron classification using spike sorting sounds easy, but may turn out difficult when only using one electrode. After all, if neurons are at the same distance from the electrode, it may be difficult to correctly sort them.

Electrophysiology: From Molecule to Cognition, from Animal to Human

Therefore, often an array of four electrodes is used, a so-called tetrode. Arranging the four electrodes three dimensionally will lead to optimal spike sorting (Buzsaki 2004) and thus enable us to record activity from multiple neurons simultaneously with good precision, which is highly advantageous compared to the single-unit recording.

As described above, multi-unit recordings provide the opportunity to measure action potentials. A similar electrode, or even the same one (e.g., Burns et al. 2010), can also record local field potentials. How is this possible? When doing a single-unit or a multi-unit recording, the specific aim is to record an action potential, an activity that has a frequency of 200–300 Hz (i.e., 200–300 spikes per second). In order to detect a spike, data including signals up to 10 KHz are generally recorded. In comparison, local field potentials have far lower frequencies. Therefore, filters typically exclude frequencies higher than 100–200 Hz, so that the spikes are not part of the output.

As opposed to the multi-unit recordings that display the output of a brain region, local field potentials assess the input to this same region. They represent the slow local current sinks (negative charges) and sources (positive charges) that are generated by the synchronous activity of a population of neurons (Wickenden 2014). Those sinks and sources are due to ions crossing cellular membranes via the ion channels. If, for instance, negatively charged ions move into the neuron, a net positivity, the current source, is left behind outside, which is picked up by the electrode tip. Because the filter settings in a local field recording only include frequencies up to 100–200 Hz, the electrode tip will pick up the postsynaptic signals.

If all neurons close to the electrode have the same orientation, and thus geometric configuration, they consist of an open field eliciting relatively strong signals close to the electrode tip. For example, the sinks and sources will be similarly aligned, far enough apart, and will be summed and conducted to the electrode. If the neurons are not nicely oriented in parallel, the sinks and sources of various neurons are much closer, which will cancel activity from the various neurons out. In other words, no clear signal reaches the electrode tip. An example of an open and closed field can be seen in ▢ Fig. 9.2. In reality, neurons in the various brain areas are not as neatly oriented as in the figure, yielding different variations in summation and cancellation of signals.

Multi-unit recordings are relevant for research similarly to the single-unit recordings described above, with the advantage of assessing action potentials from several neurons at once. It was, for instance, using such setups

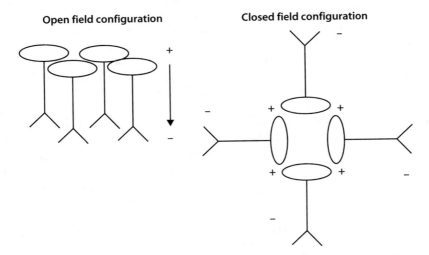

▢ **Fig. 9.2** Layout of an open and a closed field configuration. The + shows the positive sources, the − depicts negative sinks. The arrow points to the direction of the signal, going from the soma (circle) to the axon terminals (lines)

that Nobel Prize winner John O'Keefe and his colleagues were able to discover place cells, a type of neurons in the hippocampus that can memorize particular environmental locations for a longer period of time (e.g., Lever et al. 2002; O'Keefe 1976). Later, they found that the experience-dependent increase in hippocampal activity depends on the autophosphorylation of the alpha-isoform of the calcium/calmodulin-dependent protein kinase II (Cacucci et al. 2007), a molecular process crucial for spatial memory formation.

Like multi-unit recordings, local field potentials also can be recorded while an animal is performing a certain activity. We (Sambeth et al. 2007) injected rats with the cholinergic antagonist scopolamine and assessed their memory performance using an object-recognition task. While the animals' memory was impaired, theta activity (see more about this type of output in the next section) was increased in the dorsal hippocampus. Next, we examined whether nicotine and the acetylcholinesterase inhibitor, donepezil, were able to reverse these effects. On the behavioral level, this did occur, with animals showing normal memory again. However, only nicotine was able to reverse the theta change, whereas donepezil did not decrease the power back to baseline. This means that the link between the behavior and electrophysiology is sometimes difficult to determine. What we do know is that activity in the theta band recorded from the rat hippocampus is related to novelty (e.g., Sambeth et al. 2009) and subsequent memory performance (for a review, see Colgin 2016).

> In sum, multi-unit local field potential recordings can be performed with the same electrode, the first measuring output, and the latter input into the system. It is the filter used that determines which activity is picked up by the electrode tip.

9.2.4 Electroencephalography

Electroencephalography, or EEG, records slow postsynaptic signals, similarly to the local field potential. One key difference is that EEG is measured from the scalp, making it noninvasive. Another difference lies in the fact that the specificity of the signal is less compared to local field potentials. While the latter are recorded from electrodes in particular regions of the brain, EEG is only able to record activity from thousands of postsynaptic currents of cortical structures. Thirdly, due to the positioning of the electrode, only signals from cells that are oriented perpendicularly, so vertically, to the electrode can be detected. To visualize, when assuming the scalp being on top of the neurons in ⬛ Fig. 9.2, the ones in the open field are oriented optimally to be picked up by the electrode. Of those neurons in the closed field, only activity of two of them can be picked up, the ones oriented vertically. For the two horizontal ones, the positive and negative charges will cancel each other out.

When recording EEG, a large variety of output signals can be detected. First of all, by filtering the signal using particular band-pass filters, we can observe neural oscillations in different frequency bands. Delta activity, the activity between 1 and 4 Hz, is of the slowest frequency and most prominent when somebody is in deep sleep or resting state. Theta activity, the signal between 4 and 7 Hz, has in humans been related to various cognitive processes, among attention (Hong et al. 2020; Meyer et al. 2019) and working memory (Alekseichuk et al. 2017). As described above, the animal theta frequency in the hippocampus has been linked to memory performance (see also Buzsaki and Moser 2013). The frequency between 8 and 13 Hz is called alpha and can in humans best be detected if somebody closes their eyes. It is also seen as a measure of rest, and alpha desynchronization is commonly seen post-stimulus when a task is executed (Klimesch 2012). Beta oscillations have a frequency between 13 and 30 Hz, whereas gamma oscillations represent the activity between 30 and 80 Hz. Gamma oscillations have often been linked to cognitive processes, with stronger gamma responses being related to attention or memory (Bosman et al. 2014), similarly to the role of theta.

Given the overlap in activities being related to certain frequency bands, much research has been done on the coupling of frequency bands. This showed, for instance, that the ratio between theta and beta activities is highly relevant to the reduced cognitive processing capacity seen in attention deficit hyperactivity disorder (Picken et al. 2020). Cross-frequency coupling also takes place, a situation in which one frequency is phase-locked to the occurrence of the other frequency. For instance, it was shown that this coupling takes place between gamma and theta activities in rats that are exploring their environment (Buzsaki and Wang 2012) or navigating in a memorized space (Buzsaki and Moser 2013).

EEG frequency bands may be abnormal in various psychiatric disorders such as depression, anxiety, or schizophrenia (see Newson and Thiagarajan 2019 for review). For drug research, this is highly relevant, as the normalization of a particular frequency band by a certain drug could mean that symptoms are decreasing. For example, Leuchter et al. (2017) treated 146 depressed patients with the selective serotonin re-uptake inhibitor (SSRI) escitalopram and recorded EEG prior to and after one week of treatment. Delta-theta power (i.e., the squared amplitude of delta and theta activity) increased and alpha power decreased in the patients treated with escitalopram as compared to those who were administered with placebo (48 other patients). Moreover, the delta-theta/alpha ratio specifically predicted the likelihood of remission after seven weeks of treatment. Leuchter et al. (2017) concluded that this ratio might function as biomarker for remission of major depressive disorder. Similarly, it was shown that EEG-arousal regulation was able to differentiate responders from non-responders to SSRI treatment of obsessive compulsive disorder (Dohrmann et al. 2017).

Definition

Event-related potential (ERP). The ERP characterizes changes in the EEG signal in response to a definable event such as a sensory stimulus or a self-paced motor response. An ERP consists of a number of components with a particular amplitude and latency.

EEG records the spontaneous fluctuations in electrical activity of the brain, visualized as different neuronal oscillations described above. However, when a particular stimulus is presented, a direct response to that stimulus is added on top of the spontaneous background EEG. If this same stimulus, or a very similar one, is presented many times, and if the segments of EEG in response to this stimulus are averaged, the spontaneous EEG responses will be cancelled out and the one to the stimulus remains. A number of distinct positively and negatively charged components are then visible and together they are what we call an event-related potential (ERP). A large number of different ERP components exist, of which I will only present a few as example.

One of the most widely examined ERP components is the P300, which is a positive deflection occurring around 300 ms after stimulus onset and which is found when presenting both auditory and visual stimuli. It can be divided into a somewhat earlier P3a and later P3b deflection. Generally, the P3a is said to reflect the detection of a novel stimulus and early attention toward this stimulus (Polich 2007). Exactly which cognitive process the P3b relates to is still a matter of debate, but as early as 1981, Donchin suggested that the P3b reflects the process whereby mental schemas are revisited and contexts updated. Johnson (1986) proposed that information transmission, subjective probability of a particular stimulus, and the meaning a stimulus has for the participant. Polich (2007), on the other hand, suggested that the P3b is associated with attention and subsequent memory processing. What is clear is that stimulus probability and expectance, task relevance, and task difficulty affect attention and working memory (Kok 2001), which will help to categorize one stimulus among the train of other stimuli. This may all affect the P3b amplitude (Sambeth 2004).

Huang and co-workers (2015) posed the dual transmitter hypothesis of the P300, stating that the P3a is related to frontal focal attention and mediated by the dopaminergic system. P3b, on the other hand, is a temporally-parietally generated component where dense noradrenaline inputs are found.

However, this is a too simplistic view on the matter, as changes in, for instance, cholinergic levels also affect P3a and P3b amplitudes (Caldenhove et al. 2017; Evans et al. 2013; Kenemans and Kähkönen 2011).

Another vastly examined ERP component is the P50, a component representing basic auditory processing. The P50 is the component of interest in sensory gating, which is normally recorded in double-click paradigms. In such a paradigm, two auditory stimuli are presented within 500 ms. Both stimuli elicit the P50, but the P50 amplitude is dramatically decreased to the second stimulus. Reduced P50 gating, thus less suppression to the second stimulus in comparison to the first, is one of the key ERP deficiencies seen in patients with schizophrenia and has therefore been used as biomarker for psychosis (see also separate box). However, this phenomenon can, for instance, also be seen in patients with Alzheimer's disease (Cancelli et al. 2006) or bipolar disorder (Cheng et al. 2016). This seems logical, as a variety of neurotransmitters have been shown to play a role in this process, among dopamine, noradrenaline, serotonin, and acetylcholine (Kenemans and Kähkonen 2011).

The final ERP component I would like to introduce is another one receiving much interest in schizophrenia research, namely the mismatch negativity (MMN). This component is typically elicited in a passive oddball paradigm, in which irrelevant standard stimuli are interspersed with relevant deviants that differ from the standards in pitch, duration, loudness, or perceived location. The deviants elicit a negative wave around 100–150 ms after stimulus onset, the MMN. MMN is somewhat reduced to pitch and more so to duration deviants in chronic schizophrenia patients, which might pose this component to also serve as biomarker (Umbricht and Krljes 2005). However, a meta-analysis showed that the picture looks different in first episode patients, in whom only a small-to-medium reduction was found to duration deviants, but not to pitch deviants (Haigh et al. 2017). Similarly to the P50 gating deficiency, the MMN impairments are also not specific for schizophrenia, as abnormalities were, for instance, shown in Alzheimer's disease (Horvath et al. 2018), depression (Bissonnette et al. 2020), and autism (Chen et al. 2020). One key difference to P50 gating abnormalities, though, is that MMN can be detected with many more different types of stimuli and some patients may show abnormalities in one stimulus type, whereas others have reduced MMN to another. Once the differentiation in abnormalities between disorders is clearer, MMN might be used as biomarker or surrogate marker in the future.

The NMDA glutamate receptor has been said to play a major role in the generation of the MMN. Moreover, due to the various findings using glutamatergic drugs, it has even been proposed that the MMN serves as index of the integrity of the NMDA receptor system (Michie et al. 2016). Dopamine and serotonin do not seem to be relevant for the MMN generation (Todd et al. 2013), neither has manipulation of the cholinergic system shown reliable results (Klinkenberg et al. 2013a, b; Caldenhove et al. 2017).

Definition

Event-related (de)synchronization (ERD/ERS). A decrease (ERD) or increase (ERS) in amplitude of a neural oscillation in response to the presentation of a stimulus.

ERPs reflect time- and phase-locked changes to the onset of distinct stimuli. The key assumption here is that a stimulus, and each of them when being repeatedly presented, resets the phase of neural oscillations, in other words locks the phase. However, time-locked, yet non-phased locked, responses also occur in which the resetting of the phase does not take place. Those responses will cancel out if the regular ERPs are calculated. Event-related (de)synchronization (ERD/ERS) is the method by which those non-phased locked signals are visualized. Classically, first all event-related trials are band-pass filtered. Next, the amplitudes are squared to obtain the power of each sample, after which an average of power samples across trials is made. Finally, averaging over time samples takes place to smooth the data and reduce variability (Pfurtscheller 2001). The result is a timeline of power fluctuations for a particular frequency band in response to the stimulus.

Abnormalities in one or more frequencies have been found in various disorders using the ERD/ERS methodology. For instance, impaired theta oscillations generated in the right frontal lobe have shown to be correlated to autistic symptoms (Larrain-Valenzuela et al. 2017). An abnormal increase in phase synchronization in beta oscillations in a fronto-parietal network during working memory performance might indicate a compensatory mechanism for impaired cognitive function in major depression (Li et al. 2017). A final example relates to Parkinson's disease, in which mu-rhythms are commonly studied. Those are rhythms falling in the 7–12 Hz range, similarly to the alpha wave. The distinction between the two is the brain area generating the wave. Whereas the alpha wave is produced in the occipital lobe, thus the visual system, the mu-rhythm is generated in the motor cortex. Heida et al. (2014) found that impaired motor learning in patients with Parkinson's disease could be related to a deficit in mu-rhythm desynchronization that is normally found when participants are performing actions with their hands.

As compared to recording of basic EEG frequencies or calculating ERPs, the use of ERD/ERS in pharmacology has been relatively scarce so far. Given that this latter method provides detailed information on the exact changes in EEG signal that take place when a subject performs a particular task, future research should look into the usefulness of ERD and ERS in the field of drug discovery.

> To conclude, an EEG recording offers the opportunity to assess various EEG frequencies, event-related potentials, and ERD/ERS. These three types of measurements may guide early drug development, as deficiencies found in brain disorders often lead to particular deficits in one or more of these output signals.

9.3 Electrophysiology in Humans and Animals

In the previous sections, I discussed the different types of electrical output signals generated by the brain. Part of these recordings is only possible using slices of the mouse or rat hippocampus. Some measurements, though, can be done in living animals and, consequently, also in humans. In this section, I will try to compare animal to human electrophysiology in order to shed more light on the translational potential of certain measures.

Almost all studies in humans record EEG, as this is a noninvasive and easily implemented measurement. In comparison, there is no single measurement in animals that is noninvasive, as electrodes always need to be implanted into the brain. This difference has an effect on the assessment of brain regions of interest. While in humans only cortical structures can be targeted, we can assess any brain structure we like in animals. This also has an influence on the amplitude of the signal being detected. As the electrode tip in a rodent study is much closer to the individual neurons, the signal detected is much larger. After all, the meninges, skull, and scalp form a safe barrier, and thus cause a reduced amplitude, in humans. For comparison, ERP components have amplitudes up to around 15 μV in humans, although components can easily reach 100–200 μV in rodent recordings.

Some differences in the manner in which recordings are done mainly relate to drug studies. For instance, one dissimilarity pertains to the state of consciousness while performing a recording. Many researchers interested in the rodent brain acquire their data while the animal is anesthetized, although by default humans are awake. Anesthetizing is often done in animals to prevent excessive movement that may disturb the signal or to provide the opportunity of injecting the drug directly into a target tissue. If not injected directly into the brain, the route of administration still differs between humans and animals, as intraperitoneal or subcutaneous injections are commonly given in animals, while drugs are administered orally in humans. All these differences in recordings in animals vs humans should be taken into account when trying to compare their results.

Comparability of the human and animal ERP has received much attention in the last few decades. A first example is the signal found in sensory gating, the P50 component. This positive component, in humans occurring around 50 ms after stimulus onset, is heavily reduced in response to the second as compared to the first click in the double click paradigm, as described previously. In a review on the translational utility of rodent hippocampal gating, Smucny et al. (2015) argued that rodent hippocampal gating comes closest to human P50 gating measured on the scalp, the rodent P20-N40 wave being highly predictive of human effects. Indeed, we also found hippocampal gating in components with similar latencies in our drug studies examining the role of the cholinergic system (Klinkenberg et al. 2013a, b) and of PDE signaling pathways (Reneerkens et al. 2013) in gating.

The earlier, sensory-related ERP components seem to behave in an analogous way in humans and animals, even though the latencies of those early components occur 1.8 times earlier in rats than in humans (Sambeth et al. 2003). They habituate, thus their amplitudes decreases due to learning taking place in the brain, similarly when simple auditory or visual stimuli are presented (Sambeth et al. 2004; Hauser et al. 2019), respond to task-related stimuli in comparable ways (Sambeth et al. 2003), and as hypothesized by Smucny et al. (2015), drug effects are also related. Later, more cognitively related ERP components, though, share far less overlap. While P3-like components may in some rat studies be found that behave similarly to the human P3b component (Sambeth et al. 2003), also regarding drug effects (Ahnaou et al. 2018), other cognitive paradigms failed to elicit such components all together (Sambeth and Maes 2006). This might be related to the fact that the cognitive components are elicited by sophistically organized cortical structures in humans in combination with the fact that rodents have a smooth cortex, thus far less cortical surface. For drug research, therefore, it is wise to administer basic sensory paradigms that have shown sufficient analogy between humans and rodents.

Drug effects on EEG frequencies have often been recorded in both humans and rodents. Similarly to the ERP recordings, a straightforward comparison is difficult to make. For instance, alpha, the "resting" frequency in humans, is not detectable clearly in rodents. Furthermore, whereas theta seems to be the most prominent hippocampal frequency eliciting clearly visible oscillations in rats and mice, these oscillations usually only become detectable after analysis on the signal was performed in humans. This, in combination with the obvious differences in human vs animal electrophysiology as mentioned above, led Blokland et al. (2015) to conclude that the most suitable comparison between humans and rodents seems to be that of resting EEG. When actually comparing studies on the cholinergic system, a high variability in effects was not only found between humans and animals but also within species (Blokland et al. 2015). The translational validity of animal EEG seems to depend on the psychoactivity and/or neural substrates examined (Drinkenburg et al. 2015). To conclude, this asks for caution when using EEG frequencies as translational factor, until research has elucidated the cause of large differences between studies.

> In sum, even though the various species highly differ in anatomy, genetics, and the manner in which recordings can be done, some important overlap in electrophysiological activity has been found in various paradigms. For drug research, it is wise to administer basic sensory paradigms as they showed the best analogy between humans and rodents.

Example: Electrophysiology in Schizophrenia

Schizophrenia is a brain disorder that is accompanied by a variety of symptoms, including positive symptoms such as the occurrence of hallucinations, negative symptoms like being unable to plan, and cognitive symptoms such as a deficit in working memory. It is well known that at least some of the symptoms, such as hallucinations, are caused by an overactivation of the dopaminergic system. In more recent years, the role of the glutamate system has started to receive more attention as well, and as in schizophrenic patients, problems may exist in the function of the NMDA receptors (Snyder and Gao 2019). Much effort has been paid to developing biomarkers or surrogate markers of schizophrenia in order to more effectively test newly developed antipsychotic drugs. A number of electrophysiological markers will be discussed below.

The sensory systems of schizophrenic patients are impaired, especially the attentional and informational components (Javitt and Freedman 2015). This may explain certain ERP deficits that have previously been found. For instance, it has long been clear that abnormalities in P50 gating exist in schizophrenic patients (Light and Braff 1998). The inhibition in response to the second click in the double-click paradigm that should occur in healthy individuals is almost absent in schizophrenic individuals. This is usually explained as an inability to inhibit redundant sensory input.

Likewise, an imbalance between excitation and inhibition can also be shown using the auditory steady-state response (ASSR). The ASSR is a response entrained to both the frequency and phase of rapidly presented auditory stimuli. In the EEG, it elicits gamma oscillations if the frequency in which the stimuli are presented is in the gamma range; usually 30–50 Hz is used. ASSR is significantly reduced in schizophrenia, likely reflecting an inhibitory deficit (Tada et al. 2020).

Not only inhibition of irrelevant information is disturbed but also attention toward what matters may be disrupted. MMN is a measure of sensory memory, and in the ERP section, it was already described that the MMN is disrupted in schizophrenia. However, how the MMN exactly behaves in this disorder is complex, probably due to the complexity of the disorder. Responses to changes in duration and pitch of a stimulus reliably reduce the MMN in chronic patients, but only duration deviants cause this reliable effect in first-episode patients (Umbricht and Krljes 2005; Haigh et al. 2017). Responses to other stimulus types like the loudness of a tone do not seem to be impaired in these patients. Research has shown that siblings of schizophrenic patients also share some of the disturbances (Sevik et al. 2011), as do individuals at ultra-high risk for the development of schizophrenia (Lavoie et al. 2018). Given that the MMN is said to reflect NMDA receptor activity, the disturbances seen in MMN amplitude in schizophrenia can be seen as evidence for the glutamate hypothesis of schizophrenia (see also Nagai et al. 2017).

Although the entrained gamma oscillations are decreased in schizophrenia (Tada et al. 2020), spontaneous gamma waves recorded in resting state seem to be increased. It has been suggested that this abnormality relates to impaired GABAergic (McNally and McCarley 2016) and glutamatergic (Uhlhaas and Singer 2015) neurotransmission seen in schizophrenia.

In sum, various EEG markers of impairment in schizophrenia can be found. Future research should combine various methodologies to provide a more detailed view on which particular stimulus (for instance, the MMN paradigm) elicits which disturbance in which situation. This way, EEG may become a key biomarker for schizophrenia.

> Conclusions

In this chapter, I tried to provide an over-view of the most commonly utilized electrophysiological techniques in science in general, and how these may be used in drug discovery. Given that in early stages of drug development, animal models are highly important, but in later stages, human clinical trials are needed, I also looked into the translational potential of some of the measurements.

Patch-clamp techniques are especially valuable for studies on the mechanism of action of drugs in the early stages of development and are done *in vitro*. When using this technique, the influence of drugs on individual ion channels or on how ions change their behavior within a neuron may be examined. In light of the recent technological advances that now offer the opportunity to record more than a thousand patches per day, this technique is becoming more relevant in the future. The IonFlux was already introduced previously, which has shown stable recordings of the cholinergic system (Yehia and Wei 2020). Another system is the Patchliner, one of the smallest patch-clamp workstations that simplifies experimental procedures (Farre et al. 2009), a key asset for scientists. One of the reasons why these (medium-throughput) systems are so relevant for early stages of drug discovery is that they are used to "filter out" false positive hits from primary screening campaigns (using, for example, high throughput voltage-gated dyes) and to select confirmed hits that can be further processed.

Next, single- and multiple-unit recordings were introduced, which reliably assess indirect action potentials. This is relevant when one is interested in drug-induced changes in neuronal output. An advantage of this measure as opposed to a whole-cell patch-clamp recording, which also measures ionic flow of an entire neuron and could thus detect an action potential, is a single-unit recording can be done in living animals, though often under anesthesia. The combination of single- or multi-unit recordings with the measurement of local field potentials, which on the other hand record the input into neurons, provides a clear picture of drug effects on neuronal communication. The key contribution in drug discovery so far has been the characterization of functional effects or potential efficacy of drugs not only to examine pharmacodynamic properties such as sedation but also to study toxicity (Drinkenburg et al. 2015).

Finally, the use of EEG and the various output parameters that may be determined from it were discussed. Not only the frequency oscillations and ERPs were introduced, but also ERD and ERS. The latter combines the stimulus-related changes traditionally examined using ERPs with non-phase locked changes in different frequency bands. These three types of measurements may guide early drug development, as deficiencies found in brain disorders often lead to particular deficits in EEG activity. The potential of newly discovered drugs to reverse these EEG deficits is relevant to explain improved behavioral outcomes or to clarify why certain symptoms are reduced and others remain when being treated with a substance.

In the second part of this chapter, I discussed the translational potential of animal electrophysiological findings to human outcomes. Recorded local fields potentials in animals share many similarities with human EEG. However, due to certain methodological constraints, experiments identical between humans and, for instance, rodents cannot be developed. This could be just one of the reasons why large variations in responding have often been found between the different species. Other factors limiting translational value may be anatomical and genetic differences between the various species of interest. Nevertheless, due to overlap in some key electrophysiological responses such as theta frequency or the animal P20-N40 components, suggested to be equivalent of the human P50, future research should continue to use rodents in early drug development. In parallel, however, more effort must be put into optimizing the animal methodology so that the translational value increases.

9

References

Ahnaou A, Biermans R, Drinkenburg WHIM (2018) Cholinergic mechanisms of target oddball stimuli detection: the late P300-like event-related potential in rats. Neural Plast 2018:4270263. https://doi.org/10.1155/2018/4270263

Alekseichuk I, Pabel SC, Antal A, Paulus W (2017) Intrahemispheric theta rhythm desynchronization impairs working memory. Restor Neurol Neurosci 35(2):147–158

Amirabadi S, Pakdel FG, Shahabi P, Naderi S, Asalou MA, Cankurt U (2014) Microinfusion of bupropion inhibits putative GABAergic neuronal activity of the ventral tegmental area. Basic Clin Neurosci 5(3):182–190

Bissonnette JN, Francis AM, Hull KM, Leckey J, Pimer L, Berrigan LI, Fisher DJ (2020) MMN-indexed auditory change detection in major depressive disorder. Clin EEG Neurosci 51(6):365–372. https://doi.org/10.1177/1550059420914200

Blokland A, Prickaerts J, van Duinen M, Sambeth A (2015) The use of EEG parameters as predictors of drug effects on cognition. Eur J Pharmacol 759:163–168

Bosman CA, Lansink CS, Pennartz CM (2014) Functions of gamma-band synchronization in cognition: from single circuits to functional diversity across cortical and subcortical systems. Eur J Neurosci 39(11):1982–1999

Burns SP, Xing D, Shapley RM (2010) Comparisons of the dynamics of local field potential and multiunit activity signals in macaque visual cortex. J Neurosci 30(41):13739–13749

Buzsaki G (2004) Large-scale recording of neuronal ensembles. Nat Neurosci 7(5):446–451

Buzsaki G, Moser EL (2013) Memory, navigation and theta rhythm in the hippocampal-entorhinal system. Nat Neurosci 16(2):130–138

Buzsaki G, Wang XJ (2012) Mechanisms of gamma oscillations. Ann Rev Neurosci 35:203–225

Cacucci F, Wills TJ, Lever C, Giese KP, O'Keefe J (2007) Experience-dependent increase in CA1 place cell spatial information, but not spatial reproducibility, is dependent on the autophosphorylation of the alpha-isoform of the calcium/calmodulin-dependent protein kinase II. J Neurosci 27(29):7854–7859

Caldenhove S, Borghans LGJM, Blokland A (2017) Sambeth A. role of acetylcholine and serotonin in novelty processing using an oddball paradigm. Behav Brain Res 331:199–204

Cancelli I, Cadore IP, Merlino G, Valentinis L, Moratti U, Bergonzi P, Gigli GL, Valente M (2006) Sensory gating deficit assessed by P50/Pb middle latency event related potential in Alzheimer's disease. J Clin Neurophysiol 23(5):421–425

Chen TC, Hsieh MH, Lin YT, Chan PS, Cheng CH (2020) Mismatch negativity to different deviant changes in autism spectrum disorders: a meta-analysis. Clin Neurophysiol 131(3):766–777

Cheng CH, Chan PS, Liu CY, Hsu SC (2016) Auditory sensory gating in patients with bipolar disorders: a meta-analysis. J Affect Disord 203:199–203

Colgin LL (2016) Rhythms of the hippocampal network. Nat Rev Neurosci 17(4):239–249

De Graaf TA, Hsieh PJ, Sack AT (2012) The 'correlates' in neural correlates of consciousness. Neurosci Biobehav Rev 36(1):191–197

Dohrmann AL, Strengler K, Jahn I, Olbrich S (2017) EEG-arousal regulation as predictor of treatment response in patients suffering from obsessive compulsive disorder. Clin Neurophysiol 128(1):1906–1914

Drinkenburg WHIM, Ruigt GSF, Ahnaou A (2015) Pharmaco-EEG studies in animals: an overview of contemporary translational applications. Neuropsychobiology 72:151–164

Evans DE, Maxfield ND, van Rensburg KJ, Oliver JA, Jentink KG, Drobes DJ (2013) Nicotine deprivation influences P300 markers of cognitive control. Neuropsychopharmacology 38(12):2525–2531

Farre C, Haythornthwaite A, Haarmann C, Stoelzle S, Kreir M, George M, Bruggemann A, Fertig N (2009) Port-a-patch and patchliner: high fidelity electrophysiology for secondary screening and safety pharmacology. Com Chem High Throughput Screen 12(1):24–37

Haigh SM, Coffman BA, Salisbury DF (2017) Mismatch negativity in first-episode schizophrenia: a meta-analysis. Clin EEG Neurosci 48(1):3–10

Hauser MFA, Wiescholleck V, Colitti-Klausnitzer J, Bellebaum C, Manahan-Vaughan D (2019) Event-related potentials evoked by passive visuospatial perception in rats and humans reveal common denominators in information processing. Brain Struct Funct 224(4):1583–1597

Heida T, Poppe NR, de Vos CC, van Putten MJ, van Vugt JP (2014) Event-related mu-rhythm desynchronization during movement observation is impaired in Parkinson's disease. Clin Neurophysiol 125(9):1819–1825

Heinricher MM (2004) 2 Principles of extracellular single-unit recording, p 8–13

Hong X, Sun J, Wang J, Li C, Tong S (2020) Attention-related modulation of frontal midline theta oscillations in cingulate cortex during a spatial cueing Go/NoGo task. Int J Psychophysiol 148:1–12

Horvath A, Szucs A, Csukly G, Sakovics A, Stefanics G, Kamondi A (2018) EEG and ERP biomarkers of Alzheimer's disease: a critical review. Front Biosci 23:183–220

Huang WJ, Chen WW, Zhang X (2015) The neurophysiology of P300 – an integrated review. Eur Rev Med Pharmacol Sci 19(8):1480–1488

Javitt DC, Freedman R (2015) Sensory processing dysfunction in the personal experience and neuronal machinery of schizophrenia. Am J Psychiatry 172(1):17–31

Johnson R (1986) a triarchic model of P300 amplitude. Psychophysiology 23:367–383

Kenemans JL, Kähkönen S (2011) How human electrophysiology informs psychopharmacology: from bottom-up driven processing to top-down control. Neuropsychopharmacology 36(1):26–51

Klimesch W (2012) α-band oscillations, attention, and controlled access to stored information. Trends Cogn Sci 16(12):606–617

Klinkenberg I, Blokland A, Riedel WJ, Sambeth A (2013a) Cholinergic manipulation of auditory processing, sensory gating, and novelty detection in human participants. Psychopharmacology (Berl) 225(4):903–921

Klinkenberg I, Sambeth A, Blokland A (2013b) Cholinergic gating of hippocampal auditory evoked potentials in freely moving rats. Eur Neuropsychopharmacol 23(8):988–997

Kok A (2001) On the utility of P3 amplitude as a measure of processing capacity. Psychophysiology 38(3):557–577

Larrain-Valenzuela J, Zamorano F, Soto-Icaza P, Carrasco X, Herrera C, Daiber F, Aboitiz F, Billeke P (2017) Theta and alpha oscillation impairments in autistic spectrum disorder reflect working memory deficit. Sci Rep 7(1):14328

Lavoie S, Jack BN, Griffiths O, Ando A, Amminger P, Couroupis A, Jago A, Markulev C, McGorry PD, Nelson B, Polari A, Yuen HP, Whitford TJ (2018) Impaired mismatch negativity to frequency deviants in individuals at ultra-high risk for psychosis, and preliminary evidence for further impairment with transition to psychosis. Schizophr Res 191:95–100

Leblois A, Pouzat C (2017) Multi-unit recording. In: Neurobiology of motor control: fundamental concepts and new directions. Wiley & Sons, Hoboken

Leuchter AF, Hunter AM, Jain FA, Tartter M, Crump C, Coook IA (2017) Escitalopram but not placebo modulates brain rhythmic oscillatory activity in the first week of treatment of major depressive disorder. J Psychiatr Res 84:174–183

Lever C, Wills T, Cacucci F, Brugess N, O'Keefe J (2002) Long-term plasticity in hippocampal place-cell representation of environmental geography. Nature 416:90–94

Li Y, Kang C, Wei Z, Qu X, Liu T, Zhou Y, Hu Y (2017) beta oscillations in major depression – signalling a new cortical circuit for central executive function. Sci Rep 7(1):18021

Light GA, Braff DL (1998) The 'incredible shrinking' P50 event-related potential. Biol Psychiatry 43(12):918–920

Liu C, Li T, Chen J (2019) Role of high-throughput electrophysiology in drug discovery. Curr Protoc Pharmacol 87(1):e69

McNally JM, McCarley RW (2016) Gamma band oscillations: a key to understanding schizophrenia symptoms and neural circuit abnormalities. Curr Opin Psychiatry 29(3):202–210

Meyer M, Endedijk HM, van Ede F, Hunnius S (2019) Theta oscillations in 4-year-olds are sensitive to task engagement and task demands. Sci Rep 9(1):6049

Michie PT, Malmierca MS, Harms L, Todd J (2016) The neurobiology of MMN and implications for schizophrenia. Biol Psychol 116:90–97

Motoi H, Jeon J-W, Juhasz C, Miyakoshi M, Nakai Y, Sugiura A, Luat AF, Sood S, Asano E (2019) Quantitative analysis of intracranial electrocorticography signals using the concept of statistical parametric mapping. Sci Rep 9:17385

Nagai T, Kirihara K, Tada M, Koshiyama D, Koike S, Suga M, Araki T, Hasimoto K, Kasai K (2017) Reduced mismatch negativity is associated with increased plasme level of glutamate in first-episode psychosis. Sci Rep 7(1):2258

Newson JJ, Thiagarajan TC (2019) EEG frequency bands in psychiatric disorders: a review of resting state studies. Fron Hum Neurosci 12:1–24

Noda Y (2020) Toward the establishment of neurophysiological indicators for neuropsychiatric disorders using transcranial magnetic stimulation-evoked potentials: a systematic review. Psychiatry Clin Neurosci 74(1):12–34

Noh MC, Stemkowski PL, Smith PA (2019) Long-term actions of interleukin-1β on K+, Na+, and Ca2+ channel currents in small, IB4-positive dorsal root ganglion neurons; possible relevance to the etiology of neuropathic pain. J Neuroimmunol 332:198–211

O'Keefe J (1976) Place units in the hippocampus of the freely moving rat. Exp Neurol 51:78–109

Pfurtscheller G (2001) Functional brain imaging based on ERD/ERS. Vis Res 41:1257–1260

Picken C, Clarke AR, Barry RJ, McCarthy R, Selikowitz M (2020) The theta/beta ratio as an index of cognitive processing in adults with the combined type of attention deficit hyperactivity disorder. Clin EEG Neurosci 51(3):167–173

Polich J (2007) Updating P300: an integrative theory of P3a and P3b. Clin Neurophysiol 118(10):2128–2148

Pourbadi HG, Naderi N, Delavar HM, Hosseinzadeh M, Mehranfard N, Khodagholi F, Janahmadi M, Motamdei F (2017) Decrease of high voltage Ca2+ currents in the dentate gyrus granule cells by entorhinal amyloidopathy is reversed by calcium channel blockade. Eur J Pharmacol 794:154–161

Quian Quiroga R, Reddy L, Kreiman G, Koch C, Fried I (2005) Invariant visual representation by single neurons in the human brain. Nature 435:1102–1107

Reneerkens OA, Sambeth A, Blokland A, Prickaerts J (2013) PDE2 and PDE10, but not PDE5, inhibition affect basic auditory information processing in rats. Behav Brain Res 250:251–256

Rizzolatti G, Fabbri-Destro M, Caruana F, Avanzini P (2018) System neuroscience: past, presence, and future. CNS Neurosci Ther 24(8):685–693

Sambeth A (2004) Studies on the effects of learning on the event-related potential. A between species comparison. PhD dissertation, Radboud University Nijmegen

Sambeth A, Maes JHR (2006) A comparison of event-related potentials of humans and rats elicited by a serial feature-positive discrimination task. Learn Motiv 37(3):269–288

Sambeth A, Maes JH, van Luijtelaar G, Molenkamp IB, Jongsma ML, van Rijn CM (2003) Auditory event-related potentials in humans and rats: effects of task manipulation. Psychophysiology 40(1):60–68

Sambeth A, Maes JH, Quian Quiroga R, Coenen AM (2004) Effects of stimulus repetitions on the event-related potential of humans and rats. Int J Psychophysiol 53(3):197–205

Sambeth A, Riedel WJ, Smits LT, Blokland A (2007) Cholinergic drugs affect novel object recognition in rats: relation with hippocampal EEG? Eur J Pharmacol 572:151–159

Sambeth A, Meeter M, Blokland B (2009) Hippocampal theta frequency and novelty. Hippocampus 19: 407–408

Sevik AE, Anil Yagcioglu AE, Yagioglu S, Karahan S, Gurses N, Yildiz M (2011) Neuropsychological performance and auditory event-related potentials in schizophrenia patients and their siblings: a family study. Schizophr Res 130:195–202

Smucny J, Steven KE, Olincy A, Tregellas JR (2015) Translational utility of rodent hippocampal auditory gating in schizophrenia research: a review and evaluation. Transl Psychiatry e587:5

Snyder MA, Gao WJ (2019) NMDA receptor hypofunction for schizophrenia revisited: perspectives from epigenetic mechanisms. Schizophr Res 217:60–70. https://doi.org/10.1016/j.schres.2019.03.010

Tada M, Kirihara K, Koshiyama D, Fujioka M, Usui K, Uka T, Komatsu M, Kunii N, Araki T, Kasai K (2020) Gamma-band auditory steady-state response as a neurophysiological marker for excitation and inhibition balance: a review for understanding schizophrenia and other neuropsychiatric disorders. Clin EEG Neurosci 51(4):234–243. https://doi.org/10.1177/1550059419868872

Todd J, Harms L, Schall U, Michie PT (2013) Mismatch negativity: translation the potential. Front Psych 4:171

Uhlhaas PJ, Singer W (2015) Oscillations and neuronal dynamics in schizophrenia: the serach for basic symptoms and translational opportunities. Biol Psychiatry 77(12):1001–1009

Umbricht D, Krljes S (2005) Mismatch negativity in schizophrenia: a meta-analysis. Schizophrenia Res 76:1–23

Wang Y, Chen AQ, Xue Y, Liu MF, Liu C, Liu YH, Pan YP, Diao HL, Chen L (2019) Orexins alleviate motor deficits via increasing firing activity of pallidal neurons in a mouse model of Parkinson's disease. Am J Physiol Cell Physiol 317(4):C800–C812

Wickenden AD (2014) Overview of electrophysiological techniques. Curr Protoc Pharmacol 64:11.1.1–111.17

Yehia A, Wei H (2020) Studying nicotinic acetylcholine receptors using the IonFlux microfluidic-based automated patch-clamp system with continuous perfusion and fast solution exchange. Curr Protoc Pharmacol 88(1):e73

MRI in CNS Drug Development

Mitul A. Mehta

Contents

10.1 Introduction – 150

10.2 Functional Markers of Drug Development with MRI – 152
10.2.1 Perfusion Imaging – 153
10.2.2 Task-Based fMRI – 154
10.2.3 Timeseries Correlations and Resting State BOLD – 155
10.2.4 Magnetic Resonance Spectroscopy (MRS) – 156
10.2.5 PhMRI and the Constraints of Fast and Slow Pharmacokinetics – 156

10.3 The Development of Neuroimaging Biomarkers for Treatment Response – 158

10.4 Limitations and Challenges – 159
10.4.1 The Reliability of MRI Methodologies Is Under-studied – 159
10.4.2 The Challenge of Haemodynamic Outcomes to Index Brain Activity – 159
10.4.3 What Is the Best Way to Characterise Non-specific Effects of Drugs – 160
10.4.4 Is It Too Soon to Use MRI as a Stratification Tool? – 160
10.4.5 There Is a Gap Between Modulation of Brain Systems and Clinical Outcome – 160

References – 162

© The Author(s), under exclusive license to Springer Nature Switzerland AG 2021
R. Schreiber (ed.), *Modern CNS Drug Discovery*, https://doi.org/10.1007/978-3-030-62351-7_10

There are three ways in which magnetic resonance imaging is typically used in drug development: to validate the role of drugs and neuropharmacological systems in terms of their effects on the brain, the development of brain markers to evaluate drugs by – these can be associated with particular disorders or symptoms – and the evaluation of compounds with neuroimaging markers. MRI is effectively multimodal. Markers of brain function are sensitive to drug modulation and include perfusion and blood flow, functional imaging with tasks, brain connectivity, tracking of activity with rapidly changing blood levels from drug administration and spectroscopy for brain metabolite levels. Fast pharmacokinetics can be tracked, and drugs with slower pharmacokinetics can be assessed for context-dependent effects and quantitative effects on blood flow and brain metabolite levels. These methods can be applied to healthy volunteers, combined with models of dysfunction, or in patients to translate evidence in experimental animals, validate theory and provide early indicators of potential efficacy. Significant challenges exist with MRI methodology such as the limited evaluation of reliability, the modelling of confounds, stratification of individuals to understand response and non-response and bridging the gap between the modulation of brain systems and clinical outcome. Numerous examples exist to demonstrate the success and future potential of MRI methodologies in the drug development process, which, if correctly applied, can provide a principled basis for some treatment options over others, accelerating the delivery of novel therapeutics for those in need.

🔄 Learning Objectives

- To understand the relevant brain function assessments available for assessment of drugs with MRI
- To understand the basics of the MRI methodology for each main method
- To appreciate the utility of functional MRI methodology
- To understand the use of models combined with MRI
- To appreciate limitations and challenges
- To be aware of future prospects for MRI methodology in drug development

10.1 Introduction

Magnetic resonance imaging is a non-invasive methodology. Modern scanners can show sensitivity to different tissue properties by the implementation of different acquisition sequences. When applied to the brain, the broad categories describing these different sequences are structural, functional and neurochemical imaging. Neurochemical imaging is used to determine the concentration of metabolites of neuronal function, metabolic activity and neurotransmission and is usually limited to small brain regions (voxels) with spectroscopy, but can be applied to larger regions such as slices or even the whole brain. Structural neuroimaging includes T1-weighted and T2-weighted scans sensitive to the contrast between grey and white matter and cerebrospinal fluid, diffusion-weighted imaging, used for the construction of white matter pathways. Functional neuroimaging refers to acquisition sequences that fluctuate according to intrinsic (e.g. thought) or extrinsic (e.g. stimulus) variations. Functional neuroimaging is realised almost exclusively with blood oxygen-dependent level (BOLD) sensitive methods, but also includes perfusion imaging (our chapter) and methods with emerging evidence of sensitivity such as diffusion imaging (Hofstetter et al. 2013).

> **Definition**
>
> T1-weighted imaging is scanning methodology that relies on longitudinal relaxation of proton spins (spin-lattice relaxation). The spins align to the external field of the scanner, and they are put into a different plane by a radiofrequency pulse. The 'relaxation' back to the background field occurs over a different time for different tissues. By taking a snapshot at a fixed time, then different tissues can be contrasted.

> **Definition**
>
> Blood Oxygen-Level Dependent (BOLD) contrast is the standard method used in functional magnetic resonance imaging. Local distortions in the magnetic field through changes in the concentration of the paramagnetic deoxyhaemoglobin and the effects on the MRI signal are used to register changes in brain activity.

> **Definition**
>
> Voxel: A three-dimensional pixel forming the element of images produced by MRI.

Magnetic resonance imaging differs from invasive techniques such as single photon emission computerised tomography (SPECT) and positron emission tomography (PET; reference to ▶ Chap. 11) which require the injection of radiotracers. MRI has better spatial resolution than surface-based methods such as electroencephalography and near infrared spectroscopy, and MRI is able to measure from both deep and superficial structures, which nonetheless may have a role.

There are a number of roles for MRI in drug development:

1. The role of neuropharmacological systems can be validated and further understood in terms of their impact on brain structure and function.
2. The development and validation of markers associated with brain dysfunction as targets for the evaluation of drugs.
3. Pharmacodynamic investigations of novel compounds with neuroimaging markers as outcomes.

MRI can be applied at any stage of the drug development pathway. Pre-clinical imaging has been used to understand the nature of drug effects on brain structure (Vernon et al. 2014) and function (Bifone and Gozzi 2012), and has particular strengths when used in combination with animal models of dysfunction. There are some clear benefits of pre-clinical evaluation, including the testing of multiple doses of drugs into exposure ranges

that may not be safe in human studies. The development of drug models linking pharmacokinetics and pharmacodynamics (PK-PD models) may reveal non-linear patterns such as inverted-U-shaped functions defining an optimum dose range. Pre-clinical MRI can also be combined with other measurements such as intracortical recording, gene editing, DREADDS and others to provide comprehensive assessments and control for potential confounds (Coimbra et al. 2013; Jonckers et al. 2015). Despite these strengths, examples of translational viability of pre-clinical MRI with animal models into successful human trials are limited.

> **Definition**
>
> Pre-clinical imaging is usually used to refer to imaging in non-human primates and rodents. The term can sometimes be confusing because pre-clinical research includes research where the subjects are not a clinical population and can include healthy volunteers.

In humans, drugs can be tested very early in the developmental pathway. Proof of principle studies during phase I, when safety and tolerability are assessed, can be used to provide translational validation of the principles of drug use – so-called proof of principle studies (Browning et al. 2019). While healthy volunteers are the typical population tested in phase I for CNS drugs, small patient studies can be included (often referred to as phase IIa when validation in the target population is initiated). Testing a drug in an impaired system may be required to provide the necessary evidence of the modulation of the system towards normalisation (Reed et al. 2019).

Testing a compound aimed at improving a function, such as emotional processing, modulates the brain regions, and circuits involved in the target function gives an early indicator of potential efficacy prior to the drug being tested in trials with patients. Other examples during these early development phases include testing that a compound reverses the effects of a drug known to model a component of a disorder, therefore providing evidence of a specific pharmacological mechanism in vivo and testing the effects on specific brain circuits in patients with known impairment in such systems to provide an early indicator of the potential for efficacy in the target population.

If neuroimaging markers demonstrate utility in stratification of patients more likely to be responders, it may be applied at later stages of development, but as yet there are no examples of this. With PET imaging there is a clear case for stratification with markers of neurodegenerative disease such as beta-amyloid and tau (Beaurain et al. 2019; Jelistratova et al. 2020), whereas with MRI there is substantial work required to produce valid stratification markers, particularly for functional imaging.

> MRI can be used to assess drugs across the developmental pathway from basic science experimental to test theories and translate evidence from experimental animals to humans, as well as testing drug treatment to reveal their mechanisms of action.

10.2 Functional Markers of Drug Development with MRI

◘ Figure 10.1 depicts the five major methods used to evaluate drug action in the brain with MRI. Theoretically these methods can be deployed in single study sessions effectively offering multimodal capabilities with a single method of assessment. Typically, an imaging session is tailored to the specific hypotheses under investigation and may be influenced by features of the drugs such as the pharmacokinetic profile. For example, an injected compound which reaches peak plasma levels within minutes and rapidly clears from the brain will not be suitable to methods requiring averaging over several minutes such as task-based functional MRI (fMRI). These methods can also be timed to capture the desired window within the pharmacokinetic profile (De Simoni et al. 2013; Paloyelis et al. 2016). There are notable limitations. For spectroscopy the major method is MRS. While MRS is usually limited to individual brain regions, the application of newer methods such as chemical exchange saturation transfer (CEST) allow more widespread evaluation (Knutsson et al. 2018; Luna et al. 2018), although they require high-field MRI, which is not widely available and presents additional challenges (Ladd et al. 2018). These challenges are not fully resolved for high-throughput evaluation of drugs. All the other methods in ◘ Fig. 10.1 are based on successful neurovascular coupling (Mathias et al. 2018), which is the process through which brain regions more metabolically and electrically active (e.g. glutamate signalling) are accompanied by a cascade of changes which result in more blood

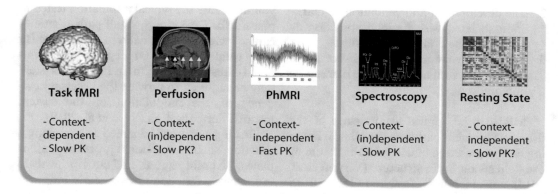

◘ **Fig. 10.1** Five principal MRI sequences used to measure drug action in the brain. The images are chosen to illustrate the methods only. Context-dependent refers to whether a specific cognitive task or behavioural paradigm to provide context. For task-based fMRI, the context is the task. Tasks can also be used with perfusion imaging, but this is uncommon due to the poor sensitivity. MRS can be used with tasks to examine brain metabolites during different task conditions, although its utility in drug development has not been determined

flow being delivered to the active area. Thus, compounds affecting neurovascular coupling may give false signals for these MRI markers of drug effect (Wise and Preston 2010). Possible solutions are included in the summary of each method below.

10.2.1 Perfusion Imaging

The principle of perfusion imaging is based on measurement of the effect of a contrast agent diffusible in blood. Exogenous contrast agents can be utilised such as gadolinium, but the endogenous contrast agent allows fully non-invasive data acquisition, and it is this approach that has yielded numerous pharmacological imaging studies (Zelaya et al. 2015).

Arterial spin labelling (ASL) uses water in the blood as the endogenous contrast agent. Two types of images are acquired, label and control. The label image is acquired after 'tagging' or encoding the inflowing blood, and the control image has different encoding and the blood is effectively untagged. The differences between these two images is regionally weighted by the perfusion in the brain tissue, often referred to as the regional cerebral blood flow (rCBF). A separate image with the same acquisition parameters but sensitive to the proton density (an Mzero image) is required for quantification (Wong et al. 1997). The two main forms of ASL are pulsed and continuously labelled. Pulsed ASL (or pASL) requires the labelling of arterial blood by inverting signal over a relatively wide volume (~100 mm) below the area of data acquisition, typically in the vicinity of the carotid arteries. This is achieved with tailored pulses (Wong et al. 1997). After a short delay, the imaging data are collected. If this is implemented with a single-shot acquisition (where data from each volume are rapidly acquired), such as echo-planar imaging, then the time for volume acquisition can be similar to functional magnetic resonance imaging. The control image can be collected using double inversion of the arterial blood and collected interleaved with the label images. Robust perfusion maps require many pairs of control-label images to provide a robust calculation of rCBF. These control-label pairs can theoretically be treated as a timeseries and thus be used to measure the effects of compounds with a rapid PK profile (minutes). In reality, single pairs are insufficient to provide a robust and reliable perfusion map to track drug effects with second-level temporal resolution (Zelaya et al. 2012).

Continuously labelled ASL has different implementations referred to as continuously labelled ASL (CASL) or 'pseudo-continuously' or 'pulsed-continuously' labelled ASL (pCASL). This method has the advantage over pASL of superior labelling efficiency, but at the loss of the potential to create a time series of second resolution (although this has been noted to be of poor robustness). This is achieved by arterial blood being labelled by flow-driven inversion of arterial blood in the presence of a magnetic field gradient, applied in the direction of blood flow into the brain. A labelling plane is used which leads to a rotation of the longitudinal magnetisation vector (the magnetisation of the blood in line with the background field produced by the scanner) about the effective field, which in turn changes the orientation of the proton spins by 180 degrees as they cross the labelling plane (Dai et al. 2008). The signal carried by the inverted proton spins degrades over time and is thus detected in the acquisition volume after a short delay. This method is extremely effective at labelling a substantial amount of arterial blood, which allows for the collection of image data using multi-shot acquisition techniques allowing whole brain coverage.

The main advantage of ASL is that is provides quantification of rCBF allowing direct comparisons across sessions. This is therefore suitable for the evaluation of the context-independent drug effects (i.e. not during a task) for acute and prolonged treatment as well as dose-response effects. Because the use of control-label pairs markedly reduces sensitivity to the sources of low frequency noise (e.g. scanner temperature fluctuations), it can be used to mark brain states such as affective states and pain states which may be prolonged.

In addition, CBF mapping can be combined with BOLD imaging, the most common technique for functional brain imaging to 'calibrate' BOLD imaging. This is because BOLD is non-quantitative and dependent on the cerebral metabolic rate of oxygen utilisation, blood flow and volume.

Importantly, ASL has very good test-retest reliability (intra-class correlation coefficients >0.80) (Li et al. 2018) making a stable measurement, useful for within-subjects studies, where a drug can be compared to placebo or other active controls. There have been attempts to harmonise ASL methods across scanner manufacturers with recommendations for multi-site studies (Alsop et al. 2015), and data sharing is beginning to demonstrate the potential to study shared and distinct mechanisms for the functional effects of drugs (Duff et al. 2015).

> Arterial spin labelling is a non-invasive method to quantify regional cerebral flow with excellent test-retest reliability and known sensitivity to drug modulation. Harmonisation across different scanner manufacturers is required to combine data across multiple sites.

10.2.2 Task-Based fMRI

MRI is a powerful methodology to identify brain regions and networks associated with specific functions. These functions could be a simple to implement as visual stimulation or hand movements or complex functions such as decision-making. Functional imaging of tasks is almost exclusively applied using blood oxygen-level dependent imaging or BOLD (Huettel et al. 2014). Alternative methods include ASL as described above, although the signal-to-noise ratio is lower than that of BOLD (Berard et al. 2020). If the impacts of the drug intervention are predicted to cause remodelling of synapses, as predicted for the novel antidepressants ketamine and psilocybin (Witkin et al. 2019), then novel approaches to detect the effects of altered synaptic function will be required (e.g. Hofstetter et al. 2013).

Functional MRI with BOLD remains the dominant methodology and this method relies on functional hyperaemia, whereby the arterial delivery of oxygenated blood at a level that exceeds the need in the target region. The result of this is that the venous outflow from an active region will have a higher level of deoxygenated haemoglobin, but because of the hyperaemia leads to a paradoxical decrease in the relative concentration of deoxyhaemoglobin. The effect of this is to increase the fMRI or BOLD contrast. This is because deoxyhaemoglobin is paramagnetic leading to local distortions in the magnetic field, but with a reduction in its concentration in an active region, there will be a reduced effect of the local distortions. Specifically, MRI signal that is sensitive to spin-spin relaxation times (T2* and T2*-weighted signal) is increased. T2*-weighted imaging is usually combined with a fast readout giving a typical temporal resolution of ~2 s for the whole brain with voxels of approximately 3 mm isotropic.

The advantages of the BOLD method are as follows:
- The rapid acquisition of whole brain images.
- Sensitivity to deep brain structures (compare to surface-based techniques such as magnetoencephaolography or EEG).
- Standardised processing pipelines that are publicly available (Esteban et al. 2020).
- A range of pre-existing tasks validated for the assessment of a variety of functions.

The challenges of the BOLD contrast are as follows:
- Following neuronal activity, the signal change peaks a number of seconds later before slowly returning to baseline. This is known as the haemodynamic response function and needs to be factored into any data modelling for task-based fMRI.
- fMRI with BOLD is non-quantitative. That is, there are properties of the signal which means that the units are arbitrary including low frequency signal drift of the signal. The impact of this is that task-based fMRI with BOLD requires a contrast with another comparison or control condition for each task.

- The influence of movement can be complex and unpredictable and must be accounted for in the analysis pipeline.
- Susceptibility artefacts result from areas of the imaging field of view that alter the local gradients in the magnetic field leading to loss of contrast. Air-tissue boundaries produce loss of contrast in orbitofrontal cortex and the temporal pole, often extending into the basal forebrain. This can result in additional loss of signal in the ventral portions of the striatum and the amygdala.

This potential for artefacts and data loss emphasises the need for excellent quality control processes in the data processing pipeline to account for scans with loss of contrast, or problematic movement. In some cases, scan volumes or entire scan runs will need to be removed from the analysis. Building such controls into the analysis pipeline is crucial to ensure decisions about data quality, and inclusion is made independent of the actual analysis (Schwarz et al. 2011).

> BOLD imaging is based on the concentration of deoxyhaemoglobin in the venous blood. BOLD imaging is non-quantitative and thus requires internal comparators. This is usually achieved through different time periods being associated with different cognitive components which are then contrasted. BOLD imaging signal can also be correlated across brain regions to index connectivity.

10.2.3 Timeseries Correlations and Resting State BOLD

The acquisition of BOLD timeseries without the use of task seemed to go against the principles of BOLD imaging, where the signal is non-quantitative and requires a contrast with other condition(s). When BOLD acquisition is acquired without a task, it is known as 'resting state' BOLD. Participants typically are asked to lie still in the scanner with their eyes open or closed and not to engage in any particular cognitive activity. When a time-series of many hundreds of data points over tens of thousands of voxels is collected, there is a multitude of processing and analysis options (Chen et al. 2020). Most commonly connectivity metrics are derived from the data which describe the correlation or coherence of signal variation across brain regions. Correlated regions are considered to be functionally connected. The connectivity described can be simple metrics between each region and every other region, can be a summary of the entire shared timeseries across defined regions, or index dynamic variations in connectivity over time. An advantage of resting state methods is the simplicity with which they can be implemented, requiring little or no training of the subject or experimenters. A challenge is the potential for artefacts in the correlation of timeseries, such as from noise in the BOLD contrast timeseries from the scanner (e.g. low frequency fluctuations) or from the subject (e.g. physiological noise from respiration or the pulsatile nature of blood flow). A method to account for these artefacts must be included in the data analysis.

Connectivity analysis can also be applied during the performance of cognitive tasks. The correlation of the timeseries BOLD data can be hypothesised to alter during changes in cognitive demands (Friston et al. 1997) and may be a viable marker of context dependent brain dysfunction.

Resting state methods have shown sensitivity to different diagnoses and to different drugs the variety of methodologies. For example, schizophrenia has been associated with impaired hippocampal connectivity (Samudra et al. 2015; Artigas et al. 2017), and hippocampal connectivity is also altered in neurochemical models of psychosis in rodents and humans (Artigas et al. 2017). Thus, these same circuits can act as a target for analysis of putative therapeutics. Despite this clear utility, defining the precise nature of connectivity difference in a disorder or change due to a drug can be problematic as it is difficult to compare findings and combine results across studies. This is principally due to the variety of acquisition and analysis methods employed. Harmonisation of acquisition methods and data sharing will be important to ensure replicability and robustness of effects and the development of connectivity-based markers of brain dysfunction and treatment response (Hong et al. 2020).

> The collection of a timeseries of data from across the whole brain can be used to calculate the associated signal across brain regions. This metric is used to index connectivity, and it is open to different methods of analysis to describe connections and network properties.

10.2.4 Magnetic Resonance Spectroscopy (MRS)

Spectroscopy is the method by which the chemical make-up of the target (which could be as varied as brain tissue or distant galaxies) can be revealed. In brain tissue, spectroscopy methods include CEST and MRS. CEST is a more recent technique (Mueller et al. 2020), the potential of which is yet to be realised in drug development. MRS is used to study metabolite levels in brain tissue (for review, see Buonocore and Maddock 2015). While protons can align along the applied magnetic field, 31P and 13C MRS can also utilised, and 23Na and 19F are options for MRS investigation, but additional equipment may be needed. At the core of MRS is the fact that the magnetic field a particular atom experience is affected by its local chemical environment, specifically the magnetic field from nearby motion of electrons. The consequence of experiencing slightly different applied fields due to the varied chemical environment is that the atoms resonate at slightly different frequencies. This 'chemical shift' effect can be used to detect different environments of relevant nuclei. It is the chemical shift that gives rise to a MR frequency spectrum consisting of nuclei which resonate at different frequencies, depending on their local environment. The frequencies are not related to the exact concentration of nuclei and depend on the exact magnetic field strength. Therefore, MRS peaks are usually expressed in dimensionless units (parts per million, ppm), with reference to a specific molecule. For proton MRS (1H-MRS), water is commonly used as a reference at 4.7 ppm. Proton MRS is of particular interest in neuroscience because it can be used to measure the important central nervous system amino acids glutamate and glycine with good reliability (van Veenendaal et al. 2018). The main inhibitory neurotransmitter GABA can also be assessed with more elaborate acquisition and analysis and reasonable reliability (van Veenendaal et al. 2018).

> MRS is the principle measure used to index brain metabolites. The largest signals are from neuronal markers, but the neurotransmitters glutamate and GABA can be resolved using specialised sequences.

10.2.5 PhMRI and the Constraints of Fast and Slow Pharmacokinetics

Drugs with fast pharmacokinetics (e.g. rapid distribution and clearance), typical of intravenously injected compounds, are difficult to study with the MRI techniques described thus far, that is, those requiring any form of averaging of response over time. For example, task-based fMRI can involve the averaging of BOLD acquisition over many minutes to obtain sufficient samples of brain activity during the task in order to construct a robust and reliable map. If drug levels are changing rapidly during such acquisition, then the meaning of such averaging is questionable.

Drugs with faster pharmacokinetics have been studied using a method referred to as pharmacological or phMRI, where changes in the fMRI timeseries are analysed following drug administration (Breiter et al. 1997; Stein et al. 1998). Any timeseries data acquisition method can be combined with compound administration in phMRI. In rodents, the use of a blood-pool contrast agents such as superparamagnetic iron oxide particles can be used to collect cerebral blood volume timeseries data (Mandeville 2012). These use the same

T2*-imaging sequences and fast acquisition as BOLD imaging. In humans, the application of phMRI has almost exclusively used BOLD contrast. The method first acquires imaging data when the participants are at rest prior to any drug delivery and data acquisition continues after drug delivery until sufficient volumes have been acquired to define a post-drug response. There is no break in the data acquisition. The expectation of phMRI is that following drug administration a rapid change in MRI signal can be detected. Due to practical considerations with pharmacokinetics and difficulties with multiple administrations of a drug in quick succession, this technique is only used for describing one timeseries per participant per session.

The description of the drug effect on the signal can be the simple difference pre- and post-drug or it can utilise an input function. This can be derived from the profile of the drug plasma levels (assuming these match brain levels), or a subjective or behavioural response to the drug (measured during the scan or in a separate cohort). These methods will identify brain areas where the temporal profile of the neuroimaging signal matches that of the plasma drug level over many minutes, or the temporal profile of the subjective/behavioural effects. Mathematical models of the profile of predicted drug effect derived from pilot or independent investigations are useful when other input functions are not available or considered inappropriate (De Simoni et al. 2013).

> PhMRI tracks the rapid (seconds to minutes) signal change during the injection or infusion of a drug. The signal change can be compared between drugs, be blocked and be related to temporal profiles in drug levels or subjective experience/behaviour.

Case Study

PhMRI of Ketamine Applied to Drug Development in Schizophrenia

Dysfunctional neurotransmission of glutamate has long been hypothesised to have a role in schizophrenia (see Javitt et al. (2018) in relation to this case study). One crucial piece of evidence was the demonstration that antagonists of the glutamate NMDA receptor exacerbated symptoms of schizophrenia in patients and produced symptoms which were schizophrenia-like (psychotomimetic) in healthy volunteers. Ketamine can be safely used to study the effects of an acute hyperglutamatergic state in volunteers with neuroimaging. The establishment of a robust and reliable signature using phMRI (De Simoni et al. 2013) allowed the design of studies to attenuate the effects of ketamine. Imaging was preferred over behavioural signals. Despite proving to be reliable they were less robust, with different participants experiencing different symptoms and thus creating a clear drug response that was sparse over some participants and symptoms. PhMRI provided continuous signatures of ketamine responses for each participant. Initially, we used two compounds which were expected to attenuate the effects of ketamine by virtue of their known mechanisms of action, and these studies constituted a validation of the approach (Doyle et al. 2013a, b; Joules et al. 2015). We subsequently used the same method to validate drug activity of two novel glutamate-modulating compounds against ketamine (Mehta et al. 2018). This study was able to provide evidence that the proposed mechanisms of the novel compounds translated to human dosing and also suggested effective doses of each drug in humans.

10.3 The Development of Neuroimaging Biomarkers for Treatment Response

The sensitivity of MRI neuroimaging to drugs is established, even though the process of developing specific markers of drug effects is an ongoing process. Exactly which markers are utilised will depend on numerous factors, such as the drug pharmacokinetics, the relevant context(s), dose ranging and neurophysiological considerations. These choices are constrained by relevant pharmacological factors but need to be cognisant of the translational application of any discoveries. That is, what is the deficit being addressed in the brain of patients. Ever since neuroimaging with MRI became available, it has been employed as a tool of discovery for the brain regions and systems that mark disorders. These markers may be the level of particular metabolites in a specific brain region, a task-related activation, or connectivity change, or a localised difference in perfusion. For example, glutamate is elevated in patients with schizophrenia (Schwerk et al. 2014; Merritt et al. 2016), and MRS would be a viable methodology to test compounds for their ability to reduce glutamate levels in vivo. Hyperperfusion has been noted and replicated in patients at risk of psychosis (Allen et al. 2016) and is therefore a candidate target for interventions to reduce such risks. Amygdala reactivity to negative affective stimuli is elevated in studies of patients with depression and a reduction in this response has been used as a target to study antidepressants (Godlewska et al. 2012). This work is crucial to the use of neuroimaging in drug development as it establishes the specific neuroimaging targets for assessing a treatment and the target effect sizes required to 'normalise' such impairments. However, examples of the shifts in these signals being accompanied by a shift in clinical presentation are limited (an indication of the potential is given by Godlewska et al. 2016). While this is very useful and is applied to MRI in studies for drug development, such a conclusion hides an array of complexities that represent real challenges extending beyond just MRI methodology.

First, the clinical diagnoses are formed from the presentations in patients and are thus probabilistic – a terms often used in neurology – or made independent of underlying biological models, which themselves may not be known. Indeed, the definition of a patient is a well-established challenge (Insel and Cuthbert 2015). Undoubtedly the functioning of the brain will be associated with the presentation of symptoms, but we must be mindful of the roles of other factors, such as social and societal factors in the expression or exacerbation of symptoms, which may limit the use of brain measurement in isolation.

Second, the presentation of multiple symptoms may be due to multiple manifestations of a 'core' deficit or multiple deficits. For example, the presentation of various motor symptoms in Parkinson's disease is due to dopaminergic pathology. However, dopaminergic pathology is not unitary and there is a gradient of pathology as well as interactions with other brain systems affected by the progressive nature of the disease. In schizophrenia, a 'core' deficit in the glutamatergic system and the dopaminergic systems have been proposed, but how these can lead to the multitude of symptoms is yet to be determined.

Third, the presentation of symptoms shared across disorders may represent a single transdiagnostic target for treatment or be a manifestation of different underlying deficits. Cognitive deficits, for example, in a particular domain such as working memory (holding information online for an action proximal in time) may be related to altered brain activity within specific regions of the brain (Owen et al. 2005; Holiga et al. 2018), connections between these regions, connections between these areas and other areas outside of the core working memory network, or even the inputs into the network (from sensory areas or internal brain systems). A drug treatment may be directed at dysfunction in one system may not ameliorate deficits across all cause impairments.

Fourth, the deficits examined cross-sectionally may be insufficient to understand a dysfunction which has developed over time and the manifestation of which may, in turn, affect brain dysfunction further. These are critical issues exemplified by developmental disorders. The treatment of children with stimulant drugs and antidepressants are known to be effective and produce measurable changes in brain circuits associated with their symptoms. Critical questions here include how these drug effects interact with development or these same brain systems? How do the drug effects alter the environment of the individual which may well be impacting on the symptoms and experiences of the same individual? These are examples where technologies such as neuroimaging alone cannot provide a clear answer, even if we rise to the challenge of asking the right questions.

> Neuroimaging biomarkers for treatments depend on knowledge of the manifestation of dysfunction in these markers. There are complexities in defining being a patient, but symptoms can also be used to define individuals, sometimes across diagnostic categories. These are potential targets for modulation by drugs.

10.4 Limitations and Challenges

10.4.1 The Reliability of MRI Methodologies Is Under-studied

Reliable measurement is of importance in neuroimaging for two main reasons. First, if we wish to establish a link between an external measure (such as symptoms) and brain imaging, both assessments need to be reliable. Second, reliable measurements of function are required for pragmatically powered studies aimed at assessing the change in measures of function. Some studies do demonstrate good reliability of functional MRI measurement. ASL performs excellently when used to quantify blood flow at rest (Hodkinson et al. 2013; Yang et al. 2019). Functional imaging with ASL performs poorly (Yang et al. 2019), and for fMRI with BOLD some studies indi-

cate good reliability measures for tasks such as the n-back (Caceres et al. 2009; Plichta et al. 2012; Holiga et al. 2018), but recent work has cast doubt on using these measures for individual-differences research (Elliott et al. 2020). Even less studied is the reliability of drug effects on the MRI signal. This is particularly relevant for models of drug attenuation to validate mechanisms where a reliable effect of drug enhances sensitivity to its modulation, and phMRI where it is the drug effect itself which is modelled. Where the reliability of drug effects has been studied moderate reliability has been demonstrated, but it can vary across brain regions (Zelaya et al. 2012; De Simoni et al. 2013).

> Reliability refers to the stability of a response across multiple timepoints of assessment. Reproducibility is the ability to demonstrate findings in a study again.

10.4.2 The Challenge of Haemodynamic Outcomes to Index Brain Activity

The principal functional neuroimaging markers of BOLD contrast and ASL indexed perfusion are complicated in drug studies as the drug could be influencing one or more of the elements within the neurovascular coupling cascade. A drug could induce a change in baseline blood flow, vascular signalling or vascular responsiveness, which could lead to a change in BOLD signal or ASL indexed perfusion in the absence of underlying neural activity, or even lead to a neural response being attenuated or undetected by reducing, for instance, vascular reactivity (Iannetti and Wise 2007). This is exemplified by the effects of the methylxanthine, caffeine which increases blood flow and reduces BOLD (Pelligrino et al. 2010), one of the reasons it is withdrawn in pharmacological neuroimaging studies. Solutions include the measurement of the effects on blood flow to 'correct' the BOLD contrast, measurement of the vascular reactivity. These methods require careful validation across multiple drug mechanisms and within experimental models of changes in neurovascular coupling, blood flow and vascular reactivity.

10.4.3 What Is the Best Way to Characterise Non-specific Effects of Drugs

Another solution to the potential direct haemodynamic effects of drug is to characterise the effects of drugs on a 'low-level' baseline condition. For example, the effects of a drug on fMRI can be tested on simple visual and motor paradigm (Harvey et al. 2018). Such approaches can be useful to characterise drug effects on the shape and timing of responses, and may even be important for defining baseline abnormalities in patient groups (Smith et al. 2008). How any effects relate to brain systems outside of those involved in 'low-level' processes and precisely how these methods are implemented remain to be fully determined.

10.4.4 Is It Too Soon to Use MRI as a Stratification Tool?

While the testing of treatment effects is typically a primary goal of clinical trials, there is variability of response. MRI can be utilised to characterise the variability of the cohort and the associated response, although challenges remain in understanding the key dimensions of stratification and their generalisability across cohorts (Berard et al. 2020). In generalised anxiety disorder and major depressive disorder, patients may have an attentional bias towards threat – a potential treatment target to reduce symptoms. Bijsterbosch et al. (2018) demonstrated potential stratification markers with fMRI across both diagnoses. Higher connectivity was linked to depression scores within the

limbic network and patients with GAD or MDD with lower limbic connectivity performed poorly on an attention-to-threat task and no such effect was shown for healthy control participants. While this finding must be seen as preliminary, it clearly demonstrates the potential utility of fMRI to select patients to inform intervention studies and this may even cut across diagnostic boundaries. Those whose fMRI-related abnormalities are farthest away from that in healthy controls may also be the ones who have greatest degree of normalisation potential. Whether this translates to predicting the degree of treatment response remains to be seen.

10.4.5 There Is a Gap Between Modulation of Brain Systems and Clinical Outcome

While there are many studies showing relationships between brain imaging measures and clinical diagnosis, symptoms and even prediction of outcome, there is less evidence that reversing, attenuating or normalising such imaging measures is directly causal in improvement in patients. Though such studies are in their infancy, examples are beginning to emerge, including elevation of hippocampal activity with antiepileptics improving cognition (Bakker et al. 2012) and alterations in emotional processing after short-term use of antidepressants predicting later clinical response (Godlewska et al. 2016).

> Challenges in the use of MRI in drug development include better understanding of the reliability of the method and the reliability and reproducibility of drug effects, the ability to resolve neuronal and haemodynamic effects, the use of MRI to select those most likely to benefit from a drug and understand how changes in MRI signals translates to improvements in symptoms and everyday life.

Case Study

Leviracetam Treatment Normalises Hippocampal Dysfunction and Improves Cognition

In a group of 17 individuals presenting with mild cognitive impairment, neuroimaging supported impaired function (Bakker et al. 2012). During an fMRI paradigm, participants were presented with pictures of everyday objects, some of which were new (not presented before), some were old (presented before) and some were lures (similar to a picture already presented). These lures were critical for the study as they served to assess the function of the dentate gyrus, part of the hippocampus, in pattern separation. The MCI individuals showed excessive hippocampal activity in the dentate gyrus region as predicted by the authors. An antiepileptic drug which lowers excessive hippocampal activity normalised dentate gyrus activity in the MCI group and improved detection of lures, although the correlation between these two effects was weak in this small group.

This study demonstrates the importance of identifying a precise neuroimaging marker of function to measure after an intervention and represents a promising approach.

Conclusion

Neuroimaging with MRI can be used to investigate different aspects of illnesses and diseases, and the differences described can be used as outcome measures to evaluate potential therapeutic interventions. The versatility of MRI allows for assessment of different aspects of pharmacological modulation in the brain, including changes in task-related activation, functional connectivity between brain regions and network topology, metabolite levels, drug-related activation as assessed using phMRI and quantitative perfusion.

These powerful techniques present with some important challenges when utilised for use in drug development. These include the need to fully understand the reliability of measurement which varies from poor to very good across studies; the choice and definition of the target population, or subpopulation; the limitations of measurement dependent on the pharmacokinetic profile of the drug (e.g. phMRI for fast-acting drugs, versus ASL for slower-acting drugs); the influence of direct haemodynamic effects of drugs and how to measure and control for them; characterising non-specific effects of drugs; and the necessary tension between using optimally suited tools for the question versus optimally validated tools as these are often not the same measurements in an evolving area such as neuroimaging.

Despite these challenges, there are examples of robust, evidence-based neuroimaging markers in use for testing drug effects with MRI. As examples, many patients with schizophrenia show altered brain connectivity during working memory tasks, and differential activity during reward processing (Leroy et al. 2020); many patients diagnosed with major depression show hyper-reactivity in the amygdala and altered activity in the anterior cingulate and frontal cortex that may be predictive of treatment response (Leppänen 2006), and ketamine phMRI has been validated for use in testing novel compounds for potential efficacy (Javitt et al. 2018; Mehta et al. 2018).

Overall, the MRI methods described in this chapter can be deployed to index drug effects by themselves, or together within the same imaging sessions, to address the nature of impairment in patient groups, at risk groups, subgroups and those with shared symptomatology. These methods can be used to examine the immediate effects of existing and novel compounds and predict the later effects in clinical outcomes. They complement other methods described in this book such as emission tomography which can be used to confirm brain penetra-

tion and dose-occupancy relationships to specific targets, and can extend or even substitute for these methods when ligands are not available or multiple targets are involved in the desired effects providing intermediate markers of functional outcome.

Acknowledgements   MAM is an employee of King's College London and is partly supported by the National Institute for Health Research (NIHR) Biomedical Research Centre at South London and Maudsley NHS Foundation Trust. The views expressed are those of the author(s) and not necessarily those of the NHS, the NIHR or the Department of Health.

References

Allen P, Chaddock CA, Egerton A et al (2016) Resting hyperperfusion of the hippocampus, midbrain, and basal ganglia in people at high risk for psychosis. Am J Psychiatry 173:392–399. https://doi.org/10.1176/appi.ajp.2015.15040485

Alsop DC, Detre JA, Golay X et al (2015) Recommended implementation of arterial spin-labeled perfusion mri for clinical applications: a consensus of the ISMRM perfusion study group and the European consortium for ASL in dementia. Magn Reson Med 73:102–116. https://doi.org/10.1002/mrm.25197

Artigas F, Schenker E, Celada P et al (2017) Defining the brain circuits involved in psychiatric disorders: IMI-NEWMEDS. Nat Rev Drug Discov 16:1–2. https://doi.org/10.1038/nrd.2016.205

Bakker A, Krauss GL, Albert MS et al (2012) Reduction of hippocampal hyperactivity improves cognition in amnestic mild cognitive impairment. Neuron 74:467–474. https://doi.org/10.1016/j.neuron.2012.03.023

Beaurain M, Salabert A-S, Ribeiro MJ et al (2019) Innovative molecular imaging for clinical research, therapeutic stratification, and nosography in neuroscience. Front Med (Lausanne) 6:268. https://doi.org/10.3389/fmed.2019.00268

Berard JA, Fang Z, Walker LAS et al (2020) Imaging cognitive fatigability in multiple sclerosis: objective quantification of cerebral blood flow during a task of sustained attention using ASL perfusion fMRI. Brain imaging Behav 14(6):2417–2428. https://doi.org/10.1007/s11682-019-00192-7

Bifone A, Gozzi A (2012) Neuromapping techniques in drug discovery: pharmacological MRI for the assessment of novel antipsychotics. Expert Opin

Drug Discov 7:1071–1082. https://doi.org/10.1517/17460441.2012.724057

Bijsterbosch JD, Ansari TL, Smith S et al (2018) Stratification of MDD and GAD patients by resting state brain connectivity predicts cognitive bias. Neuroimage Clin 19:425–433. https://doi.org/10.1016/j.nicl.2018.04.033

Breiter HC, Gollub RL, Weisskoff RM et al (1997) Acute effects of cocaine on human brain activity and emotion. Neuron 19:591–611. https://doi.org/10.1016/s0896-6273(00)80374-8

Browning M, Kingslake J, Dourish CT et al (2019) Predicting treatment response to antidepressant medication using early changes in emotional processing. Eur Neuropsychopharmacol 29:66–75. https://doi.org/10.1016/j.euroneuro.2018.11.1102

Buonocore MH, Maddock RJ (2015) Magnetic resonance spectroscopy of the brain: a review of physical principles and technical methods. Rev Neurosci 26:609–632. https://doi.org/10.1515/revneuro-2015-0010

Caceres A, Hall DL, Zelaya FO et al (2009) Measuring fMRI reliability with the intra-class correlation coefficient. NeuroImage 45:758–768. https://doi.org/10.1016/j.neuroimage.2008.12.035

Chen K, Azeez A, Chen DY, Biswal BB (2020) Resting-state functional connectivity: signal origins and analytic methods. Neuroimaging Clin N Am 30:15–23. https://doi.org/10.1016/j.nic.2019.09.012

Coimbra A, Baumgartner, R. Schwarz, AJ (2013) Pharmacological fMRI in Drug Discovery and Development. In: Garrido, L. & Beckmann, L. (eds.) New Applications of NMR in Drug Discovery and Development. Royal Society of Chemistry

Dai W, Garcia D, de Bazelaire C, Alsop DC (2008) Continuous flow-driven inversion for arterial spin labeling using pulsed radio frequency and gradient fields. Magn Reson Med 60:1488–1497. https://doi.org/10.1002/mrm.21790

De Simoni S, Schwarz AJ, O'Daly OG et al (2013) Test-retest reliability of the BOLD pharmacological MRI response to ketamine in healthy volunteers. NeuroImage 64:75–90. https://doi.org/10.1016/j.neuroimage.2012.09.037

Duff EP, Vennart W, Wise RG et al (2015) Learning to identify CNS drug action and efficacy using multi-study FMRI data. Sci Transl Med 7:274ra16 https://doi.org/10.1126/scitranslmed.3008438

Doyle OM, Ashburner J, Zelaya FO et al (2013a) Multivariate decoding of brain images using ordinal regression. NeuroImage 81:347–357. https://doi.org/10.1016/j.neuroimage.2013.05.036

Doyle OM, De Simoni S, Schwarz AJ et al (2013b) Quantifying the attenuation of the ketamine pharmacological magnetic resonance imaging response in humans: a validation using antipsychotic and glutamatergic agents. J Pharmacol Exp Ther 345:151–160. https://doi.org/10.1124/jpet.112.201665

Esteban O, Ciric R, Finc K et al (2020) Analysis of task-based functional MRI data preprocessed with fMRIPrep. Nat Protoc 15:2186–2202. https://doi.org/10.1038/s41596-020-0327-3

Friston KJ, Buechel C, Fink GR et al (1997) Psychophysiological and modulatory interactions in neuroimaging. NeuroImage 6:218–229. https://doi.org/10.1006/nimg.1997.0291

Godlewska BR, Norbury R, Selvaraj S et al (2012) Short-term SSRI treatment normalises amygdala hyperactivity in depressed patients. Psychol Med 42:2609–2617. https://doi.org/10.1017/S0033291712000591

Godlewska BR, Browning M, Norbury R et al (2016) Early changes in emotional processing as a marker of clinical response to SSRI treatment in depression. Transl Psychiatry e957:6. https://doi.org/10.1038/tp.2016.130

Harvey J-L, Demetriou L, McGonigle J, Wall MB (2018) A short, robust brain activation control task optimised for pharmacological fMRI studies. PeerJ 6:e5540. https://doi.org/10.7717/peerj.5540

Hodkinson DJ, Krause K, Khawaja N et al (2013) Quantifying the test-retest reliability of cerebral blood flow measurements in a clinical model of on-going post-surgical pain: a study using pseudo-continuous arterial spin labelling. Neuroimage Clin 3:301–310. https://doi.org/10.1016/j.nicl.2013.09.004

Hofstetter S, Tavor I, Tzur Moryosef S, Assaf Y (2013) Short-term learning induces white matter plasticity in the fornix. J Neurosci 33:12844–12850. https://doi.org/10.1523/JNEUROSCI.4520-12.2013

Holiga Š, Sambataro F, Luzy C et al (2018) Test-retest reliability of task-based and resting-state blood oxygen level dependence and cerebral blood flow measures. PLoS One 13:e0206583. https://doi.org/10.1371/journal.pone.0206583

Hong S-J, Xu T, Nikolaidis A, et al (2020) Toward a connectivity gradient-based framework for reproducible biomarker discovery. bioRxiv. https://doi.org/10.1101/2020.04.15.043315

Huettel SA, Song AW, Mccarthy G (2014) Functional magnetic resonance imaging, 3rd edn. Sunderland, Sinauer Associates

Iannetti GD, Wise RG (2007) BOLD functional MRI in disease and pharmacological studies: room for improvement? Magn Reson Imaging. 25(6):978-88. https://doi.org/10.1016/j.mri.2007.03.018.

Insel TR, Cuthbert BN (2015) Medicine. Brain disorders? Precisely. Science 348:499–500. https://doi.org/10.1126/science.aab2358

Javitt DC, Carter CS, Krystal JH et al (2018) Utility of imaging-based biomarkers for glutamate-targeted drug development in psychotic disorders: a random-

ized clinical trial. JAMA Psychiat 75:11–19. https://doi.org/10.1001/jamapsychiatry.2017.3572

Jelistratova I, Teipel SJ, Grothe MJ (2020) Longitudinal validity of PET-based staging of regional amyloid deposition. Hum Brain Mapp 41(15):4219–4231. https://doi.org/10.1002/hbm.25121

Jonckers E, Shah D, Hamaide J et al (2015) The power of using functional FMRI on small rodents to study brain pharmacology and disease. Front Pharmacol 6:231. https://doi.org/10.3389/fphar.2015.00231

Joules R, Doyle OM, Schwarz AJ et al (2015) Ketamine induces a robust whole-brain connectivity pattern that can be differentially modulated by drugs of different mechanism and clinical profile. Psychopharmacology 232:4205–4218. https://doi.org/10.1007/s00213-015-3951-9

Knutsson L, Xu J, Ahlgren A, van Zijl PCM (2018) CEST, ASL, and magnetization transfer contrast: how similar pulse sequences detect different phenomena. Magn Reson Med 80:1320–1340. https://doi.org/10.1002/mrm.27341

Ladd ME, Bachert P, Meyerspeer M et al (2018) Pros and cons of ultra-high-field MRI/MRS for human application. Prog Nucl Magn Reson Spectrosc 109:1–50. https://doi.org/10.1016/j.pnmrs.2018.06.001

Leppanen JM (2006) Emotional information processing in mood disorders: a review of behavioral and neuroimaging findings. Curr Opin Psychiatry 19:34–39. https://doi.org/10.1097/01.yco.0000191500.46411.00

Leroy A, Amad A, D'Hondt F et al (2020) Reward anticipation in schizophrenia: a coordinate-based meta-analysis. Schizophr Res 218:2–6. https://doi.org/10.1016/j.schres.2019.12.041

Li Z, Vidorreta M, Katchmar N et al (2018) Effects of resting state condition on reliability, trait specificity, and network connectivity of brain function measured with arterial spin labeled perfusion MRI. Neuroimage 173:165–175. https://doi.org/10.1016/j.neuroimage.2018.02.028

Luna A, Martín Noguerol T, Mata LA (2018) Fundamentals of functional imaging II: emerging MR techniques and new methods of analysis. Radiologia 60(Suppl 1):23–35. https://doi.org/10.1016/j.rx.2018.03.001

Mandeville JB (2012) Iron FMRI measurements of CBV and implications for BOLD signal. Neuroimage 62:1000–1008. https://doi.org/10.1016/j.neuroimage.2012.01.070

Mathias EJ, Kenny A, Plank MJ, David T (2018) Integrated models of neurovascular coupling and BOLD signals: responses for varying neural activations. NeuroImage 174:69–86. https://doi.org/10.1016/j.neuroimage.2018.03.010

Mehta MA, Schmechtig A, Kotoula V et al (2018) Group II metabotropic glutamate receptor agonist prodrugs LY2979165 and LY2140023 attenuate the functional imaging response to ketamine in healthy subjects. Psychopharmacology 235:1875–1886. https://doi.org/10.1007/s00213-018-4877-9

Merritt K, Egerton A, Kempton MJ et al (2016) Nature of glutamate alterations in schizophrenia: a meta-analysis of proton magnetic resonance spectroscopy studies. JAMA Psychiat 73:665–674. https://doi.org/10.1001/jamapsychiatry.2016.0442

Mueller S, Stirnberg R, Akbey S et al (2020) Whole brain snapshot CEST at 3T using 3D-EPI: aiming for speed, volume, and homogeneity. Magn Reson Med 84(5):2469–2483. https://doi.org/10.1002/mrm.28298

Owen AM, McMillan KM, Laird AR, Bullmore E (2005) N-back working memory paradigm: a meta-analysis of normative functional neuroimaging studies. Hum Brain Mapp 25:46–59. https://doi.org/10.1002/hbm.20131

Paloyelis Y, Doyle OM, Zelaya FO et al (2016) A spatio-temporal profile of in vivo cerebral blood flow changes following intranasal oxytocin in humans. Biol Psychiatry 79:693–705. https://doi.org/10.1016/j.biopsych.2014.10.005

Pelligrino DA, Xu H-L, Vetri F (2010) Caffeine and the control of cerebral hemodynamics. J Alzheimers Dis 20(Suppl 1):S51–S62. https://doi.org/10.3233/JAD-2010-091261

Plichta MM, Schwarz AJ, Grimm O et al (2012) Test-retest reliability of evoked BOLD signals from a cognitive-emotive fMRI test battery. NeuroImage 60:1746–1758. https://doi.org/10.1016/j.neuroimage.2012.01.129

Reed JL, Nugent AC, Furey ML et al (2019) Effects of ketamine on brain activity during emotional processing: differential findings in depressed versus healthy control participants. Biol Psychiatry Cogn Neurosci Neuroimaging 4:610–618. https://doi.org/10.1016/j.bpsc.2019.01.005

Samudra N, Ivleva EI, Hubbard NA et al (2015) Alterations in hippocampal connectivity across the psychosis dimension. Psychiatry Res 233:148–157. https://doi.org/10.1016/j.pscychresns.2015.06.004

Schwerk A, Alves FDS, Pouwels PJW, van Amelsvoort T (2014) Metabolic alterations associated with schizophrenia: a critical evaluation of proton magnetic resonance spectroscopy studies. J Neurochem 128:1–87. https://doi.org/10.1111/jnc.12398

Schwarz AJ, Becerra L, Upadhyay J et al (2011) A procedural framework for good imaging practice in pharmacological FMRI studies applied to drug development #1: processes and requirements. Drug Discov Today 16:583–593. https://doi.org/10.1016/j.drudis.2011.05.006

Smith EE, Vijayappa M, Lima F et al (2008) Impaired visual evoked flow velocity response in cerebral amyloid angiopathy. Neurology 71:1424–1430. https://doi.org/10.1212/01.wnl.0000327887.64299.a4

Stein EA, Pankiewicz J, Harsch HH et al (1998) Nicotine-induced limbic cortical activation in the human brain: a functional MRI study. Am J Psychiatry 155:1009–1015. https://doi.org/10.1176/ajp.155.8.1009

van Veenendaal TM, Backes WH, van Bussel FCG et al (2018) Glutamate quantification by PRESS or MEGA-PRESS: validation, repeatability, and concordance. Magn Reson Imaging 48:107–114. https://doi.org/10.1016/j.mri.2017.12.029

Vernon AC, Crum WR, Lerch JP et al (2014) Reduced cortical volume and elevated astrocyte density in rats chronically treated with antipsychotic drugs-linking magnetic resonance imaging findings to cellular pathology. Biol Psychiatry 75:982–990. https://doi.org/10.1016/j.biopsych.2013.09.012

Wise RG, Preston C (2010) What is the value of human FMRI in CNS drug development? Drug Discov Today 15:973–980. https://doi.org/10.1016/j.drudis.2010.08.016

Wong EC, Buxton RB, Frank LR (1997) Implementation of quantitative perfusion imaging techniques for functional brain mapping using pulsed arterial spin labeling. NMR Biomed 10:237–249. https://doi.org/10.1002/(sici)1099-1492(199706/08)10:4/5<237::aid-nbm475>3.0.co;2-x

Witkin JM, Martin AE, Golani LK et al (2019) Rapid-acting antidepressants. Adv Pharmacol 86:47–96. https://doi.org/10.1016/bs.apha.2019.03.002

Yang FN, Xu S, Spaeth A et al (2019) Test-retest reliability of cerebral blood flow for assessing brain function at rest and during a vigilance task. NeuroImage 193:157–166. https://doi.org/10.1016/j.neuroimage.2019.03.016

Zelaya FO, Zois E, Muller-Pollard C et al (2012) The response to rapid infusion of fentanyl in the human brain measured using pulsed arterial spin labelling. MAGMA 25:163–175. https://doi.org/10.1007/s10334-011-0293-4

Zelaya FO, Fernandez-Seara MA, Black KJ et al (2015) Perfusion in pharmacologic imaging. In: Bammer R (ed) MR & CT perfusion imaging: clinical applications and theoretical principles. Lippincott Williams & Wilkins, Philadelphia

10

Positron Emission Tomography in Drug Development

Frans van den Berg and Eugenii A. (Ilan) Rabiner

Contents

11.1   Introduction – 166

11.2   Principles of PET – 166

11.3   CNS Radiotracer Development – 168

11.4   Building the "Three Pillars": Imaging in Early-Phase Drug Development – 170

11.4.1   Pillar 1: Drug Exposure at Target Site – Biodistribution – 170

11.4.2   Pillar 2: Target Engagement – 173

11.4.3   Pillar 3: Pharmacologic Activity – 176

References – 180

The original version of the chapter has been revised. A correction to this chapter can be found at https://doi.org/10.1007/978-3-030-62351-7_20

Central nervous system (CNS) drug development poses some unique challenges for drug developers in terms of drug delivery, assessment of target engagement, and observation of drug's pharmacological effects. These challenges are reflected in the huge costs and the high attrition rate of CNS drugs. Positron emission tomography (PET) has become an indispensable tool in the development of drugs for CNS disorders. PET imaging allows developers to measure tissue exposure, target engagement, and pharmacological activity, providing development teams with a clear strategy to establish the "three pillars of survival" and allowing them to make crucial decisions about their drug's progression. This chapter will focus on each of the three "pillars" and how PET imaging can support CNS drug development.

Learning Objectives
- Understanding the principles of PET imaging
- Understanding key requirements for radiotracers
- Understanding the concept of biodistribution in PET imaging
- Understanding the concept of target engagement in PET imaging
- Using PET imaging in assessing pharmacological activity

11.1 Introduction

The concept of emission and transmission tomography, as proposed by Kuhl, Chapman, and Edwards in the late 1950s, evolved into what we know today as positron emission tomography (PET). PET is, and remains, the premier tool for imaging neurochemistry in living humans, has become one of the key enabling technologies in translational medicine, and is an invaluable tool in the drug development process.

As our knowledge and understanding of CNS pathologies grow, we are able to identify new targets for therapeutic intervention. However, only 15% of central nervous system (CNS) drug candidates progress from phase 1

to market approval. When coupled with a mean research and development investment of $1336 million per drug (Wouters et al. 2020), it should come as no surprise that pharmaceutical companies are increasingly looking at ways to optimize decision-making processes in drug development. Drug development in the CNS faces inherent challenges due to the presence of the blood–brain barrier, which limits drug access to the CNS. In addition, measuring drug presence in the brain and observing the biochemical/physiological consequences of that presence are considerably more complex than for other systems. PET imaging is uniquely positioned to inform us about the underlying mechanisms of CNS disease and identify new targets for therapeutic intervention, as well as characterize the pharmacokinetics (PK) of CNS drugs (drug biodistribution) and their molecular selectivity. PET can provide accurate quantification of drug-target engagement and pharmacological activity of CNS drugs in the living human brain. These capabilities make PET imaging indispensable in the early triage of candidate molecules, minimizing late-stage attrition in CNS drug development.

11.2 Principles of PET

PET is a noninvasive imaging technique that allows for true molecular imaging. PET provides the visualization and quantification of specific molecular targets, enabling the monitoring of physiological processes by utilizing high-affinity and high-selectivity radioactively labeled compounds, administered in tracer concentrations (i.e., at doses that do not produce any meaningful pharmacological activity). These radiotracers must possess the physiochemical characteristics that enable them to be labeled with a short-lived positron emitting isotope (or radionuclide). ^{11}C, with a half-life of 20.38 min, and ^{18}F, with a half-life of 109.8 min, are the most commonly used radionuclides, though shorter- and longer-lived ones (^{15}O, half-life of 2 min; ^{89}Zr, half-life of 78.5 h) are utilized for specific indications (see ■ Table 11.1 for a list of common PET radionuclides).

PET imaging usually involves injection of the radiotracer into a peripheral vein, followed by its distribution across the body. The decay of a PET radionuclide leads to an emission of a positron (effectively a positively charged electron). On emission, the positron will encounter an electron (this typically happens within a few millimeters in the body), and this resultant matter–anti-matter "annihilation event" generates two collinear photons (i.e., the photons travel at 180 degrees to each other) with an energy of 511 keV each. These photons are detected almost simultaneously by the ring of detectors of the PET camera (coincident detection) allowing great accuracy in the location of the initial photon-electron annihilation event. A 3D image is reconstructed from all the coincident events (after correction for signal scatter and attenuation), providing for an accurate quantification of the radioactivity (and hence radionuclide concentration) in a given volume of tissue (typically reported as kBq/ml of tissue). The whole process is depicted in ◘ Fig. 11.1. PET images vary over time as the distribution of radiotracer, and hence the concentration of annihilation events, changes and can therefore often have limited anatomical definition. For this reason, PET data are often combined with a companion image of high anatomical resolution (magnetic resonance imaging, MRI, or computed tomography, CT). The resulting images show the quantitative distribution of radioactivity in the body over time, mapped onto the anatomy. It is important to note that it is the distribution of the radionuclide, not the radiotracer molecule that is directly measured in a PET scanner, while the biological interpretation of the scan results is based on the distribution of the radiotracer. Therefore, useful interpretation of the results of a PET study requires a thorough understanding of the metabolic fate of the labeled radiotracer, and the contribution of the parent radiotracer, as well as that of any radiolabelled metabolites, to the total radioactivity concentration measured at any particular time in any particular tissue.

◘ **Table 11.1**   Radionuclides (Anand et al. 2009)

Isotope	Half life	Positron energy (MeV)
C-11	20 min	0.385
N-13	10 min	0.492
O-15	2 min	0.735
F-18	110 min	0.250
K-38	8 min	1.216
Cu-62	10 min	1.315
Cu-64	12.7 h	0.278
Ga-68	68.1 h	0.836, 0.352
Rb-82	1.3 min	1.523, 1.157
I-124	4.2 days	1.691, 7.228,1.509,1.376

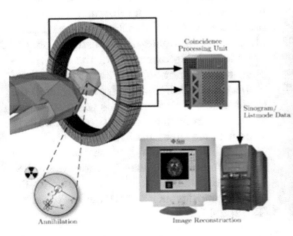

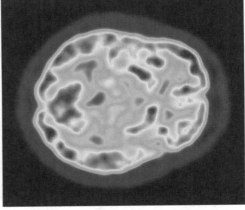

◘ **Fig. 11.1**   PET scanner – obtaining an image

PET imaging is an exquisitely sensitive technique and modern-day PET scanners are able to detect picomolar (10^{-12}) and even femtomolar (10^{-15}) concentrations of the radiotracer, providing a sensitivity nine orders of magnitude higher than MRI. However, PET suffers in comparison to MRI in terms of spatial and temporal resolution. Whereas an MRI scan can yield a spatial resolution of <1 mm, the best current PET scanners offer resolution of 1–2 mm, in the center of the field of view that effectively translates to a general resolution of 4–5 mm. The hard limit of PET spatial resolution is the so-called "positron range" (the distance a positron travels following emission, before encountering an electron). While the positron range varies for different radionuclides, being inversely related to the energy of the emitted positron, the smallest average positron range is ~ 1 mm (0.8 mm for ^{18}F and 1.3 mm for ^{11}C), providing the theoretical limit to effective spatial resolution for PET. PET also has very low temporal resolution, this being determined by the physiological distribution and pharmacokinetics of the labeled radiotracer. As such, the biological information derived from PET data is of the order of tens of seconds (for blood flow and blood volume measurements) to tens of minutes for the quantification of the density of specific molecular targets (such as neuroreceptors). PET can thus be best conceptualized as a high-sensitivity molecular imaging tool that allows the accurate quantification of a concentration of a specific molecule over a period of ~ 1–2 h.

Other practical limitations in PET imaging include the need for complex technology, the high cost involved, and the need for a sophisticated research team to perform and interpret the imaging data.

> In summary, PET provides the ability to determine, noninvasively, the concentration of a molecule in a volume of tissue to femtomolar concentrations. Such information provides the basis for biological applications of PET imaging.

11.3 CNS Radiotracer Development

A molecule (such as an investigational drug) can be radiolabeled and its distribution in the body can be monitored over time. Such a radiotracer can provide information on the concentration of the molecule itself, over a particular time-span, but typically cannot provide any information on the molecular targets it is interacting with, as its signal is dependent almost entirely on the "free" radiotracer concentration and a low-affinity/high-capacity "non-specific" interaction with the body tissues.

A small subset of radiotracers (known as radioligands) provide a signal from which a displaceable component (reporting on the high-affinity/low-capacity interaction with specific molecular targets, such as neurotransmitters and enzymes of transporters) can be extracted and separated from the "free" and "non-specific" components. These radioligands can provide quantitative information on the concentration of molecules of interest and as such can be used to track such concentrations over time (e.g., measure the concentration of a neuroreceptor over the course of a disease) or measure the binding of unlabeled drugs to such targets (e.g., relating the dose of a novel drug to the blockade of its target enzyme), in order to estimate the drug dose required to provide meaningful occupancy of its molecular target.

PET imaging is entirely dependent on the availability of suitable tracer compounds, and the low number of suitable tracers remains a bottleneck in the use of PET in CNS research, in general, and drug development process, in particular. The radiolabeling of a particular molecule to evaluate its biodistribution can provide useful information for the drug developer, but it is not as useful as that obtained with a proper radioligand (see examples in the sections below); however, it is technically easier, as the molecule needs to only fulfill the structural requirements in ◻ Table 11.2.

On the other hand, development of a radioligand is significantly more complex, leading to a high attrition rate due to the many properties such a molecule has to satisfy (◻ Table 11.2).

□ **Table 11.2** Desired attributes for successful CNS PET ligands (Zhang and Villalobos 2016)

Structural requirements	*PK properties*
Structural handle for ^{18}F or ^{11}C labeling	Brain permeable
Amenable for late-stage radiolabeling	No brain permeable radiolabeled metabolites
Pharmacology	
Low non-specific binding (NSB)[a]	
High affinity[b]	
High selectivity versus competing targets[c]	

[a]The acceptable levels of NSB are going to depend on the levels of displaceable binding. Typically, the ratio of displaceable signal-to-NSB should be >1.5, with ratios between 3 and 5 considered to be optimal
[b]The affinity required for a usable signal is a function of target density, as the PET signal will be proportional to the ratio of target density and ligand affinity for the target (B_{max}/K_d). Typically, low single-digit nM to sub-nM affinity is required as a minimum
[c]The required affinity ratio between any competing targets will be dependent on the ratio of the densities of the targets in a region of interest (ROI), as the PET signal is dependent on both factors (see note b above)

The National Institute of Mental Health (NIMH) (NIMH – Approved CNS radiotracers n.d.) currently lists about 160 CNS radiotracers that have been advanced for human use, covering a range of neuroreceptor targets including metabotropic (i.e., adenosine A1, D1/D2), ionotropic (NMDA, P2X7), transporters (DAT, SERT), synaptic proteins (SV2A), channel like targets (TSPO), enzymes (AChE, Cox-1), and aggregated proteins (β-amyloid, Tau, α-synuclein). There remain, however, thousands of proteins in the CNS that may be used as biomarkers for disease detection, progression monitoring, and patient phenotyping and constitute viable targets for drug developers. The discovery and development of CNS tracers to utilize this wealth of targets continue to be a time-consuming and costly undertaking.

The key challenge that faces CNS PET radioligand development is creating a molecule with an adequate signal-to-noise ratio (SNR). The SNR will be a function of the displaceable-to-NSB ratio (see □ Table 11.2) as well as the within-subject variability of this ratio. Adequate quantification of the density of the molecular target of the radioligand requires the evaluation of both the accumulation and washout of the radioligand from the brain (with a few notable exceptions). An important parameter is therefore the reversibility of the radioligand tissue kinetics within the time-scale of the PET scan (typically 90–120 min).

In order to reach CNS targets, the radioligand should be able to cross the blood–brain barrier (BBB). While compounds with higher lipophilicity will generally have better BBB penetration, the magnitude of NSB is often proportional to the lipophilicity. Empirically, an optimal window for lipophilicity appears to be between LogD of 1 and 4, though successful radioligands with LogD outside this window are available.

Developing a radioligand for novel targets is therefore a complicated endeavor and may take 1–2 years from start to successful human characterization, even when no significant hurdles are encountered. With significant effort and resources required for such development and the high rate of failure, optimization of the radioligand development process is highly desirable. Past efforts have focused on the characteristics of the candidate molecule; however, as we saw above, factors inherent in the molecular target (such as target density and anatomical distribution) play an important role in the success of any radioligand. Biomathematical approaches, that utilize known ligand and target parameters, allow the triage of potential radioligand candidates and reduce the risk of failure (Guo et al. 2009, 2012). A choice of better ligand candidates, combined with a well-defined iterative development pathway pipeline (see □ Fig. 11.2 for an example), provides a significant improvement in the chances of successful novel radioligand development.

Small animal PET is increasingly being used to study drug PK and PD, before conducting human studies (Chatziioannou 2002).

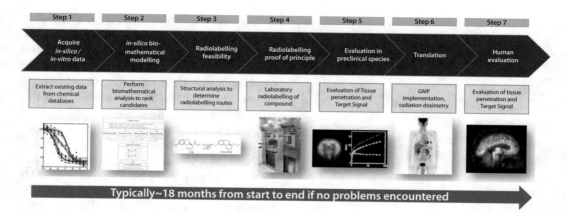

■ **Fig. 11.2**   An illustration of a well-defined development pathway pipeline. Copyright Invicro, used with permission

Relevant animal models provide valuable information concerning the absorption, distribution, metabolism, and elimination (ADME) of candidate compounds. Small animal PET also allows us to obtain tissue biopsy samples to determine if radiolabeled metabolites contribute to the PET signal in tissue, a procedure that is not possible in human PET studies (Hicks et al. 2006). The caveat to drawing conclusions based on small animal PET data is that despite genetic similarities, animal models do not always represent the full phenotype of human physiology, and they may not necessarily provide the same information nor necessarily reflect the process that they purport to evaluate (Hicks et al. 2006). However, the ability to extrapolate from animals to human studies makes PET imaging a logical technique for both preclinical testing of new drugs and for the validation of new tracers.

> The quantification of radiotracer concentration by PET, when combined with a suitable biomathematical model, provides information on the kinetics of the molecule in human tissue. Some radiotracers that possess suitable physiochemical and pharmacological qualities can be used to provide meaningful information on the concentration of endogenous molecules (receptors/enzymes/transporters) that can be used to index biological processes in vivo.

11.4   Building the "Three Pillars": Imaging in Early-Phase Drug Development

PET imaging in early drug development obtains information that is critical in triaging drug candidates to ensure that molecules with the best chance of success are advanced to proof-of-concept stage. These characteristics, collectively referred to as the "three pillars of survival" (Morgan et al. 2012), provide an integrated understanding of a drug candidate's pharmacokinetic/pharmacodynamic characteristics, namely, drug exposure at the target site, quantitative information on the relationship between drug exposure and target binding, and confirmation of relevant pharmacological activity related to drug binding to its molecular target.

11.4.1   Pillar 1: Drug Exposure at Target Site – Biodistribution

The pharmacological activity of any CNS drug depends directly on the free drug concentration in brain tissue, which in turn depends on the drug's ability to cross the blood–brain barrier (BBB). While various approaches are used to predict BBB permeability based on drug physiochemical characteristics (Zhao et al. 2003; Vilar et al. 2010), and in vitro assays, ultimately in vivo tests are required to provide reliable estimates. PET can provide an accurate measure of BBB penetration, and higher species tend to be very good predictors of humans. The BBB functions as a filter, with hydrophilic drugs generally having low BBB penetration by passive diffusion, and therefore tend to be excluded from the CNS unless specific transport mechanisms are present. The BBB also contains efflux transporters (such as the P-glycoprotein (PGP)) that actively remove drugs from the CNS. Establishing the free drug concentration in the brain is often a critical step in early-phase drug development and is often difficult to establish with conventional methods that predict BBB permeability and liability to efflux transporters based on the drug's physicochemical properties. PET imaging is the only practical method that enables

direct measurement of drug kinetics across the human BBB.

Where possible, drug exposure in the brain would be determined indirectly, during a target occupancy study designed to evaluate direct drug-target interactions (see section below). However, such a study requires the availability of a suitable radioligand for the molecular target in question. Where such a radioligand is not available, the drug molecule may be labeled with a PET radionuclide, and information about its brain exposure obtained in the course of a biodistribution study. See ◘ Fig. 11.3 below for a summary of differences between useful radioligands and other radiolabeled molecules.

When a prospective drug molecule can be labeled with a PET radioisotope without changing the molecule's structure, PET pro-

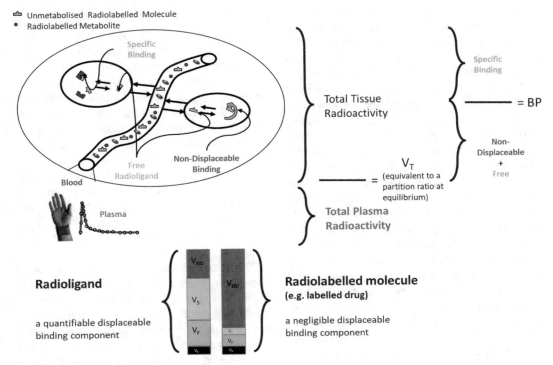

◘ **Fig. 11.3** Components of a PET signal: This figure represents the components of the measurable PET signal. While the acquired signal cannot be broken down to its components due to limitations in the anatomical resolution of the PET camera, a combination of PET and ancillary data within a suitable biomathematical model allows the determination of the components of the total PET signal. V_{ND} represents the non-displace-

able binding component, V_S represents the specific binding component, V_f represents the free fraction, and V_B represents the fractional blood volume component. The distinction between a radioligand and a radiotracer lies in the presence of a quantifiable displaceable component in the case of a radioligand. Copyright Invicro, used with permission

vides a means to directly follow the drug molecule's delivery, distribution and clearance in the living body, and retention and kinetics in the tissue of interest over time. These biodistribution studies can provide data on the delivery of drugs to the brain, influx and efflux rates, and free drug concentration (C_{FT}) in brain. Such studies often employ only tracer or microdoses of a compound and can, in theory, take place before phase 1 studies (they are sometimes known as phase 0 studies), giving developers the opportunity to gain early insights into their drug's characteristics and to discontinue development in the case of negative findings (e.g., poor BBB permeability).

Small molecules are typically labeled directly with ^{11}C or ^{18}F (if a fluorine is present in the structure), while large molecules may be labeled directly with ^{89}Zr or other longer lived isotopes, or pre-targeted with ^{18}F; however, large molecules are not usual in CNS drug development and will not be discussed further here. During the radiolabelling process, it is important to maintain the authenticity of the compound, as any chemical modification may alter the behavior of the molecule. In the past, such studies may have just evaluated the accumulated radioactivity in the brain at a given time-point following intravenous administration. Such a study design may provide a qualitative estimate of a particular molecule's tendency to cross the BBB, but it is susceptible to several confounding factors. The measured PET signal would represent the total accumulation of radiolabeled species at a particular time point, but would not provide any information about the nature of the molecules accumulating (i.e., parent molecule and/or any radiolabeled metabolites). In addition, little information on the transport rates of the molecule would be available, and only the total drug concentration would be estimated (a sum of the free, the non-specifically bound, a small specifically bound component, as well as the drug within the vascular lumen in the brain).

A more sophisticated design (see ◻ Table 11.3) would involve radiolabeling

◻ **Table 11.3**   CNS biodistribution study design parameters (Suridjan et al. 2019)

Four to six healthy subjects
T1-weighted MRI scan to provide a structural image (and exclude confounding pathology)
Unlabeled oral dose of study drug (optional)
Dynamic PET acquisition following injection of a microdose of ^{11}C or ^{18}F labeled drug
Arterial blood samples to correct for radiolabeled metabolites
Tracer kinetics analysis including blood volume correction

an appropriate candidate drug and administrating a tracer dose intravenously (usually <10 micrograms) to a healthy subject. The radiotracer's distribution would be examined over time (typically ~1-2 h post-injection) by collecting dynamic PET data. Serial blood samples would be obtained from the radial artery, allowing the quantification of the time-course of total radioactivity in blood and plasma, as well as the fraction of unmetabolized drug in plasma. Combining the PET data and the arterial blood data in an appropriate kinetic model, a number of parameters can be derived, including total concentration of radioactivity in the brain over time, the rate of drug entry from the blood across the BBB (K_1), and the equilibrium partition coefficient of the free and non-specifically-bound drug between brain and plasma (volume of distribution of the non-displaceable compartment – V_{ND}).

While the parameters above can be useful in assessing the kinetics of a compound in the brain, further information can be obtained by combining the V_{ND} with the free fraction of the drug in plasma (f_p, defined as the fraction of the total drug concentration not bound to plasma proteins, in the absence of any specific binding) and the free fraction of the drug in tissue (f_{ND}, defined as the fraction of the free + non-displaceable drug concentration in the brain). V_{ND} is defined as the ratio (at equilibrium) of the total drug concentration in tissue (C_T) to that in plasma (C_p):

$$V_{ND} = \frac{C_T}{C_P}$$

It can also be shown that under the conditions of passive diffusion, the free drug concentration (i.e., unbound drug) will be the same on both sides of the BBB at equilibrium. Therefore,

$$C_P f_p = C_T f_{ND}$$

and

$$V_{ND} = \frac{f_p}{f_{ND}}$$

This relationship between V_{ND}, f_p, and f_{ND} has been demonstrated to hold for a large number of molecules that cross the BBB by passive diffusion (Gunn et al. 2012), and conversely, a significant deviation from this relationship indicates a concentration gradient across the BBB at equilibrium. Where $V_{ND} < f_p/f_{ND}$, an interaction of a drug with an efflux transporter, such as P-glycoprotein (PGP), is present, and conversely $V_{ND} > f_p/f_{ND}$ indicates the effects of an influx transporter. While V_{ND} and f_p can be obtained during the course of a PET study, f_{ND} has to be obtained from a separate in vitro experiment. Equilibrium dialysis is a widely available method to estimate tissue-free fractions, and when combined with PET data can provide valuable information on whether the drug is subject to any transporter limitations with regard to its BBB penetration (Ridler et al. 2014).

A further extension of data obtained in the course of a biodistribution PET study is the use of the free tissue concentration (C_{FT}) to estimate dose-related target occupancy. Using the measured plasma concentration and controlling for V_{ND} and f_{ND}, C_{FT} can be calculated.

$$C_{FT} = f_{ND} V_{ND} C_p$$

Target occupancy may then be estimated by combining C_{FT} with the known drug affinity (K_d)

$$\text{Fractional occupancy} = \frac{C_{FT}}{C_{FT} + K_d}$$

The approach above can provide extensive information on drug brain penetration and target interaction; however, it has a number of caveats. First, this approach is relatively inefficient when applied to a development program, as the process of radiolabeling and setup of manufacturing to GMP requirements will be needed for every new candidate. Second, the estimation of target occupancy will depend significantly on the K_d estimate used. As this will be derived from in vitro experiments, and/or conducted in different species, significant differences can occur from the true human in vivo K_d. For these reasons, biodistribution studies are not recommended as the approach of choice in early-phase drug development, when a true radioligand for a particular target is available. When such a ligand is available, a target engagement or a "target occupancy study" is recommended, and the next section will describe these.

11.4.2 Pillar 2: Target Engagement

The fundamental principle in play here is that target engagement (or target occupancy) is a prerequisite for pharmacological activity. The ability to demonstrate drug molecule interaction with its intended target (usually a receptor, enzyme, or transporter) provides evidence of drug distribution into that tissue, and in the case of CNS drugs, confirms BBB penetration. Occupancy studies provide direct information on, and enable the quantification of, the magnitude of target engagement. These studies can thus provide valuable dosing information, and the relationship between drug dose level and the percentage of receptor occupancy achieved is information critical to the development of any CNS drug candidate. Once a suitable PET ligand for a target is available, it can be used to evaluate and compare multiple alternative candidate molecules without further extensive radiochemistry, making this approach very efficient and cost-effective.

Target occupancy studies require the existence or development of a PET ligand appropriate for the molecular target. The characteristics of such a radioligand were discussed in the previous section, so the only additional point worth mentioning here is that in case of a de novo ligand development being required, this process should be planned well in advance, so that the ligand is ready for use as part of the First-Time-In-Human (FIH) study (the process can take 12–24 months). Radioligand development often utilizes similar molecules to the drug candidate and, therefore, can be done efficiently in parallel with the drug discovery program.

A target engagement (or an occupancy) study typically involves the administration of a baseline PET scan designed to quantify the density of available target molecules in the brain, followed by dosing of the study drug and repeated PET scans to quantify the

reduction in the available target density due to occupancy of the target by the drug. Due to the relatively high cost of PET studies and the ethical imperative to minimize human exposure to novel chemical entities and additional ionizing radiation, drug developers are encouraged to use adaptive study designs when designing a target occupancy study. Adaptive study designs (◼ Fig. 11.4) allow drug developers to measure the occupancy of a particular dose of a drug in the first few individuals and adjust the doses used in subsequent cohorts based on the results obtained (Zhang and Fox 2012; Zamuner et al. 2010). Such a design means that the information acquired from each scan is used optimally and study size and duration are minimized. A mathematical model is then used to combine occupancy and plasma concentration of the drug to understand the relationship between plasma exposure and target occupancy.

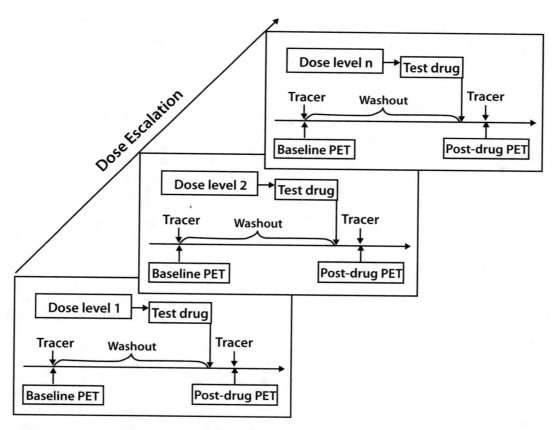

◼ **Fig. 11.4** Design of receptor occupancy studies using positron emission tomography imaging to evaluate a test drug (Zhang and Fox 2012)

The value of information derived from an occupancy study is highest at the very beginning of the development program. Knowledge from this kind of study can be used to formulate key decisions in drug development such as refining the clinically relevant doses for later-phase efficacy studies. For example, the efficacious levels of dopamine D_2 receptor occupancy for antipsychotics (Farde et al. 1989) and relationship of D_2 receptor occupancy to incidence of adverse events (Kapur et al. 2000), can be used to determine the desired dose range of novel antipsychotic medication (te Beek et al. 2012), thereby reducing the size, cost, and complexity of phase II efficacy trials. Therefore, occupancy studies are best performed during or just after the first time in human single ascending dose (FTIH SAD) studies.

A question that is often raised is whether the occupancy study should be conducted following a single or repeat dosing of the study drug? In addition, should such studies be conducted in healthy volunteers or patients from the target disease population? As the aim of such studies is to estimate the clinically meaningful doses to be used in patients, who would be receiving repeat doses of the drug, it would seem natural to perform occupancy studies in patients following repeat dose administration. However, such a conclusion would be false for a number of reasons. While the use of patients may seem to be a natural option, it is complicated logistically and ethically. We recommend that occupancy studies be conducted as early as possible, ideally concurrently with the FTIH SAD studies. However, at this early stage, the available safety information is generally not sufficient for repeat dose studies in patients and, therefore, such an occupancy study would be delayed to the later stages of the development program, reducing its value. In addition, studies in patient populations are more complicated logistically (both in terms of subject recruitment and study design) and may require the washout of patients' standard medication, which may be ethically difficult. Finally, there are very few situations, where a study in patients will provide differing information from that in healthy volunteers. The fractional occupancy of a drug depends on two parameters only – the free concentration of the drug and the affinity of the drug for its target. While patients may have differences from healthy volunteers in the total density of a target, this parameter is not relevant for the determination of fractional occupancy. Similarly, the brain access of a drug and its distribution in the brain are typically unaffected by a particular disease status, and hence the relationship between measured plasma exposure and free brain concentration is generally similar in patients and healthy individuals. For these reasons, occupancy studies conducted in healthy volunteers will generally provide the same information as those conducted in patients. The exceptions would be disease conditions where (1) the molecular target is not expressed in healthy individuals, (2) the molecular target is structurally different in patients, leading to a difference in affinity between patients and healthy controls, and (3) the brain entry of a drug is different in patients (e.g., due to an active transporter being present, that is significantly affected by the disease process).

Similarly, the use of repeat-dosing designs would seem to be indicated if we want to estimate the doses required for clinical application. Here again though, a combination of logistic and scientific considerations combine to propose a different conclusion. The use of repeat-dose design is not possible during a SAD study due to safety considerations and would necessitate a delay in obtaining occupancy information, reducing its value. A more serious objection is based on the fact that target up- or downregulation following repeated dosing of a drug is a common feature of pharmacology. As occupancy studies rely on the knowledge of the total number of available targets (obtained in a baseline PET scan) to estimate fractional occupancy, a change in this density will lead to a bias in the measured occupancy. As information on the repeat-dose effects of a new drug on its molecular target is generally not available, this is a risk that cannot be easily discharged. A solution that has been developed is to conduct an occupancy study that examines target occupancy over an extended time period following a single dose of the drug, evaluating occupancy and drug plasma concentration on at least two occasions post-dose (Abanades et al. 2011). The relationship between plasma concentration and target occupancy will be either "direct" (i.e., the same plasma concentration will lead to the same target occupancy) or "indirect" (i.e., the relationship between plasma concentration and target occupancy will depend on the time post-dose). In the case of "direct" kinetics, the occupancy relationship estimated in a single-dose study will hold true for a repeat-dose study – something true for > 80% of drugs in our experience. "Indirect" kinetics, where there is a hysteresis between plasma drug concentration, typically arises in cases of slow drug target kinetics (e.g., irreversible binding being the extreme case) or slow clearance of free drug from the brain due to a BBB transport limitation. An example of the different types of kinetics can be seen in data obtained when comparing the well-known μ-opioid receptor inhibitor naltrexone, with the experimental compound GSK1521498, conducted using the μ-opioid receptor radioligand [¹¹C]carfentanil (Rabiner et al. 2011) (see ◻ Fig. 11.5).

Occupancy studies conducted as part of a phase I program and employing optimized adaptive study designs have become an essential component of CNS drug development, reducing the size and duration of early-phase clinical trials, as well as allowing the discontinuation of poor drug candidates, leading to a significant saving in resources.

11.4.3 Pillar 3: Pharmacologic Activity

The use of PET imaging in estimating the activity of pharmacological agents as part of the drug development process is not well established. The relative lack of specificity in linking drug administration and changes in target activity has limited the use of pharmacodynamic PET imaging for critical go-no-go decision-making. However, the value of pharmacodynamic imaging in certain clinical settings must not be underestimated. Imaging can help to detect early disease, monitor disease progression, and distinguish between responders and non-responders in patients undergoing therapeutic intervention. The use of PET to detect pharmacological activity of an administered drug can be divided into these broad categories:

- Change in brain activity and/or metabolism
- Change in molecular target expression
- Change in molecular target affinity (allosteric modulation)
- Change in neurotransmitter release

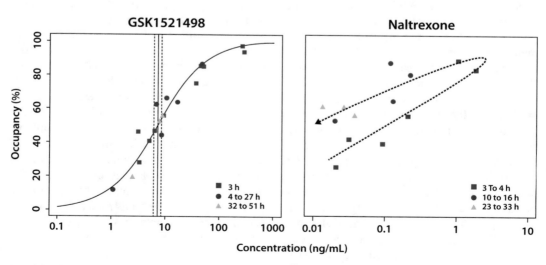

◻ **Fig. 11.5** Relationship between μ-opioid receptor occupancy and plasma concentration for two antagonists examined using [¹¹C]carfentanil PET (Rabiner et al. 2011). GSK1521498 is an example of direct kinetics, whereas naltrexone represents indirect kinetics

Positron Emission Tomography in Drug Development

Today, [^{18}F]-flurodeoxyglucose (FDG) is the oldest radiotracer that is being used in the brain, and it has been used to index brain metabolic activity in Alzheimer's disease (AD) patients. FDG PET may be useful in evaluating treatments for AD (Cummings et al. 2013), or it may act as a viable surrogate endpoint in the evaluation of potential disease-modifying therapy. The use of FDG PET in drug development, however, remains limited due to the relative lack of selectivity of the change in FDG PET signal. More recently, PET occupancy studies have successfully incorporated functional MRI to provide a pharmacodynamic response linked to target occupancy by a drug. A demonstration of differences in the functional effects of μ-opioid receptor occupancy by two μ-opioid receptor antagonists, linked to their occupancy of the target, is demonstrated in ◘ Fig. 11.6 (Rabiner et al. 2011). The utilization of MRI in drug development lies outside the scope of this chapter and will be discussed elsewhere.

Change in molecular target expression, either as a consequence of the disease process or as a result of therapeutic intervention, is well suited for evaluation using PET. The accumulation of misfolded protein species such as beta-amyloid (βA), tau protein (τ-protein), alpha-synuclein (α-SYN), and others has become recognized to be a key characteristic of neurodegenerative disorders. A number of PET radioligands for the quantification of βA deposits have been well characterized (e.g., the ^{18}F labeled Amyvid and Neuraceq). The use of these ligands in the diagnosis of AD, and selection of subjects with AD or, more importantly, mild-cognitive disorder (MCI) for clinical trials, has become standard practice. PET has also been used to quantify the clearing of βA by novel therapies such as aducanumab and gantenerumab (Sevigny et al. 2016, 2017; Klein et al. 2019) (◘ Fig. 11.7).

More recently, research into tauopathies, a class of neurodegenerative diseases involving the aggregation of tau protein (τ) into neurofibrillary or gliofibrillary tangles (NFTs) in the human brain, has evolved rapidly, with no fewer than 12 putative tau radiotracers progressing into human studies in the last 7 years (McCluskey et al. 2020). As of now, no tau-based therapy is available, but the ability to track the tau load and assess the potential

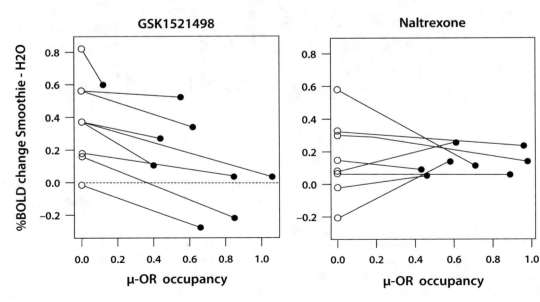

◘ **Fig. 11.6** Plots of blood oxygenation level-dependent (BOLD) activity from a functional MRI evaluation as function of μ-opioid receptor occupancy (Rabiner et al. 2011). Open circles represent baseline evaluation and filled circles represent post-dose evaluation. Lines link evaluations for the same individual

Biogen - Aducanumab

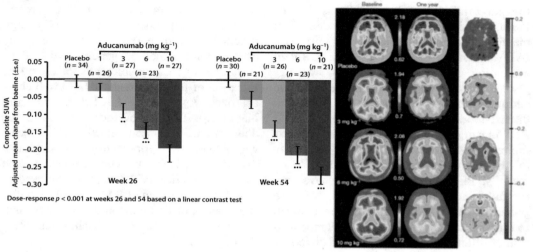

Roche - Gantenerumab

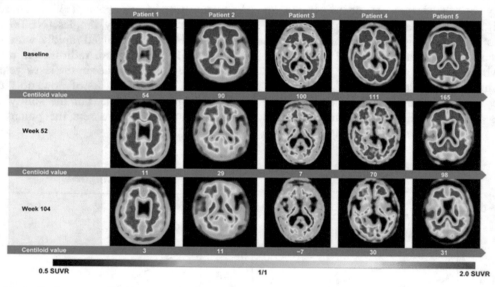

◘ **Fig. 11.7** Beta-amyloid plaque clearance from the brain measured with PET (Sevigny et al. 2016, 2017; Klein et al. 2019)

therapeutic effect would be invaluable for drug developers.

The total signal detected following the injection of a PET radioligand depends on the affinity of the radioligand for the target as much as target density (indeed it is proportional to the product of density and affinity). Therefore, drug effects that produce changes in the affinity of the molecular target for the radioligand can be detected. Such change in

affinity can occur when the radioligand binding site affinity of phosphodiesterase 10 (PDE10) is increased following the binding of cyclic nucleotides AMP and GMP (cAMP and cGMP) to allosteric binding sites. A blockade of other PDE species (in this case PDE2 and PDE4) increases the concentration of cyclic nucleotides in the striatum and leads to an increase in the signal of a PDE10-specific radioligand (Ooms et al. 2016). PET evaluating PDE10-specific signal can therefore serve as a biomarker of pharmacodynamic effects of medications that alter cyclic nucleotide concentrations in the human brain.

PET allows the measurement of changes in certain neurotransmitter concentrations in the human brain. This has been best established for dopamine using D2 receptor antagonist ligands such as [11C]raclopride (reviewed by Laruelle 2000), although the small signal change elicited by even powerful dopamine releasers such as dex-amphetamine did not encourage the evaluation of a modulation of this effect by novel drugs. More recently, agonist radioligands such as [11C]PHNO and [11C]NPA have been shown to be significantly more sensitive to dopamine fluctuations than antagonists such as [11C]raclopride (Narendran et al. 2004; Shotbolt et al. 2012). This higher signal enabled the quantification of dose-dependent modulation of dex-amphetamine-induced dopamine release, by the novel GPR139 agonist TAK-041, in the human brain (Tauscher et al. 2018). More recently, developments have enabled similar studies for the detection of changes in opioid, acetylcholine, and serotonin concentrations (Colasanti et al. 2012; Gallezot et al. 2014; Erritzoe et al. 2020), opening new avenues in detecting pharmacodynamic effects of CNS medication.

The application of PET imaging in drug development has primarily been concentrated in CNS and oncology therapeutics. As our knowledge of PET imaging and suitable targets evolve, there is an exciting potential to widen the scope of PET applications. Inflammatory disease is one such promising area with researchers keen to explore the role inflammation plays in the pathophysiology of neuropsychiatric conditions such as neurodegenerative and mood disorders. The best characterized radioligand target has been the 18 kDa translocator protein (TSPO), which is upregulated on activated microglia (Tronel et al. 2017). Earlier results with first-generation radioligands such as 11C-PK11195 were difficult to interpret due to the low signal to noise ratio of this radioligand (Marques et al. 2019). The development of second-generation radioligands (such as [11C]PBR28,

[18F]DPA-713, and [18F]FEPPA) with significantly higher signal to noise ratios was initially hampered by unexplained variability in the signal between individuals. The identification of the rs6971 single-nucleotide polymorphism in the TSPO gene, as the cause of this variability, has enabled a simple genetic test to control for such variability across the population (Owen et al. 2012). Nevertheless, problems in the interpretation of changes in TSPO availability are due to the presence of TSPO on astroglia and endothelial cells, low concentration of TSPO in certain disease states, and lack of selectivity between pro-inflammatory M1 microglia and anti-inflammatory M2 microglia (Tronel et al. 2017), which means that TSPO PET has not been widely utilized in drug development. The identification of novel inflammatory targets and the development of radioligands to quantify them promise great benefits in the development of modulators of neuroinflammation.

The advent of fully integrated human PET/MRI systems (Heiss 2009) provided researchers the opportunity to integrate data from PET occupancy studies and MRI methods such as functional MRI (fMRI) and MR spectroscopy (MRS). While PET and MR have been very successfully integrated in the past, using dedicated cameras, the simultaneous acquisition of molecular (PET) and functional (MRI) parameters may provide insights into the effect diseases have on the brain.

> ❯ The "three pillars of drug development" envisage the confirmation of a molecule entering the tissue of interest to interact with the desired molecular target and produce a pharmacological effect at doses that are clinically meaningful. The presence of such information in early-phase drug development significantly increases the probability of success in late phases, reducing the overall burden and cost of drug development.

❯ Conclusion

PET imaging allows for precision pharmacology and is an indispensable tool in the development of drugs for CNS disorders.

It enables drug developers to directly quantify drug concentration, the level of target engagement, and assess pharmacodynamic effects of drug treatment in the living brain. These three "pillars of survival" provide development teams with a clear strategy to enable critical decision-making in the difficult and expensive world of drug development. Applied appropriately, it can lead to early identification of non-viable compounds or refine subsequent later-phase studies, saving both money and time, commodities crucial in drug development.

References

Abanades S et al (2011) Prediction of repeat-dose occupancy from single-dose data: characterisation of the relationship between plasma pharmacokinetics and brain target occupancy. J Cereb Blood Flow Metab 31:944–952. https://doi.org/10.1038/jcbfm.2010.175

Anand SS, Singh H, Dash AK (2009) Clinical applications of PET and PET-CT. Med J Armed Forces India 65:353–358. https://doi.org/10.1016/s0377-1237(09)80099-3

Chatziioannou AF (2002) Molecular imaging of small animals with dedicated PET tomographs. Eur J Nucl Med Mol Imaging 29:98–114. https://doi.org/10.1007/s00259-001-0683-3

Colasanti A et al (2012) Endogenous opioid release in the human brain reward system induced by acute amphetamine administration. Biol Psychiatry 72:371–377. https://doi.org/10.1016/j.biopsych.2012.01.027

Cummings JL et al (2013) Alzheimer's disease drug development: translational neuroscience strategies. CNS Spectr 18:128–138. https://doi.org/10.1017/s1092852913000023

Erritzoe D et al (2020) Serotonin release measured in the human brain: a PET study with [11C]CIMBI-36 and d-amphetamine challenge. Neuropsychopharmacology 45:804–810. https://doi.org/10.1038/s41386-019-0567-5

Farde L, Wiesel FA, Nordstrom AL, Sedvall G (1989) D1- and D2-dopamine receptor occupancy during treatment with conventional and atypical neuroleptics. Psychopharmacology 99 Suppl:S28–S31. https://doi.org/10.1007/bf00442555

Gallezot J-D et al (2014) Evaluation of the sensitivity of the novel α4β2* nicotinic acetylcholine receptor PET radioligand 18F-(−)-NCFHEB to increases in synaptic acetylcholine levels in rhesus monkeys. Synapse 68:556–564. https://doi.org/10.1002/syn.21767

Gunn RN et al (2012) Combining PET biodistribution and equilibrium dialysis assays to assess the free brain concentration and BBB transport of CNS drugs. J Cereb Blood Flow Metab 32:874–883. https://doi.org/10.1038/jcbfm.2012.1

Guo Q, Brady M, Gunn RN (2009) A biomathematical modeling approach to central nervous system radioligand discovery and development. J Nucl Med 50:1715–1723

Guo Q, Owen DR, Rabiner EA, Turkheimer FE, Gunn RN (2012) Identifying improved TSPO PET imaging probes through biomathematics: the impact of multiple TSPO binding sites in vivo. NeuroImage 60:902–910. https://doi.org/10.1016/j.neuroimage.2011.12.078

Heiss W-D (2009) The potential of PET/MR for brain imaging. Eur J Nucl Med Mol Imaging 36:105–112. https://doi.org/10.1007/s00259-008-0962-3

Hicks RJ, Dorow D, Roselt P (2006) PET tracer development--a tale of mice and men. Cancer Imaging 6: S102–S106. https://doi.org/10.1102/1470-7330.2006.9098

Kapur S, Zipursky R, Jones C, Remington G, Houle S (2000) Relationship between dopamine D(2) occupancy, clinical response, and side effects: a double-blind PET study of first-episode schizophrenia. Am J Psychiatry 157:514–520. https://doi.org/10.1176/appi.ajp.157.4.514

Klein G et al (2019) Gantenerumab reduces amyloid-β plaques in patients with prodromal to moderate Alzheimer's disease: a PET substudy interim analysis. Alzheimers Res Ther 11:101. https://doi.org/10.1186/s13195-019-0559-z

Marques TR et al (2019) Neuroinflammation in schizophrenia: meta-analysis of in vivo microglial imaging studies. Psychol Med 49:2186–2196. https://doi.org/10.1017/s0033291718003057

McCluskey SP, Plisson C, Rabiner EA, Howes O (2020) Advances in CNS PET: the state-of-the-art for new imaging targets for pathophysiology and drug development. Eur J Nucl Med Mol Imaging 47:451–489. https://doi.org/10.1007/s00259-019-04488-0

Morgan P et al (2012) Can the flow of medicines be improved? Fundamental pharmacokinetic and pharmacological principles toward improving Phase II survival. Drug Discov Today 17:419–424. https://doi.org/10.1016/j.drudis.2011.12.020

Narendran R et al (2004) In vivo vulnerability to competition by endogenous dopamine: Comparison of the D2 receptor agonist radiotracer (−)-N-[11C]propyl-norapomorphine ([11C]NPA) with the D2 receptor antagonist radiotracer [11C]-raclopride. Synapse 52:188–208. https://doi.org/10.1002/syn.20013

https://www.nimh.nih.gov/research/research-funded-by-nimh/therapeutics/cns-radiotracer-table.shtml.

NIMH - Approved CNS radiotracers, https://www.nimh.nih.gov/research/research-funded-by-nimh/therapeutics/cns-radiotracer-table.shtml

Ooms M et al (2016) [18F]JNJ42259152 binding to phosphodiesterase 10A, a key regulator of medium spiny neuron excitability, is altered in the presence of cyclic AMP. J Neurochem 139:897–906. https://doi.org/10.1111/jnc.13855

11

Owen DR et al (2012) An 18-kDa translocator protein (TSPO) polymorphism explains differences in binding affinity of the PET radioligand PBR28. J Cereb Blood Flow Metab 32:1–5. https://doi.org/10.1038/jcbfm.2011.147

Rabiner EA et al (2011) Pharmacological differentiation of opioid receptor antagonists by molecular and functional imaging of target occupancy and food reward-related brain activation in humans. Mol Psychiatry 16:826–835. https://doi.org/10.1038/mp.2011.29

Ridler K et al (2014) An evaluation of the brain distribution of [(11)C]GSK1034702, a muscarinic-1 (M 1) positive allosteric modulator in the living human brain using positron emission tomography. EJNMMI Res 4:66. https://doi.org/10.1186/s13550-014-0066-y

Sevigny J et al (2016) The antibody aducanumab reduces Aβ plaques in Alzheimer's disease. Nature 537:50–56. https://doi.org/10.1038/nature19323

Sevigny J et al (2017) Addendum: The antibody aducanumab reduces Aβ plaques in Alzheimer's disease. Nature 546:564. https://doi.org/10.1038/nature22809

Shotbolt P et al (2012) Within-subject comparison of [11C]-(+)-PHNO and [11C]raclopride sensitivity to acute amphetamine challenge in healthy humans. J Cereb Blood Flow Metab 32:127–136. https://doi.org/10.1038/jcbfm.2011.115

Suridjan I, Comley RA, Rabiner EA (2019) The application of positron emission tomography (PET) imaging in CNS drug development. Brain Imaging Behav 13:354–365. https://doi.org/10.1007/s11682-018-9967-0

Tauscher J et al (2018) TAK-041 modulates amphetamine-induced dopamine release in the human brain: a phase −1 [^{11}C]PHNO PET study. Poster presented at the ACNP 57th Annual Meeting in Miami Dec 9–12th 2018

te Beek ET et al (2012) In vivo quantification of striatal dopamine D2 receptor occupancy by JNJ-37822681 using [11C]raclopride and positron emission tomography. J Psychopharmacol 26:1128–1135. https://doi.org/10.1177/0269881111435251

Tronel C et al (2017) Molecular targets for PET imaging of activated microglia: the current situation and future expectations. Int J Mol Sci 18:802. https://doi.org/10.3390/ijms18040802

Vilar S, Chakrabarti M, Costanzi S (2010) Prediction of passive blood–brain partitioning: Straightforward and effective classification models based on in silico derived physicochemical descriptors. J Mol Graph Model 28:899–903. https://doi.org/10.1016/j.jmgm.2010.03.010

Wouters OJ, McKee M, Luyten J (2020) Estimated research and development investment needed to bring a new medicine to market, 2009-2018. JAMA 323:844–853. https://doi.org/10.1001/jama.2020.1166

Zamuner S et al (2010) Adaptive-optimal design in PET occupancy studies. Clin Pharmacol Ther 87:563–571. https://doi.org/10.1038/clpt.2010.9

Zhang Y, Fox GB (2012) PET imaging for receptor occupancy: meditations on calculation and simplification. J Biomed Res 26:69–76. https://doi.org/10.1016/S1674-8301(12)60014-1

Zhang L, Villalobos A (2016) Strategies to facilitate the discovery of novel CNS PET ligands. EJNMMI Radiopharm Chem 1:13. https://doi.org/10.1186/s41181-016-0016-2

Zhao YH, Abraham MH, Zissimos AM (2003) Determination of McGowan volumes for Ions and correlation with van der Waals Volumes. J Chem Inf Comput Sci 43:1848–1854. https://doi.org/10.1021/ci0341114

Application of Cognitive Test Outcomes for Clinical Drug Development

Chris J. Edgar

Contents

12.1   Introduction – 184

12.2   Use of Cognition as a Pharmacodynamic/
        Exploratory Efficacy Outcome in Early Drug
        Development – 185
12.2.1   Overview and Rationale – 185
12.2.2   MoA and Preclinical Data to Guide Test Selection – 186
12.2.3   Interpretation of Trial Data: Go/No-Go
        Decision-Making – 186

12.3   Use of Cognition to Guide Decision-Making
        About Safety and Tolerability – 189
12.3.1   Overview and Rationale – 189

12.4   Use of Cognition to Assess Clinical Benefit of
        Treatment Interventions – 193
12.4.1   Overview and Rationale – 193

        References – 195

© The Author(s), under exclusive license to Springer Nature Switzerland AG 2021
R. Schreiber (ed.), *Modern CNS Drug Discovery*, https://doi.org/10.1007/978-3-030-62351-7_12

Cognitive tests are objective measures of task performance providing insights into the potential of drug treatments to impact patient function/activities of daily living. Such tests can be useful tools in all phases of clinical trials in humans. In early development, prior to phase 3 trials, they can be used to assess pharmacodynamics and identify potential efficacy and/or safety and tolerability signals. These data can be informative for go/no-go decision-making as part of a package of evidence that the drug is well tolerated, in support of a proposed mechanism of action, or for early signals of efficacy. In phase 3 pivotal trials, cognitive tests can be employed as clinical outcome assessments (COAs) to confirm evidence of clinical benefit. For each of these contexts of use, it is important to establish their practicality/utility, validity, reliability, and ability to detect change and to ensure they are fit-for-purpose.

⊙ **Learning Objectives**

— Objective, performance-based assessments of cognition are useful drug development tools, which can be used to evaluate questions of pharmacodynamics, tolerability, and efficacy.

— Cognitive tests should be established to be fit-for-purpose, i.e., valid, reliable, able to detect change, and practical for use.

— Different evidence of their fitness-for-purpose may be needed for each context in which they are used, e.g., stage of drug development and study population.

— Cognitive tests may be included in phase 1 clinical trials whose primary aim is safety and pharmacokinetics, may be a critical component of a drug development model (such as the scopolamine challenge model of cognitive impairment), or be selected as a primary outcome for efficacy in a pivotal phase 3 clinical trial.

12.1 Introduction

The terms "cognition," and "cognitive function" are commonly used to describes those functions of the brain that support our daily behaviors and actions. These brain processes, including our abilities to concentrate, to learn and remember, and to plan and execute actions, are essential to function in daily life. Assessment of cognitive function using cognitive tests is an important way of evaluating functional status and brain health more broadly. While such abilities may be assessed by self-report, observation, or interview, in drug development they are more commonly evaluated by objective tests. Cognitive tests or tasks are assessments that require the performance of a well-defined and standardized activity by an individual, with the intent of measuring some underlying cognitive process or set of processes. These tests are distinguished from motor function and sensory tests that may assess grip strength, coordination, and visual or auditory acuity, but the potential overlap between them is an important consideration. Such assessments have a long history as tools to further the scientific understanding of brain and behavior and have been central to the fields of neuropsychology, experimental psychology, and psychopharmacology (Wasserman and Kaufman 2016). In drug development, cognitive assessments have long been used in both preclinical and clinical phases to answer a range of important questions about the effect of drugs on behavior and inform us as to their safety/tolerability, potential for further development, and efficacy (Hall et al. 1990).

A theme throughout this chapter will be that cognitive tests should be valid, reliable, and able to detect relevant changes. These properties need to be established for the intended context of use within clinical drug development including, but not limited to, the type of endpoint which is being defined (i.e., pharmacodynamic, efficacy, or safety/tolerability), the study population, the trial design, and the nature of the intervention. Here, validity will be used to refer primarily to evidence that the assessment approach measures the relevant concepts of interest (content validity) and that the outcome measure assesses those aspects or domains of cognition it purports to (construct validity). Such properties can be context dependent and so may change, for example, in healthy volunteers versus a cognitively impaired patient population. This is important since cognitive tests themselves may be applicable to multiple use cases, applications, and contexts of use. A range of cognitive ability exists even in healthy individuals and many cognitive tests have been developed to discriminate performance in the normal range from impairment, for the purpose of neuropsychological evaluation. Thus, a cognitive test may be applicable to a range of ability (e.g., healthy to cognitively impaired), a range of ages (e.g., children to older adults), and a range of applications. This may be contrasted with other concepts of interest for measurement in clinical trials, where certain signs and symptoms are evident only in the clinical population, e.g., tumors, seizures, and viral load. A consequence then is that cognitive tests may be incorporated into clinical trial protocols for reasons of efficacy and safety, or with no (single) objective specified.

In the following sections, considerations for the development and selection of cognitive assessments, related trial design, and practical examples are given to cover their use in three distinct applications:

— As pharmacodynamic or exploratory efficacy assessments in early clinical drug development
— As safety/tolerability assessments throughout a clinical drug development program
— As efficacy assessments to determine clinical benefit

12.2 Use of Cognition as a Pharmacodynamic/ Exploratory Efficacy Outcome in Early Drug Development

12.2.1 Overview and Rationale

In early clinical drug development (trials in humans), e.g., phases 1a, 1b, 2a, and 2b, cognitive assessments can be applied to detect potential pharmacodynamic and early efficacy signals. Such signals, which could constitute both improved and worsened cognitive test performances, may help in the generation of information regarding questions of "ability of the drug to cross the blood–brain barrier," "ability of the drug to interact with and modulate central nervous system (CNS) function," "association between drug pharmacokinetics and pharmacodynamics (PK/PD)," and "presence of efficacy signals." The development of CNS-penetrant drugs may or may not be intended for treatment of disorders of the CNS, e.g., neurological, neuropsychiatric, and neurodevelopmental indications. For example, many sedative/hypnotic and pain medications may be used in populations without marked CNS disease or dysfunction, such as the use of zolpidem or other "z-drugs" in the treatment of short-term insomnia (Roth et al. 1995). Importantly, the presence of worsened cognition in early development and/ or healthy volunteers may be a useful signal, since it may, for example, be important evidence of CNS penetrance and target engagement, be a manageable side effect in clinical use, and/or not ultimately be present in the intended population and with the approved dose regimen.

Early clinical development, including first-in-human (FIH) studies, e.g., single ascending dose (SAD) and multiple ascending dose (MAD) studies, designed to gather data on safety and tolerability, general pharmacokinetic (PK) and pharmacodynamic (PD) characteristics, and identify the maximum tolerated dose (MTD), will typically be conducted in heathy volunteers. Thus, the context in which cognitive assessments are applied is often in clinical trials in healthy individuals and busy trial designs, primarily intended to assess safety and PK. Therefore, it is important that cognitive assessments (tests and test batteries) are brief, multiply repeatable to detect on-set, off-set of drug action and peak effect over the course of hours and/or days, and bidirectionally sensitive to improved and worsen performance in healthy individuals (Wesnes 2002). In respect of the cognitive domains and processes of interest, this will primarily be guided by considerations around the hypothesized mechanism of action (MoA) of the drug, preclinical data from various animal models, and potentially any expectations related to the target indication(s), the so-called target product profile (TPP) (Potter 2015).

> Key facts: Cognitive tests can be useful tools in early clinical drug development to evaluate pharmacodynamics and identify potential efficacy and/or safety and tolerability signals. The tests with the greatest utility will be those which are objective, brief, repeatable, stable, and with demonstrated bidirectional sensitivity to change, even in healthy adults.

12.2.2 MoA and Preclinical Data to Guide Test Selection

It is important to be cautious with respect to the potential for overinterpretation of preclinical data and the extent to which such data can inform expectations regarding target cognitive domains and tests in humans. Cognitive domains and processes may differ in important ways between species, and between species test analogues may not be exactly comparable, or other factors e.g., around

reward and motivation could influence outcomes (Al Dahhan et al. 2019). Even within species, our assumptions regarding cognitive domains may fail to account for the challenges in developing unique probes of single processes and domains, the many interactions between cognitive processes and brain regions, the ways in which such relationships might change with disordered cognition, and more basic considerations of test difficulty/cognitive load and fatigue/test position within a broader battery. While the translatability of animal models to human cognition has been challenged and trial failure rates are undoubtedly high, there are many reasons such as target engagement and dose selection that drive failure rates. There are several examples of drugs with well-established preclinical models, reliable effects in healthy volunteers, and demonstrable efficacy in the clinic, including for example, amphetamines in attention deficit hyper activity disorder (ADHD) (Sagvolden 2000; Heal et al. 2013), and acetylcholinesterase inhibitors in Alzheimer's disease (AD) (Brinkman and Gershon 1983).

12.2.3 Interpretation of Trial Data: Go/No-Go Decision-Making

An important challenge arises in the interpretation and evaluation of early clinical trial data and how to make go/no-go decisions, i.e., whether to continue studies of the drug into later phases of development. The detection of pharmacodynamic and early efficacy signals, while encouraging, is not a guarantee of success in later-phase clinical drug development. In the field of AD drug development, for example, failure rates have been notoriously high. AD may be contrasted with other fields of drug development, with 28% of drugs approved in 2017 being oncology therapies, but 0% AD therapies (Cummings et al. 2019). Although negative trials may have resulted from multiple reasons including lack of efficacy, poor safety, or tolerability concerns, the appropriate use of cognitive tests as part of the go/no-go decision-making process is critical. Key factors include the identification of converging lines of evidence and setting a threshold for a required magnitude of clinical effect size related to the likely clinical relevance of findings for later-stage development.

The presence of a signal on a single cognitive assessment outcome or parameter may provide only weak support for a go decision. Typically, batteries of cognitive tests are employed in studies and there are multiple parameters derived and analyzed, so some findings may be due to chance. Furthermore, given the possible lack of domain/process specificity of cognitive tests, combined with the importance of functional brain networks, an effect on a single parameter may be a relatively unlikely hypothesis. Therefore, the setting of a go criterion might deliberately include a need for an effect to be identified on two or more related parameters, which might be cognitive test outcomes, or different modalities. Quantitative electroencephalography (qEEG) has been used to characterize electrophysiologic features of cognition and of neuropathologic aging. qEEG recordings of AD patients show patterns of slowing, with excessive theta (slow wave) activity in early stages of the disease and reduced alpha and beta (fast wave) activities in later stages. Established treatments for AD (the acetylcholinesterase inhibitors – see ► Sect. 2.2) have been shown to shift qEEG activity patterns toward a more normal profile and such changes have been proposed as indicative of likely treatment response (Lanctôt et al. 2003). Thus, the combination of cognitive tests and qEEG can be used to increase confidence in go decision-making. For example, the experimental drug PQ912, an inhibitor of the glutaminyl cyclase enzyme, has shown evidence for both a significant reduction in theta power in the EEG frequency analysis and a significant improvement in a N-back working memory test in a phase 2a study. Along with evidence for target occupancy and effects on exploratory biomarkers, a go decision was made for a phase 2b study (Scheltens et al. 2018).

Another means of increasing confidence may be via the setting of an effect size threshold. It is now well understood that the use of p-values alone is an insufficient means of understanding clinical trial data and that metrics that inform us of the relative size of an effect are also important (Lytsy 2018). Ultimately, drugs are not approved on the basis of statistical significance alone; they must also demonstrate a sufficiently large effect, which confers a clinical benefit for patients. The concept of the minimally clinically important difference (MCID) has been extensively discussed (Coon and Cappelleri 2016). In general, this has been defined as the smallest difference that patients or clinicians are aware of, often making use of anchor scales such as Patient or Clinician Global Impression of Change to define relevant change on a target outcome measure. In practice, such approaches have rarely been used for cognitive test outcomes since there are questions around the use of and suitability of anchor scales in this context, particularly global impressions of change, since important cognitive changes may be difficult to self-report and observe. Comparisons of effect size between groups have been an important means of evaluating trial outcomes, used in comparative and meta-analytic studies, and also suggested as method to estimate MCIDs (Angst et al. 2017). Using the common measure of effect size Cohen's d, Cohen proposed that effect sizes may be interpreted as "small" (0.2), "medium" (0.5), and "large" (0.8) effect size (Cohen 1988). Cohen immediately qualified this by highlighting that effect size is a relative metric and thus in some circumstances smaller effect sizes may increase in importance and vice versa. In reviewing various MCIDs, Angst et al. 2017 identified a range of 6–10% of the total score of a given outcome, corresponding to an effect size of 0.3–0.5 (standard deviations). They state that this effect size range could be used as a general estimate for different outcome measures and settings and applied across different trials. In the review of Leucht et al. 2012, psychiatry drugs were shown to have effect sizes in the range 0.3 (fluoxetine in major depressive disorder) to 1.0

(amphetamine in attention deficit hyperactivity disorder), when considering continuous clinical outcome assessments (Leucht et al. 2012). Effect size is easily applied to different types of outcome measure including cognitive tests. Several lines evidence converge on the use of an effect size at the level >/=0.3 as a threshold for relevant effects on cognitive/behavioral assessments. Effects at or above this level might be considered as potentially clinically meaningful and worthy of further investigation, while effects below this might be irrelevant. Such a threshold can then be applied as a go/no-go criterion.

> Go/no-go decision-making is a critical component of the drug development process and describes the use of criteria to evaluate clinical trial outcomes across PK/PD, safety and tolerability, and potential efficacy of an investigational drug. Data from cognitive tests can be employed to provide convergent evidence in support of this decision-making process and go/no-go criteria should be established for the strength of evidence in terms of the consistency and size of an effect, as well as its statistical reliability.

Case Study

Practical Example: Use of the Scopolamine Challenge Model in the Development of an H3 Antagonist for Cognitive Impairment

The scopolamine challenge model refers to the use of scopolamine to induce a cognitive impairment either in animals or in humans, through the blockade of central muscarinic receptors. Muscarinic receptor blockade, for example, via atropine and scopolamine, has long been known to produce memory impairment and scopolamine has been established as a safe and reproducible model of cognitive deficits such as those seen in AD. The potential efficacy of drugs for the treatment of AD and other forms of cognitive impairment can then be evaluated in animals and healthy adults by their ability to prevent or reverse the scopolamine-induced deficits, using cognitive tests to measure these changes, typically in a crossover study design (Buccafusco 2008).

In order to explore the potential of enhancement of histaminergic neurotransmission, or histaminergic plus cholinergic neurotransmission for improving cognition in Alzheimer's disease, the novel histamine H3 receptor inverse agonist (MK-3134) was studied along with donepezil, an acetylcholinesterase inhibitor (AChE-I), with established efficacy in the scopolamine challenge model (Cho et al. 2011). The study was a double-blind, placebo-controlled, five-period, crossover study carried out in 31 healthy adult male subjects. Cognition was assessed using a battery of cognitive tests, including a hidden pathway maze learning test designed to assess executive function. The study demonstrated a robust effect of scopolamine, for example, errors on the maze learning test increased following administration of scopolamine alone, peaking at 2-hours post-dose, with a very large effect size (Cohen's d >2). This effect was attenuated by both MK-3134 and donepezil, supporting the ability of H3 and AChE-I to reverse cognitive effects of central cholinergic blockade. While MK-3134 has not continued in development, the H3 antagonist pitolisant is now in use for the treatment of excessive daytime sleepiness in narcolepsy (Kollb-Sielecka et al. 2017) (◘ Fig. 12.1).

12

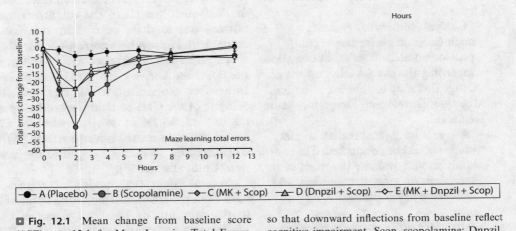

■ Fig. 12.1 Mean change from baseline score (±SE) over 12 h for Maze Learning Total Errors. Errors were multiplied by −1 for display purposes so that downward inflections from baseline reflect cognitive impairment. Scop, scopolamine; Dnpzil, donepezil; MK, MK-3134 (Cho et al. 2011)

12.3 Use of Cognition to Guide Decision-Making About Safety and Tolerability

12.3.1 Overview and Rationale

Multiple regulatory guidance documents stress the need to evaluate cognition throughout drug development, and cognition is mentioned across both early and late clinical development in the context of safety/tolerability; for example:

- ▬ EMA
  - – Extrapolation of efficacy and safety in pediatric medicine development
    - – "5.5.4. Examples where extrapolation is not recommended in diseases where there are differences in terms of neurodevelopment stages, including growth, sexual and cognitive development"
  - – Guideline on clinical investigation of medicinal products in the treatment of hypertension – Revision 4
    - – Safety aspects "8.1.5. Effects on target organ damage… Special emphasis should be placed on cognitive functions and central nervous system (CNS)-effects… especially in the elderly."
  - – Adequacy of guidance on the elderly regarding medicinal products for human use
    - – "Irritable bowel syndrome CPMP/EWP/785/97 (Mar 03). Because of the fact that IBS is not a life-threatening condition and some of the drugs for its treatment may cause adverse reactions among the elderly (e.g. cognitive symptoms), depending on the pharmacology of the product, safety studies, which specifically target the elderly, might be of interest."

(see ▶ https://www.ema.europa.eu/en/human-regulatory/research-development/scientific-guidelines)

- ▬ FDA
  - – Evaluating Drug Effects on the Ability to Operate a Motor Vehicle
    - – "Driving is a complex activity involving a wide range of cognitive, perceptual, and motor activities. Reducing the incidence of motor vehicle accidents (MVAs) that occur because of drug-impaired driving is a public health priority. A systematic effort to identify drugs that increase the risk of MVAs is a critical component of assessing drug risk and designing strategies to reduce this risk."

– Assessment of Abuse Potential of Drugs
  – "Clinical studies that evaluate cognition and performance can provide additional safety information regarding the psychoactive effects of drugs that may have abuse potential."
– Migraine: Developing Drugs for Acute Treatment
  – "A 1-year long-term pediatric safety study should be conducted. That study should evaluate the effect of treatment on growth, cognition, and endocrine development."

(see ▶ https://www.fda.gov/regulatory-information/search-fda-guidance-documents)

The use of spontaneous adverse events as well as self-report instruments, e.g., questionnaires and visual analogues scales, can be informative regarding the presence and severity of cognitive impairment. The Medical Dictionary for Regulatory Activities (MedDRA) has been used in conjunction with FDAs guidance regarding abuse potential and evaluating drug effects on the ability to operate a motor vehicle to create a list of preferred terms that represent possible effects on mental ability, e.g., "distractibility," "disturbance in attention," "somnolence," and "sedation." Several instruments such as the "Bond and Lader Visual Analogues Scales" and the "Addiction Research Center Inventory – 49 item" have been used for many years as self-ratings of drug effects (Bond and Lader 1974; Martin et al. 1971). Self-report, though important, is recognized to be insufficient for understanding drug effects due to the inability of individuals to reliably recognize relevant impairment. The FDA states that "In general, patient self-perception is not adequate for evaluating the presence or degree of driving impairment or for mitigating risk." Prior studies have shown the presence of effects on performance-based cognitive assessments that are not evident in self-ratings. For example, impairments to tests of attention (simple and choice reaction time, vigilance), tracking, and word recognition were identified at 48 hours following 4 days of dosing with haloperidol in healthy volunteers (Beuzen et al. 1999). This impairment was marked, such that six of the volunteers were advised not to drive or operate dangerous equipment. However, no corresponding effect was seen to the "Alertness" rating the from the Bond and Lader Visual Analogues Scales. In another experiment, in subjects receiving 5.25 mg of the CNS-sedating benzodiazepine midazolam, 66.7% of subjects who did not report sedation-related adverse effects (AEs) still displayed substantial cognitive dysfunction (Collie et al. 2006).

Several conceptual models of cognition have been proposed that might be applicable to the selection of cognitive domains and assessments for safety evaluation. The FDA states that the functional domains of alertness/arousal/wakefulness, attention and processing speed, reaction time/psychomotor functions, sensory-perceptual functioning, and executive functions are important for driving ability. Alternatively, an input, storage, and control model has been proposed as a "taxonomy of cognition" relevant for safety/tolerability, where input refers to perception, attention/concentration, and wakefulness; storage to working memory, long-term memory, and forgetting; and control to decision-making, planning, flexibility, and impulse control (Roiser et al. 2016). Other models that might be used to guide the selection of assessment domains include the DSM-V cognitive domains for identification of neurocognitive disorders (complex attention, executive function, learning and memory, language, perceptual-motor, social cognition) and NIH-Toolbox cognition subdomains for identification of health, success in school and work, and independence in daily functioning (executive function, episodic memory, language, processing speed, working memory, and attention (Sachdev et al. 2009; Weintraub et al. 2013).

A stepwise approach to cognitive safety/tolerability can be incorporated from FIH studies onward. This will include initial considerations around whether there is a plausible mechanism for CNS impairing effects in the MoA and/or emerging evidence in AE data that might necessitate targeted performance-based assessment. Following this, emerging evidence for CNS impairing effects in SAD/ MAD studies could prompt more detailed investigations. Investigation in these early trials can be critical as they often include higher doses than will be used in later efficacy trials and allow exploration of CNS impairing effects at higher exposures. However, there are many careful considerations in the design of such studies, as the presence and timing of effects can be hard to predict. CNS effects are intended to occur at night and cannot be assumed to be absent the following day, especially in the morning ("next-day, residual effects"); drugs or active metabolites with a long half-life can result in higher blood levels after multiple doses than after a single dose, causing greater impairment with chronic versus initial use; initial exposure may be more impairing than chronic exposure because of the development of pharmacological tolerance or habituation over time. Evaluations might also include drug-drug and drug-alcohol interaction crossover design trials with single or repeat dosing, including ethanol, sedating antihistamines, and benzodiazepine-like drugs, since interactions can lead to higher blood levels or increased pharmacodynamic effects of drugs. Additionally, the use of a positive control more generally can be important for the interpretability of any observed effects. Further on in development, the exploration of effects in potentially vulnerable populations, for example, elderly, pediatric, or target patient population(s); exploration of effects following chronic exposure; and developmental trajectory in pediatric populations can become important.

> Self-report and spontaneous adverse events are insufficient for the detection of cognitive impairment, since individuals may lack awareness of the presence and degree of impairment. Objective cognitive tests should be used to evaluate the potential for adverse cognitive effects of CNS drugs, making use of the step-wise drug development process to evaluate the impact of PK/PD, dose, exposure, population characteristics, and other relevant factors.

Case Study

Practical Example: An Early-Phase Cognitive Test Battery to Evaluate Ability to Operate a Motor Vehicle

Many classes of drug designed to act on the CNS will also disrupt cognition. One important consequence of this cognitive disruption, and a public health priority, is the potential to impair ability to operate a motor vehicle. Driving is a complex activity involving cognitive, perceptual, and motor functions. It is also well established that self-perception of impairment and ability (e.g., via patient report, interview, or spontaneous adverse event reports) is inadequate to evaluate such risks. The FDA recommend evaluating impaired driving using a tiered approach including clinical/standardized behavioral assessments with high sensitivity for detecting impairment early in development and designed to characterize clinical relevance later in development (FDA 2017). The following functional domains are considered important for driving ability by the FDA: alertness/ arousal/wakefulness; attention and processing speed; reaction time/psychomotor functions; sensory-perceptual functioning; executive functions. The proposed early-phase battery consists of four tests designed as sensitive assessments of these functional domains with well-characterized clinical relevance against established cognitive impairment and has been proposed for use in early drug development. Key properties include its brevity (approximately 15 minutes in duration), repeatability, stability, construct validity, test-retest reliability, and ability to detect relevant changes. Construct validity, the extent to which a given test outcome measures the intended domain, has been established by correlation with established tests of the same domains (Pietrzak et al. 2009; Maruff et al. 2009). Test-retest reliability, the agreement between the results of repeated measurements, is indicative of a measure's "reproducibility" (FDA 2009). The tests in the early-phase battery have previously demonstrated adequate (>0.7) to excellent reliability over various different time intervals (Falleti et al. 2006; Lim et al. 2013; Louey et al. 2014; Darby et al. 2014). Multiple studies have demonstrated the sensitivity of the early-phase battery to interventions known to impair driving ability such as fatigue, alcohol, and benzodiazepines (■ Table 12.1). This includes both highly sensitive characterization of clinical pharma-

cology in a dose- and time-dependent manner, as well as establishing effect size benchmarks for clinically relevant impairment (Falleti et al. 2003; Snyder et al. 2005; Morrison et al. 2018). In their practical application, such tests can then characterize deleterious effects on cognition of various drugs, including their magnitude and time course (■ Fig. 12.2). This information can then be used to inform considerations regarding both further development of the drug, for example, specific driving studies, and their safe use, for example, precautions regarding driving and operating heavy machinery.

■ **Table 12.1** An "early-phase" battery

Primary cognitive domains	Test paradigm	Construct validity	Ability to detect change (alprazolam 1 mg)
Reaction time/ psychomotor functions	Simple reaction time	Grooved Pegboard Dominant Hand ($r = 0.81$)	$d = 0.80$
Attention and processing speed	Choice reaction time	Trail making test part A ($r = 0.76$)	$d = 0.94$
Attention and processing speed	N-back working memory	Symbol digit modalities Test ($r = 0.81$)	$d = 0.40$
Executive functions	Hidden-pathway maze learning	NAB mazes ($r = 0.56$)	$d = 0.63$

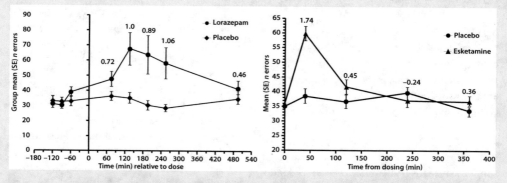

■ **Fig. 12.2** Worsening and recovery of cognitive test performance (hidden-pathway maze learning) following lorazepam and esketamine (Maruff 2019)

12

12.4 Use of Cognition to Assess Clinical Benefit of Treatment Interventions

12.4.1 Overview and Rationale

To establish clinical benefit for new therapeutic interventions, it is critical that treatment outcomes reflect those aspects of health which are clinically important and meaningful to patients. The US Food and Drug Administration (FDA)–National Institutes of Health (NIH) BEST Resource Glossary define clinical benefit as "A positive clinically meaningful effect of an intervention, i.e., a positive effect on how an individual feels, functions, or survives." As stated by Walton et al. (2015), "When clinical assessments are used as clinical trial outcomes, they are called clinical outcome assessments (COAs)." The FDA categorize COAs used for assessing clinical benefit based on the source of the data, i.e., patient (patient-reported outcome [PRO] assessments), clinician (clinician-reported outcome [ClinRO] assessments), non-clinician observer (observer-reported outcome [ObsRO] assessments], and performance by the patient of standardized, quantifiable tasks not requiring judgment by clinicians or others (performance outcome [PerfO] assessments) (Walton et al. 2015; FDA-NIH Biomarker Working Group 2018).

A PerfO assessment is a type of clinical outcome assessment (COA) that is "…based on standardized task(s) performed by a patient that is administered and evaluated by an appropriately trained individual or is independently completed." PerfO assessments include measures of physical function (e.g., mobility), sensory function (e.g., visual acuity), and cognitive function (e.g., working memory). At the time of writing, the FDA is developing a series of patient-focused drug development (PFDD) guidance documents addressing methodological approaches to facilitate and enhance the incorporation of the patient's voice in medical product development and regulatory decision-making (▶ https://www.fda.gov/drugs/development-approval-process-drugs/fda-patient-focused-drug-development-guidance-series-enhancing-incorporation-patients-voice-medical). Thus, good practices are emerging in this area, but they are yet to be fully defined. Cognitive tests then fall into the category of PerfO assessments, and when applied to establishing clinical benefit for new therapeutic interventions, i.e., as primary endpoints and/or in product labeling, they must be meaningful to patients, valid, reliable, and responsive to treatment. In addition, PerfO assessments should have a detailed manual that clearly describes the way they are to be performed and scored.

> ❯ Cognitive tests when used as a clinical outcome assessment (COA) to confirm evidence of clinical benefit fall into the category of PerfO assessments. Although good practice for their development and validation is not well established, validity, reliability, and ability to detect change as well as patient relevance are important considerations common to all COAs.

Cognitive PerfO assessments have been accepted as primary or co-primary endpoints in several pivotal phase 3 programs or included in product labeling for indications including Alzheimer's disease, cognitive impairment associated with schizophrenia, attention deficit hyperactivity disorder, heart failure, and major depressive disorder. In practice, regulatory acceptance has typically followed from the regulatory interactions that occur during protocol development and the progress from one phase of development to another. Critically, in the past 10 years or so, much greater scrutiny has been given to the extent to which COA are suited to the intended context of use (❑ Fig. 12.3), and the perception of acceptability through prior use has begun to be challenged and viewed as potentially contributing to the continued use of assessments that are not fit-for-purpose, while at the same time slowing the development of revised or novel outcomes. What is deemed acceptable to the regulator has also changed markedly, with the introduction of a qualification pathway, the 2009 PRO

Division of neurology products (DNP)				
Disease/condition	Concept	COA tool & type	COA context of use	Drug name & approval date
Alzheimer's disease (AD)	Day-to-day function	Modified Alzheimer's disease cooperative study–Activities of daily living: **ClinRO**	Adult patients with moderate to severe AD	Namenda (memantine hydrochloried) *October 16, 2003*
	Cognitive function	Severe impairment battery (SIB): **PerfO**		
	Global impression	Clinical global impression (CGI-C): **ClinRO**		

◾ **Fig. 12.3** FDA COA compendium excerpt (▶ https://www.fda.gov/drugs/development-resources/clinical-outcome-assessment-compendium, April 2020)

guidance, development of the Division of Clinical Outcome Assessment (DCOA), and the ongoing development of new guidance. Thus, significant changes and developments in the acceptability and application of cognitive PerfO can be expected in the coming years.

Case Study

Practical Example: The MATRICS Consensus Cognitive Battery as a Co-primary Endpoint for Cognitive Impairment Associated with Schizophrenia

Extensive evidence has shown cognitive impairment to be a core symptom of schizophrenia that has a negative impact on function. The April 2004 MATRICS/FDA/NIMH workshop was held in order to begin to develop guidelines for the design of clinical trials of cognitive-enhancing drugs for people with schizophrenia. One important consequence arising from this and subsequent development work was a proposed battery of performance-based outcome (PerfO) assessments (cognitive tests), the "MCCB" to assess cognitive treatment effects in people with schizophrenia (Green et al. 2004). The use of the MATRICS Consensus Cognitive Battery (MCCB) as a co-primary endpoint for efficacy has been accepted by the FDA for previous pivotal clinical trials assessing cognition in schizophrenia (e.g., EVP-6124/Encenicline: NCT01716975 and NCT01714661). However, no formal FDA guidelines exist in this area. EMA describes the battery as "acceptable but other, comparable, test batteries may also be used provided their validity is demonstrated" (EMA 2012). Inclusion of the MCCB is frequent but not universal in such clinical trials, with other neurocognitive test batteries such as BACS, CANTAB and Cogstate also commonly included (Keefe et al. 2013). Importantly, since no pivotal program using the MCCB has yet been successful, no approval has yet been granted on the basis of a cognitive test battery. The conceptual framework for the MCCB incorporates the assessment of seven cognitive domains, selected on the basis of a review of the literature and input from experts. These domains are: working memory, attention/vigilance, verbal learning and memory, visual learning and memory, reasoning and problem-solving, speed of processing, and social cognition (Kern et al. 2004). Well-established neuropsychological tests of these domains were then selected based on their test-retest reliability, utility as a repeated measure, relationship to functional outcome, potential response to pharmacologic agents, and practicality/tolerability.

Following the development of the MCCB and its use in clinical trials, significant concerns have been raised regarding patient and trial burden, particularly related to the duration of the battery at around 90 minutes, the process to select the tests themselves, and the use of survey methodologies. Recognition of the issue of assessment burden in trials is increasing and development of shorter assessments is widely viewed as important. The FDA has directly highlighted this issue in COA development funding: "COAs and endpoints for use in schizophrenia trials, including but not limited to, shortened versions of current instruments, as appropriate" (▶ https://grants.nih.gov/grants/guide/rfa-files/RFA-FD-19-006.html). In addition, researchers involved in the MATRICS initiative have stated that the evaluation of computerized/digital tests was largely discounted at the time but would be a critical consideration now.

12

▶ Conclusions

Cognition underpins human behavior and our ability to perform functions in everyday life. Drug treatments have the potential to impair cognitive function, e.g., impacting our ability to drive, and to improve cognitive function, e.g., the ability of an Alzheimer's treatment to improve patient's capacity to perform activities of daily living. Therefore, understanding the impact of a drug treatment on cognition may be particularly important. Objective cognitive tests provide insights into human behavior beyond what can be learnt from self-report and observation and can be useful tools throughout all phases of clinical drug development.

In early human trials, they can help understand and evaluate pharmacodynamics and identify potential efficacy and/or safety and tolerability signals. Here the tests with the greatest utility will be those which are objective, brief, repeatable, stable, and with demonstrable bidirectional sensitivity to change, even in healthy adults. This information can be used to support go/no-go decision-making from one phase of development of an investigational drug to the next. Data from cognitive tests may provide convergent evidence in support of this decision-making process and go/no-go criteria should be established for the strength of evidence in terms of consistency and size of an effect, as well as its statistical reliability.

When used as a clinical outcome assessment (COA) to confirm evidence of clinical benefit in a pivotal phase 3 trial, cognitive tests fall into the category of PerfO assessments. Although good practice for their development and validation is not well established, validity, reliability, and ability to detect change and patient relevance must be understood in order to draw valid conclusions regarding clinical benefit.

In conclusion, cognitive tests are useful and flexible drug development tools with many applications. Given this diversity of uses, careful planning and consideration are needed to define the research question(s)

being addressed, the context within which the tests will be used, and how the different possible outcomes from a trial should be interpreted. A lack of thoroughness and care, may lead to the selection of tests which are not fit-for-purpose and/or an inability to fully interpret trial data.

References

Al Dahhan NZ, De Felice FG, Munoz DP (2019) Potentials and pitfalls of cross-translational models of cognitive impairment. Front Behav Neurosci 13:48. https://doi.org/10.3389/fnbeh.2019.00048

Angst F, Aeschlimann A, Angst J (2017) The minimal clinically important difference raised the significance of outcome effects above the statistical level, with methodological implications for future studies. J Clin Epidemiol 82:128–136. https://doi.org/10.1016/j.jclinepi.2016.11.016

Beuzen JN, Taylor N, Wesnes K, Wood A (1999) A comparison of the effects of olanzapine, haloperidol and placebo on cognitive and psychomotor functions in healthy elderly volunteers. J Psychopharmacol 13(2):152–158. https://doi.org/10.1177/026988119901300207

Bond A, Lader M (1974) The use of analogue scales in rating subjective feelings. Br J Med Psychol 47(3):211–218. https://doi.org/10.1111/j.2044-8341.1974.tb02285.x

Brinkman SD, Gershon S (1983) Measurement of cholinergic drug effects on memory in alzheimer's disease. Neurobiol Aging 4(2):139–145. https://doi.org/10.1016/0197-4580(83)90038-6

Buccafusco J (2008) The revival of scopolamine reversal for the assessment of cognition-enhancing drugs. https://doi.org/10.1201/noe1420052343.ch17

Cho W, Maruff P, Connell J et al (2011) Additive effects of a cholinesterase inhibitor and a histamine inverse agonist on scopolamine deficits in humans. Psychopharmacology 218(3):513–524. https://doi.org/10.1007/s00213-011-2344-y

Cohen J (1988) Statistical power analysis for the behavioral sciences. Routledge

Collie A, Maruff P, Snyder PJ, Darekar A, Huggins JP (2006) Cognitive testing in early phase clinical trials: outcome according to adverse event profile in a Phase I study. Hum Psychopharmacol 21(7):481–488. https://doi.org/10.1002/hup.799

Coon CD, Cappelleri JC (2016) Interpreting change in scores on patient-reported outcome instruments. Ther Innov Regul Sci 50(1):22–29. https://doi.org/10.1177/2168479015622667

Cummings J, Feldman HH, Scheltens P (2019) The "rights" of precision drug development for Alzheimer's disease. Alzheimers Res Ther 11(1):76. https://doi.org/10.1186/s13195-019-0529-5

Darby DG, Fredrickson J, Pietrzak RH, Maruff P, Woodward M, Brodtmann A (2014) Reliability and usability of an internet-based computerized cognitive testing battery in community-dwelling older people. Comput Human Behav 30:199–205. https://doi.org/10.1016/j.chb.2013.08.009

EMA. Guideline on Clinical Investigation of Medicinal Products, Including Depot Preparations in the Treatment of Schizophrenia Guideline on Clinical Investigation of Medicinal Products, Including Depot Preparations, in the Treatment of Schizophrenia. Vol EMA/CHMP/.; 2012:40072/2010 Rev. 1

Falleti MG, Maruff P, Collie A, Darby DG, McStephen M (2003) Qualitative similarities in cognitive impairment associated with 24 h of sustained wakefulness and a blood alcohol concentration of 0.05%. J Sleep Res 12(4):265–274. https://doi.org/10.1111/j.1365-2869.2003.00363.x

Falleti M, Maruff P, Collie A, Darby D (2006) Practice effects associated with the repeated assessment of cognitive function using the CogState battery at 10-minute, one week and one month test-retest intervals. J Clin Exp Neuropsychol 28(7):1095–1112. https://doi.org/10.1080/13803390500205718

FDA (2009) Patient-reported outcome measures: use in medical product development to support labeling claims guidance for industry, p 1–39. https://doi.org/10.1111/j.1524-4733.2009.00609.x

FDA (2017) Evaluating Drug Effects on the Ability to Operate a Motor Vehicle Guidance for Industry. https://www.fda.gov/downloads/Drugs/GuidanceComplianceRegulatoryInformation/Guidances/UCM430374.pdf

FDA-NIH Biomarker Working Group (2018) FDA BEST Glossary. In: BEST (Biomarkers, EndpointS, and Other Tools) Resource

Green MF, Nuechterlein KH, Gold JM et al (2004) Approaching a consensus cognitive battery for clinical trials in schizophrenia: the NIMH-MATRICS conference to select cognitive domains and test criteria. Biol Psychiatry 56(5):301–307. https://doi.org/10.1016/j.biopsych.2004.06.023

Hall ST, Puech A, Schaffler K, Wesnes K, Gamzu ER (1990) Group report 3: early clinical testing of cognition enhancers: prediction of efficacy. Pharmacopsychiatry 23(Suppl 2):57–58; discussion 59. https://doi.org/10.1055/s-2007-1014534

Heal DJ, Smith SL, Gosden J, Nutt DJ (2013) Amphetamine, past and present - A pharmacological and clinical perspective. J Psychopharmacol 27(6):479–496. https://doi.org/10.1177/0269881113482532

Keefe RSE, Buchanan RW, Marder SR et al (2013) Clinical trials of potential cognitive-enhancing drugs in schizophrenia: what have we learned so far? Schizophr Bull 39(2):417–435. https://doi.org/10.1093/schbul/sbr153

Kern RS, Green MF, Nuechterlein KH, Deng BH (2004) NIMH-MATRICS survey on assessment of neuro-cognition in schizophrenia. Schizophr Res 72(1):11–19. https://doi.org/10.1016/j.schres.2004.09.004

Kollb-Sielecka M, Demolis P, Emmerich J, Markey G, Salmonson T, Haas M (2017) The European Medicines Agency review of pitolisant for treatment of narcolepsy: summary of the scientific assessment by the Committee for Medicinal Products for Human Use. Sleep Med 33:125–129. https://doi.org/10.1016/j.sleep.2017.01.002

Lanctôt KL, Herrmann N, LouLou MM (2003) Correlates of response to acetylcholinesterase inhibitor therapy in Alzheimer's disease. J Psychiatry Neurosci 28(1):13–26

Leucht S, Hierl S, Kissling W, Dold M, Davis JM (2012) Putting the efficacy of psychiatric and general medicine medication into perspective: review of meta-analyses. Br J Psychiatry 200(2):97–106. https://doi.org/10.1192/bjp.bp.111.096594

Lim YY, Jaeger J, Harrington K et al (2013) Three-month stability of the cogstate brief battery in healthy older adults, mild cognitive impairment, and alzheimer's disease: results from the Australian imaging, biomarkers, and lifestyle-rate of change substudy (AIBL-ROCS). Arch Clin Neuropsychol 28(4):320–330. https://doi.org/10.1093/arclin/act021

Louey AG, Cromer JA, Schembri AJ et al (2014) Detecting cognitive impairment after concussion: sensitivity of change from baseline and normative data methods using the CogSport/Axon cognitive test battery. Arch Clin Neuropsychol 29(5):432–441. https://doi.org/10.1093/arclin/acu020

Lytsy P (2018) P in the right place: revisiting the evidential value of P-values. J Evid Based Med 11(4):288–291. https://doi.org/10.1111/jebm.12319

Martin WR, Sloan JW, Sapira JD, Jasinski DR (1971) Physiologic, subjective, and behavioral effects of amphetamine, methamphetamine, ephedrine, phenmetrazine, and methylphenidate in man. Clin Pharmacol Ther 12(2):245–258. https://doi.org/10.1002/cpt1971122part1245

Maruff P (2019) Use of cognition to guide decision making about the safety and efficacy of drugs in early-phase clinical trials. In: Nomikos GG, Feltner DE (eds) Translational medicine in CNS drug development. Academic Press

Maruff P, Thomas E, Cysique L et al (2009) Validity of the CogState brief battery: relationship to standardized tests and sensitivity to cognitive impairment in mild traumatic brain injury, schizophrenia, and AIDS dementia complex. Arch Clin Neuropsychol 24(2):165–178. https://doi.org/10.1093/arclin/acp010

Morrison RL, Fedgchin M, Singh J et al (2018) Effect of intranasal esketamine on cognitive functioning in healthy participants: a randomized, double-blind, placebo-controlled study. Psychopharmacology 235(4):1107–1119. https://doi.org/10.1007/s00213-018-4828-5

Pietrzak RH, Olver J, Norman T, Piskulic D, Maruff P, Snyder PJ (2009) A comparison of the CogState schizophrenia battery and the measurement and treatment research to improve cognition in schizophrenia (MATRICS) battery in assessing cognitive impairment in chronic schizophrenia. J Clin Exp Neuropsychol 31(7):848–859. https://doi.org/10.1080/13803390802592458

Potter WZ (2015) Optimizing early Go/No Go decisions in CNS drug development. Expert Rev Clin Pharmacol 8(2):155–157. https://doi.org/10.1586/17512433.2015.991715

Roiser JP, Nathan PJ, Mander AP, Adusei G, Zavitz KH, Blackwell AD (2016) Assessment of cognitive safety in clinical drug development. Drug Discov Today 21(3):445–453. https://doi.org/10.1016/j.drudis.2015.11.003

Roth T, Roehrs T, Vogel G (1995) Zolpidem in the treatment of transient insomnia: a double-blind, randomized comparison with placebo. Sleep 18(4):246–251. https://doi.org/10.1093/sleep/18.4.246

Sachdev P, Andrews G, Hobbs MJ, Sunderland M, Anderson TM (2009) Neurocognitive disorders: cluster 1 of the proposed meta-structure for DSM-V and ICD-11. Psychol Med 39(12):2001–2012. https://doi.org/10.1017/S0033291709990262

Sagvolden T (2000) Behavioral validation of the spontaneously hypertensive rat (SHR) as an animal model of attention-deficit/hyperactivity disorder (AD/HD). Neurosci Biobehav Rev 24(1):31–39. https://doi.org/10.1016/S0149-7634(99)00058-5

Scheltens P, Hallikainen M, Grimmer T et al (2018) Safety, tolerability and efficacy of the glutaminyl cyclase inhibitor PQ912 in Alzheimer's disease: results of a randomized, double-blind, placebo-controlled phase 2a study. Alzheimers Res Ther 10(1):107. https://doi.org/10.1186/s13195-018-0431-6

Snyder PJ, Werth J, Giordani B, Caveney AF, Feltner D, Maruff P (2005) A method for determining the magnitude of change across different cognitive functions in clinical trials: the effects of acute administration of two different doses alprazolam. Hum Psychopharmacol 20(4):263–273. https://doi.org/10.1002/hup.692

Walton MK, Iii JHP, Hobart J et al (2015) Clinical outcome assessments: conceptual foundation — Report of the ISPOR clinical outcomes assessment – Emerging good practices for outcomes research task force. Value Heal 18:741–752. https://doi.org/10.1016/j.jval.2015.09.2863

Wasserman JD, Kaufman AS (2016) A history of mental ability tests and theories. In: Barr WB, Bieliauskas LA (eds) The Oxford handbook of history of clinical neuropsychology. Oxford University Press. https://doi.org/10.1093/oxfordhb/9780199765683.013.32

Weintraub S, Dikmen SS, Heaton RK et al (2013) Cognition assessment using the NIH Toolbox. Neurology 80(11 Suppl 3):S54–S64. https://doi.org/10.1212/wnl.0b013e3182872ded

Wesnes K (2002) Assessing cognitive function in clinical trials: latest developments and future directions. Drug Discov Today 7(1):29–35. https://doi.org/10.1016/S1359-6446(01)02068-2

Exciting Research in the Field of Sexual Psychopharmacology: Treating Patients with Inside Information

Paddy Janssen

Contents

13.1   General Introduction – 200

13.2   Sexual Psychopharmacology – 201

13.3   Endophenotypes in the Rat – 203

13.4   Endophenotypes in Humans – 203

13.5   Genetic Research of the Intravaginal Ejaculation
        Latency Time – 204

13.6   The Influence of Viagra (the Pharmaceutical
        Industry) – 206

13.7   Sexual Dysfunctions in Women – 206

13.8   Discussion – 207

        References – 208

© The Author(s), under exclusive license to Springer Nature Switzerland AG 2021
R. Schreiber (ed.), *Modern CNS Drug Discovery*, https://doi.org/10.1007/978-3-030-62351-7_13

In this chapter, the use of pharmacogenetics will be clarified, with a focus on sexual psychopharmacology in men and women. Multidisciplinary collaboration is crucial in sexual psychopharmacological research. As with many disciplines, the boundaries of sexual psychopharmacology are not sharply limited. Knowledge of the phenotype is essential when conducting research, applying genetics, and using psychopharmaceuticals, whereby a clearly defined (endo)phenotype and careful methodology are crucial. Furthermore, this approach reduces the ultimate risk of pharmacotherapeutic overtreatment within a certain indication and can reduce pharmacotherapeutic failure. Biological variation makes a theoretical "one-size-fits-all" strategy illogical. Practical experience so far shows that the use of psychotropic drugs within a total population results in a low success rate. Current practice is characterized by trial and error in which of the many psychotropic drugs that can be used for an indication, it is not clear in advance what will determine the chance of success at an individual level (Sinyor et al. 2010).

Recent developments show that it is possible to use pharmacotherapy at an individual level using additional phenotypic and genotypic information (Tuiten et al. 2018). The purpose of this chapter is to provide insight into this new approach using recent developments in sexual psychopharmacology.

Learning Objectives
- Insight into the field of sexual psychopharmacology
- Use new insights to optimize and personalize psychopharmacological interventions
- Use of companion diagnostics
- Insight into the predictive value of pharmacogenetics

13.1 General Introduction

From the 1980s to the present, hundreds of publications have been published describing pharmacogenetics within the field of psychopharmacology. These publications mainly describe the relationship between cytochrome P450 enzyme system (CYP450) and drug metabolism. Despite over 30 years of experience, there is an ongoing discussion about the burden of proof, and uniform clinical implementation is lacking.

In relation to the use of selective serotonin reuptake inhibitor (SSRIs) (and tricyclic antidepressants (TCAs)), the results, new insights, and applicability are limited after the introduction of pharmacogenetics. An important cause of the still limited applicability and predictability is the current focus on pharmacokinetics. The predictive value therefore mainly relates to the concentration due to the metabolism of the drug. Furthermore, the question is whether the effect of agents without a concentration–effect relationship can be predicted with a pharmacogenetic test that provides insight into the pharmacokinetics. Toxicity and adherence can be an additional argument for determining CYP metabolism.

Within other medical professions such as oncology, the focus is on pharmacodynamics, and it has been accepted in recent years that associated diagnostics are being developed.

The idea of combining drugs and diagnostics is not new. When tamoxifen was developed in 1970s for the treatment of breast cancer, data on estrogen-receptor status were correlated with the treatment outcome. Based on the phase II study, published in 1976, the investigators concluded a high degree of correlation between response and positive estrogen receptor assay suggesting the value of the diagnostic test as a means to select patients for tamoxifen treatment. Despite the fact that this conclusion was drawn more than 40 years ago, the adaptation of the drug-diagnostic co-development has been slow and it is only within the last years that it has gained acceptance in oncology. A more recent example of CDx is screening for the presence of the human epidermal growth factor receptor (HER)-2 prior to the use of trastuzumab (Herceptin®) in patients with metastatic breast cancer (Spector and Blackwell 2009). Trastuzumab treatment is effective in patients with HER-2-positive breast cancer. HER2 FISH PharmDx™ and HER2 FISH PharmDx™ kits are both FDA-approved companion diagnostics for trastuzumab (Herceptin®).

> **Definition**
>
> Companion Diagnostic: A companion diagnostic (CDx) is a diagnostic test used as a companion to a therapeutic drug to determine its applicability to a specific person.

Shifting the focus to pharmacodynamics provides new insights that can be used in personalized medicine with psychotropic drugs. With regard to psychopharmacology, first steps of companion diagnostics are introduced in 2018 within the field of sexual psychopharmacology (Tuiten et al. 2018).

13.2 Sexual Psychopharmacology

The development of sexual psychopharmacology can be roughly divided into three periods.

The first period lasted from 1919 to 1933, the period of the medical specialists.

The second period lasted from 1970 to 2000, the period of the neuroscientists.

We are now living in the third period, which started around 2000, the period of the pharmaceutical industry (Fig. 13.1).

A case of premature ejaculation has already been described in Greek mythology. In this case Athena visited Hephaestus desirous to get arms. He, being forsaken by Aphrodite, fell in love with Athena and began to pursue her; but she fled. When he got near her with much ado (for he was lame), he attempted to embrace her; but she being a chaste virgin, would not submit to him, and he dropped his seed on the leg of the goddess. In disgust, she wiped off the seed with wool and threw it on the ground. Erichthonius, a mythical king of Athens was conceived when the seed of Hephaestus fell on Gaea, the goddess of earth. Ehrentheil published this case in 1974 entitled "A case of premature ejaculation in Greek mythology." This case was the inspiration for the famous Italian painter Paris Bordone for his painting "Athena Scorning the Advances of Hephaestus."

At the beginning of the last century, the seed of sexual psychopharmacology was sown at the Institut für Sexualwissenschaft in Berlin. The endocrinologist Bernhard Schapiro and Magnus Hirschfeld developed the drug "Testifortan" for the treatment of erectile dysfunction and the drug "Präjaculin" for the treatment of hypersexuality (Dose and Herrn 2006; Hirschfeld and Testifortan 1927; Schapiro 1932). Both products were manufactured by the Hamburg-based pharmaceutical company Promonta. In 1933, these studies within the Institut fur Sexualwissenschaft came to an end. From this time until the 1970s, no fundamental research has been carried out in sexual psychopharmacology.

The second period begins in the 1970s with some publications by medical specialists, now mainly from psychiatrists, on the successful use of tricyclic antidepressants and other psychotropic drugs in premature ejaculation (Baldwin et al. 2015; Clayton et al. 2014; Eaton 1973). Attempts have also been made to treat erectile dysfunction during this time (Virag 1982).

Fundamental research also started to take off in the late 1980s (Ahlenius et al. 1980). Some psychopharmacologists will study the sexual behavior of the laboratory rat and show that the neurotransmitter serotonin plays a role in ejaculation. In the Netherlands in the late 1980s and early 1990s, this fundamental research was carried out by neuroscientists such as Prof. dr. Berend Olivier, Prof Dr. ir. Chose Slob, Dr. Jan Mos, Dr. Hemmie Berendsen, and Dr. Rik Broekkamp (Coolen et al. 1996; de Jong et al. 2005, 2006a; Gower et al. 1986; Haensel et al. 1991). Also neuroanatomical studies, as performed by Dr. Jan Veening, provided new insights into the neuronal circuits and neurotransmitters that play a role in the sexual functioning of the rat (de Jong et al. 2006b).

In the early 1990s, sexual psychopharmacology received a tremendous boost after the introduction of the selective serotonin reuptake inhibitors, the SSRIs, and the accidental discovery that this new class of antidepressants significantly delays ejaculation (Baldwin et al. 2015; Clayton et al. 2014).

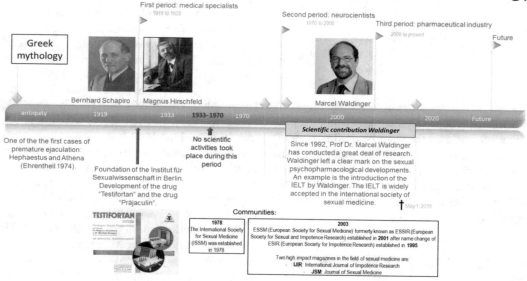

□ Fig. 13.1   Timeline with some important developments in sexual psychopharmacology

Since 1992, Prof. Dr. Marcel Waldinger has conducted a great deal of research into the ejaculation-delaying effect of SSRIs and in particular that of paroxetine (Waldinger et al. 1994, 1998a, b, 2002, 2007; Waldinger and Olivier 1998). With the findings of studies, Waldinger left a clear mark on the sexual psychopharmacological developments during this period. This period is a special period in history. Special, because during this period, an evidence-based design and methodology for drug studies were devised and carried out by independent researchers, without the intervention of the pharmaceutical industry (Waldinger et al. 2004; Waldinger 2003).

In order to measure the delay in ejaculation exactly, Waldinger asked the participating men in his studies to do this with a stopwatch. The time to measure was called the IELT, the intravaginal ejaculation latency time, i.e., the time that elapses from the start of vaginal penetration to the start of intravaginal ejaculation.

The systematic application of this simple measurement method enabled the clinical study of premature ejaculation and the study of the ejaculation-delaying effect of antidepressants to take off (Serretti and Chiesa 2009; Waldinger 2002). Without the use of this precise and therefore more objective measurement method, several new ejaculation phenomena would have been difficult to uncover.

For example, Waldinger et al. discovered that the ejaculation time in a cohort of men with premature ejaculation and in the general population is not divided according to a Gaussian curve (Waldinger et al. 2005a, 2009). Waldinger et al. also found evidence of endophenotypes in laboratory rats and Waldinger. found time and time again that SSRIs differ in the degree to which they delay ejaculation. It has been conclusively established in human and animal experiments that paroxetine causes the strongest delay in ejaculation of all antidepressants. The antidepressant fluvoxamine causes little delay in men. Animal studies have also shown that the antidepressant fluvoxamine causes little delay.

The animal experimental research of Dr. Trynke de Jong has provided indications that these differences between the SSRIs are caused by a different effect of the SSRIs on desensitization of the 5-HT1A receptor.

The studies with SSRIs and other antidepressants were done in men with the primary form of premature ejaculation, also known as lifelong premature ejaculation (LPE). These are men who have a rapid ejaculation with every woman, with every sexual contact and from the first sexual contacts in puberty or adolescence.

The stopwatch studies showed that 90% of men with a primary premature ejaculation experience ejaculation within 1 minute of penetration. Of this group of men, 80% ejaculates within 30–40 seconds, and 50% even within 15 seconds.

In the DSM-IV-TR and the American Psychiatric Association's classification system (APA), premature ejaculation is defined as ejaculation that occurs "shortly before, during, or shortly after" vaginal penetration (McMahon et al. 2008). For years, however, the question was what was actually meant by the two words "shortly after." According to patients' stopwatch measurements, this was approx. 1 minute. This figure of approximately 1 minute was included in the new definition of Lifelong Premature Ejaculation by the International Society for Sexual Medicine (ISSM) in 2008. The time criterion found by Waldinger has also been included in the newly published DSM-V (American Psychiatric Association 2000; McMahon et al. 2008; Waldinger et al. 1998c, 2005b; Waldinger and Schweitzer 2006a, b, 2008; Waldinger 2006, 2007).

13.3 Endophenotypes in the Rat

Another important finding was the existence of endophenotypes with regard to ejaculation time, the IELT. The existence of this is by Prof. dr. Dr. Berend Olivier and Prof Dr. ir. Waldinger postulated in the mid-1990s based on sexual behavior studies in laboratory rats. As an example of translational research, Oliver and Waldinger translated this finding into the humane situation (Chan et al. 2008). They argued that ejaculation time is likely to be distributed along a continuum in the general male population. That is, there is a gradual transition from men who are ejaculating very quickly, on average, or with difficulty.

These endophenotypes are determined by neurobiological and genetic, and thus biological, factors.

This hypothesis was first investigated in male rats by Dr. Tommy Pattij. In studies, various parameters of the sexual behavior of the rat are objectively measured (Mos et al. 1991; Olivier et al. 1998, 2006; Pattij et al. 2005b).

These parameters are the frequency of ejaculations, the number of ascents, and intromissions in the female, and the latency time before ejaculation occurs. In each group of rats, it appears that the ejaculation time is continuously divided with a fixed percentage of approx. 10% that ejaculates very quickly and approx. 10% that ejaculates only after a lot of effort. In short, a continuum of ejaculation time actually appeared to exist in rats (Pattij et al. 2005a).

In a pooled population of male Wistar rats (total $N > 1300$), obtained from 15 subsequent mating test experiments, male rats on either side of the Gaussian distribution were defined as "slugging" (0–1 ejaculation), "normal" (2–3 ejaculations), and "rapid" ejaculators (4–5 ejaculations).

These studies showed that rats can also suffer from "premature ejaculation."

The work of Tommy Pattij is mainly focused on male sexual function. Different groups are active in the field of sexual psychopharmacology. A group led by Jim Pfauss also examines animal models of female sexual functioning. Jim Pfauss has done a lot of animal experimental research into sexual behavior (Pfaus 2009).

13.4 Endophenotypes in Humans

Two large-scale stopwatch studies were conducted in an unselected cohort from the general male population in four European countries and in the United States. Both studies showed that ejaculation time (IELT) did not behave according to a Gaussian curve but had a skewed distribution from which a median IELT of 5.4 and 6.0 minutes could be calculated, respectively. By setting the IELT values below the 0.5 and 2.5 percentile relative to the male population, it was found that men with an IELT of 1–1.5 minutes had a statistically different value from the rest of the male population. Both cohort studies confirmed our hypothesis from the study with laboratory rats that the ejaculation time manifests itself in a non-Gaussian continuum in humans and rats with which the biological character of the ejaculation time can also be solidified. The phenotype distribution described by Waldinger was confirmed years later in large-scale population studies by Guliano and Patrick.

Within Caucasian populations, the phenotype distribution is reproducible and can be expressed with a population density formula (Janssen and Waldinger 2016, 2019a, 2019b) (⬛ Fig. 13.2).

Various population studies show that approximately 1% of the male population has an IELT of <1 min and can be defined as LPE according to the definition of Waldinger et al. The population frequency distribution of LPE men is statistically significantly different from the IELT distribution in the overall population.

> The phenotype distribution of the IELT in the general male population described by Waldinger was confirmed in different large-scale population studies and can be expressed using a probability density formula.

Approximately 1% of the male population has an IELT of <1 min and can be defined as LPE according to the definition of Waldinger et al.

13.5 Genetic Research of the Intravaginal Ejaculation Latency Time

In 1998, Waldinger postulated that the primary ejaculatio praecox, and in particular the duration of the IELT, is partly genetically determined. In a study, 237 men with lifelong PE were asked whether they were willing to ask family members about the occurrence of PE. Due to embarrassment, only 14 of them consented to ask male relatives about their ejaculation time. These 14 men found a total of 11 first-degree male relatives' available information for direct personal interview.

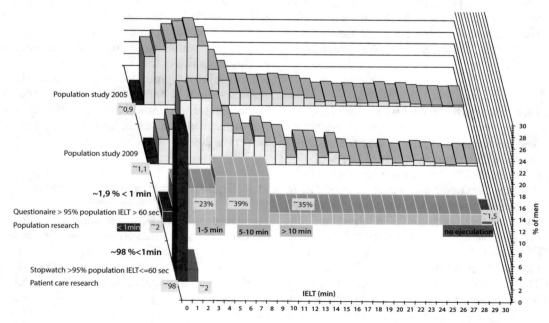

⬛ **Fig. 13.2** IELT distribution of men in the general male population. Different population studies show similar results in the general male population. Dutch-Caucasian men with lifelong premature ejaculation show a strongly deviating distribution with regard to the IELT

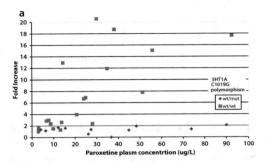

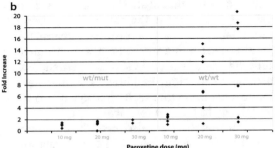

Fig. 13.3 Fold increase of the IELT as a function of the paroxeine plasma concentration **a** and paroxetine dose **b**. Within the concentration and dose data, deviation is shown based on the C-1019G polymorphism of the 5-HT1A receptor. Non-response is defined as a FI less than 2, and response is defined as an FI more than 2. Response is related to the 5-HT1A receptor C1019G polymorphism and is not related to the paroxetine concentration and/or paroxetine dose

Indeed, ten of them also ejaculated within 1 minute or less. The calculated risk in this small selected group of men to have a first-degree relative with PE was 91%. The odds of family occurrence are therefore much higher compared to a suggested population prevalence rate of 2–39%. Moreover the high odds indicate a familial occurrence of the syndrome far higher than by chance alone (Waldinger et al. 1998d).

In 2006, genetic research was started in men with primary ejaculation praecox. In the first study of 89 men, ejaculation time, IELT, was measured at home with the stopwatch. Laboratory research showed that genetic polymorphism of the 5-HT transporter has an influence on ejaculation time (Janssen et al. 2009). The activity of the serotonin transporter determines how much serotonin is present in the synapse and thus has a role in serotonergic neurotransmission. Purely and only by the very accurate measurements of the stopwatch, it could be shown that in the group of men who ejaculate within 1 minute, men with an LL genotype have a 100% faster ejaculation than men with an SS genotype. Without the use of the stopwatch, which made it possible to take measurements with precision, the difference in ejaculation time between the LL and SS genotypes would most likely have been difficult.

With regard to treatment outcome of LPE to paroxetine treatment, a study in 2014 found that there was no association between paroxetine serum concentration and paroxetine-induced fold increase (IELT after treatment divided by IELT before treatment) of the geometric mean IELT (Janssen et al. 2014d). In ◼ Fig. 13.3, the response expressed as fold increase as a function of concentration and dose is shown. It was found that serum concentrations of paroxetine are higher in men with CYP2D6 *3 and*4 variations, but as there is no serum paroxetine concentration and IELT effect relationship, this genotype for paroxetine metabolism is not relevant for paroxetine-induced ejaculation delay. With regard to treatment outcome, genetic research demonstrated that all non-responders to paroxetine treatment were heterozygote (GC genotype) for the 5-HT1A receptor C-1019G variation. In contrast, all paroxetine responders had the wild-type (CC) genotype for the 5-HT1A receptor C-1019 G variation.

Further it was found that paroxetine induced serum prolactin concentration changes in men with lifelong PE. In the paroxetine non-responders, prolactin concentrations tend to become reduced, whereas in paroxetine responders, prolactin concentrations increased. The SmPC of paroxetine describes elevation of prolactin. The study shows that prolactin change can be differentiated based on response. Without differentiation, an increase in prolactin is seen in the total population.

These findings show that precise methodological clinical outcome definition is needed in this field of research (Janssen et al. 2014a, b, c, e). In future, precise clinical definition and additional diagnostics can improve clinical outcome and at the same time prevent redundant pharmacotherapeutical treatment.

It is clear that a man who always accidentally ejaculates within 15 seconds actually has a rapid ejaculation. Waldinger's studies have made it plausible that such rapid ejaculation is very likely to have a biological cause.

However, there are also men who complain about rapid ejaculation when the ejaculation time is not fast objectively. In 2006, Waldinger addressed this issue in several publications. Waldinger has proposed to distinguish this group as a separate subtype of premature ejaculation. Rather, this is a problem in the perception of sexuality. Psychoeducation and possibly psychotherapy are more indicated in these men. Pharmacotherapy with orally taken SSRIs or TCAs is not indicated in these men.

> — Paroxetine has a strong effect on the delay of ejaculation.
> — There is no relation between paroxetin concentration and delay of ejaculation.
> — Paroxetin-induced ejaculation delay is influenced by genetic factors.

13.6 The Influence of Viagra (the Pharmaceutical Industry)

In 1998, oral drug sildenafil was first used to treat erectile dysfunction. In a few years, this has led to enormous changes in thinking about sexuality and the erection disorder in particular. It has also led to a new medical field, which has become known as sexual medicine (Goldstein et al. 1998). In this field, the medical approach and the pharmacological treatment of sexual dysfunctions are central, but attention is also paid to psychological, social, and cultural factors that play a role in most sexual dysfunctions. Partly due to the success of sildenafil, several small and large pharmaceutical companies have become interested in developing drugs to treat both male and female sexual dysfunctions over the past decade.

Definition
Sexual medicine: In this field, the medical approach and the pharmacological treatment of sexual dysfunctions are central.

13.7 Sexual Dysfunctions in Women

Low sexual desire and/or arousal are the most common sex-related complaints reported by women (West et al. 2008). They often result in sexual dissatisfaction, which in turn impacts psychological well-being and can result in severe personal distress (Brotto et al. 2016; Davison et al. 2009; Shifren et al. 2008). These complaints are classified in the Diagnostic and Statistical Manual for Mental Disorders, edition 5 (DSM-5), as female sexual interest/arousal disorder (FSIAD). The disorder is likely caused by a complex interaction of psychological and neurobiological factors and is prevalent among women of all ages and ethnicities (Kingsberg et al. 2015).

The researchers Jos Bloemers and Kim van Rooij describe that there are at least two HSDD/FSIAD subgroups, namely one subgroup in which the patients are treated due to their low sensitivity to sexual stimuli and one subgroup in which the patients are treated due to their excessive activity of sexual-inhibiting mechanisms – this gives not one but two therapeutic indication areas.

With this knowledge, more selective pharmacotherapeutic or psychological interventions can then be developed, instead of a "one-size-fits-all" approach. This was one of the foundations of Emotional Brain's drug development program, which has led to two drug on-demand treatments for HSDD/FSIAD.

The first is the combined administration of sublingual testosterone (0.5 mg) with sildenafil (50 mg) for women with HSDD/FSIAD with low sensitivity to sexual stimuli and, second, the combined administration of sublingual testosterone (0.5 mg) with buspirone (10 mg) for women with HSDD/FSIAD with increased activity of sexual inhibition.

Exciting Research in the Field of Sexual Psychopharmacology: Treating Patients...

207

13

According to the model, HSDD is correlated with low sensitivity to sexual cues or by overactivation of sexual inhibitory mechanisms. Subjects with high sensitivity for sexual cues will be more sensitive for positive sexual stimuli and experiences, which can lead to a hedonic sex life. However, high-sensitivity subjects are also more vulnerable to negative sexual experiences, and as a result more susceptible for learning a negative association with sex. Exposure to stimuli that make an appeal to the sexual motivational system can then automatically elicit an inhibitory response (e.g., a phasic increase in serotonin activity in the dorsolateral prefrontal cortex [PFC]) to diminish conscious or subconscious negative affective state induced by the undesired sexual motivational state. The strength of the inhibitory response might be a function of the sensitivity of the sexual brain system and the duration and severity of negative sexual experiences. Low sensitivity for sexual cues might be the result of a low sensitive androgenic receptor system in the brain, and/or a low level of intracellular androgenic activity, and/or tonic high serotonergic activity in particular areas in the PFC (which may function as a filter for emotional positive and negative stimuli). Subjects with a low sensitive system will have decreased levels of activation of sexual excitatory mechanisms, resulting in low sexual desire. This state of low sensitivity can be interpreted as a biological trait and is unlikely to be caused by sexual experiences. Different combinations of various levels of sensitivity and inhibition are possible, but high inhibition is more likely to occur in sexually high sensitive subjects (resulting in HSDD). Subjects with HSDD and a low sensitive sexual brain are less likely to suffer from high inhibition because they already have low propensity to respond to sexual stimuli; they have less need to inhibit their sexual response.

A method to distinguish the subgroups from each other has been published (Tuiten et al. 2018). This method should also be easy and manageable for the clinician, who should ultimately determine which of the two drugs to prescribe.

The published method describes scores to predict drug efficacy in subtypes of female sexual interest/arousal disorder. The resulting prediction score was validated and shown to effectively and reliably predict which women would benefit from which on-demand drug. The 16 single-nucleotide polymorphisms (SNPs) used in the model all 16 affect pharmacodynamics. The genotype scores predict drug efficacy in subtypes of female sexual interest/arousal disorder. The SNPs predicting kinetics including CYP did not influence the predictive value of the model.

In the end, the phenotype prediction score demarcation formula had the following characteristics: accuracy, sensitivity, specificity, positive predictive value, and negative predictive value were all near 0,8.

> Companion diagnostics can improve pharmacotherapeutical treatment outcome of FSIAD.

13.8 Discussion

The shift from "trial and error" to personalized medicine requires additional efforts. Despite the fact that many psychotropic drugs have been on the market for decades, the effect of many drugs is hardly known. Furthermore, with the current use of psychotropic drugs, it is still almost unclear in which patient a desired effect will occur. The approach described in this chapter may contribute in the future to a more personalized approach with predictive value (Tuiten et al. 2018; Belsky and Israel 2014; Demirkan et al. 2011; Yang et al. 2011). A possible bottleneck within psychopharmacology is the low pricing of the different psychotropics. In the case of the very cheap psychotropic drugs, a companion diagnostic will result in additional costs. In the case of expensive medicines such as oncolytics, immediate financial gains can be made by using a companion diagnostic. When using psychotropic drugs, other costs and social considerations not directly related to the drug should also play a role in deploying additional diagnostics. In addition, a personalized approach can help prevent overprescription.

▶ Conclusion

Human sexual behavior is regulated by very complex, and often difficult to investigate, biological mechanisms in the central and peripheral nervous systems. Moreover, this area of research is still surrounded by various taboos. Recent years have shown that sexual psychopharmacological research also needs new, sometimes uncommon, objectifying research methods. Partly because of all these factors, this type of research is a thrilling and an exciting affair.

Translational research provides an opportunity to gain a better understanding of the underlying neurobiological mechanisms of sexual functions and dysfunctions. Translational research contributes to an ever better understanding of the pathogenesis of sexual dysfunction and, on the other hand, to better treatment strategies for patients.

Combination of animal experimental and human sexual psychopharmacological research makes it possible to develop psychotropic drugs without sexual side effects as well as (sex-promoting) pro-sexual drugs.

It is clear that at the very beginning of the scientific study of sexuality, i.e., at the beginning of the twentieth century, efforts have been made with medicines to treat sexual dysfunctions. Developing this type of medicine is therefore certainly not new. The recent developments presented in this chapter show that it is possible to make use of personalized psychopharmacological pharmacotherapy in the future.

References

Ahlenius S, Larsson K, Svensson L (1980) Further evidence for an inhibitory role of central 5-HT in male rat sexual behavior. Psychopharmacology 68:217–220

American Psychiatric Association (2000) Diagnostic and statistical manual of mental disorders, 4th edn. Text Revision (DSM-IV-TR). American Psychiatric Association, Washington

Baldwin DS, Manson C, Nowak M (2015) Impact of antidepressant drugs on sexual function and satisfaction. CNS Drugs 29(11):905–913. https://doi.org/10.1007/s40263-015-0294-3. Review

Belsky DW, Israel S (2014) Integrating genetics and social science: genetic risk scores. Biodemography Soc Biol 60:137–155

Brotto L, Atallah S, Johnson-Agbakwu C et al (2016) Psychological and interpersonal dimensions of sexual function and dysfunction. J Sex Med 13:538–571

Chan JS, Olivier B, de Jong TR, Snoeren EM, Kooijman E, van Hasselt FN, Limpens JH, Kas MJ, Waldinger MD, Oosting RS (2008) Translational research into sexual disorders: pharmacology and genomics. Eur J Pharmacol 585(2–3):426–435

Clayton AH, El Haddad S, Iluonakhamhe JP, Ponce Martinez C, Schuck AE (2014) Sexual dysfunction associated with major depressive disorder and antidepressant treatment. Expert Opin Drug Saf 13(10):1361–1374. https://doi.org/10.1517/14740338.2014.951324. Epub 2014 Aug 22. Review

Coolen LM, Peters HJ, Veening JG (1996) Fos immunoreactivity in the rat brain following consummatory elements of sexual behavior: a sex comparison. Brain Res 738:67–82

Davison SL, Bell RJ, LaChina M et al (2009) The relationship between self-reported sexual satisfaction and general wellbeing in women. J Sex Med 6:2690–2697

de Jong TR, Pattij T, Veening JG, Waldinger MD, Cools AR, Olivier B (2005) Effects of chronic selective serotonin reuptake inhibitors on 8-OH-DPAT-induced facilitation of ejaculation in rats: comparison of fluvoxamine and paroxetine. Psychopharmacology (Berl) 179:509–515

de Jong TR, Snaphaan LJ, Pattij T, Veening JG, Waldinger MD, Cools AR, Olivier B (2006a) Effects of chronic treatment with fluvoxamine and paroxetine during adolescence on serotonin-related behavior in adult male rats. Eur Neuropsychopharmacol 16:39–48

de Jong TR, Veening JG, Waldinger MD, Cools AR, Olivier B (2006b) Serotonin and the neuro-biology of the ejaculatory threshold. Neurosci Biobehav Rev 30:893–907

Demirkan A et al (2011) Genetic risk profiles for depression and anxiety in adult and elderly cohorts. Mol. Psychiatry 16:773–783

Dose R, Herrn R (2006) Verloren 1933: Bibliothek und Archiv des Instituts für Sexualwissenschaft in Berlin. In: Zeitschrift für Bibliothekswesen und Bibliographie Sonderhefte, vol 88. Vittorio Klostermann, Frankfurt am Main Sonderheft, pp 37–51

Eaton H (1973) Clomipramine in the treatment of premature ejaculation. J Int Med Res 1:432–434

Goldstein I, Lue TF, Padma-Nathan H, Rosen RC, Steers WD, Wicker PA (1998) Oral sildenafil in the treatment of erectile dysfunction. Sildenafil Study Group. N Engl J Med 338:1397–1404

Gower AJ, Berendsen HH, Broekkamp CL (1986) Antagonism of drug-induced yawning and penile erections in rats. Eur J Pharmacol 122:239–244

Haensel SM, Mos J, Olivier B, Slob AK (1991) Sex behavior of male and female Wistar rats affected by the serotonin agonist 8-OH-DPAT. Pharmacol Biochem Behav 40:221–228

Hirschfeld M, Testifortan SB (1927) Therapie der Potenzstorungen (Prospekt). Chemische Fabrik Promonta G.m.b. H, Hamburg

Janssen PK, Waldinger MD (2016) The mathematical formula of the intravaginal ejaculation latency time (IELT) distribution of lifelong premature ejaculation differs from the IELT distribution formula of men in the general male population. Investig Clin Urol 57(2):119–126

Janssen PKC, Waldinger MD (2019a) Use of a confirmed mathematical method for back-analysis of IELT distributions: ejaculation time differences between two continents and between continents and men with lifelong premature ejaculation (part 1). Int J Impot Res 31(5):334–340

Janssen PKC, Waldinger MD (2019b) Men with subjective premature ejaculation have a similar lognormal IELT distribution as men in the general male population and differ mathematically from males with lifelong premature ejaculation after an IELT of 1.5 minutes (part 2). Int J Impot Res 31(5):341–347

Janssen PKC, Bakker SC, Rethelyi J, Zwinderman AH, Touw DJ, Olivier B, Waldinger MD (2009) Serotonin transporter promoter region (5-HTTLPR) polymorphism is associated with the intravaginal ejaculation latency time in Dutch men with lifelong premature ejaculation. J Sex Med 6:276–284

Janssen PKC, Van Schaik R, Olivier B, Waldinger MD (2014a) Re: gene mapping of serotoninergic system polymorphisms provides insight on pathology and treatment of men with lifelong premature ejaculation. Asian J Androl 16(4):643

Janssen PK, Olivier B, Zwinderman AH, Waldinger MD (2014b) Measurement errors in polymerase chain reaction are a confounding factor for a correct interpretation of 5-HTTLPR polymorphism effects on lifelong premature ejaculation: a critical analysis of a previously published meta-analysis of six studies. PLoS One 9(3):e88031

Janssen PK, Rv S, Olivier B, Waldinger MD (2014c) The 5-HT2C receptor gene Cys23Ser polymorphism influences the intravaginal ejaculation latency time in Dutch Caucasian men with lifelong premature ejaculation. Asian J Androl 16(4):607–610

Janssen PK, Touw DJ, Schweitzer H, Waldinger MD (2014d) Non-responders to daily paroxetine and another SSRI in men with lifelong premature ejaculation: a pharmacokinetic dose-escalation study for a rare phenomenon. Korean J Urol 55(9):599–607

Janssen PK, van Schaik R, Zwinderman AH, Olivier B, Waldinger MD (2014e) The 5-HT1A receptor C(1019)G polymorphism influences the intravaginal ejaculation latency time in Dutch Caucasian men with lifelong premature ejaculation. Pharmacol Biochem Behav 121:184–188

Kingsberg SA, Clayton AH, Pfaus JG (2015) The female sexual response: current models, neurobiological underpinnings and agents currently approved or under investigation for the treatment of hypoactive sexual desire disorder. CNS Drugs 29:915–933

McMahon CG, Althof S, Waldinger MD, Porst H, Dean J, Sharlip I, Adaikan PG, Becher E, Broderick GA, Buvat J, Dabees K, Giraldi A, Giuliano F, Hellstrom WJ, Incrocci L, Laan E, Meuleman E, Perelman MA, Rosen R, Rowland D, Segraves R (2008) An evidence-based definition of lifelong premature ejaculation: report of the International Society for Sexual Medicine (ISSM) ad hoc committee for the definition of premature ejaculation. J Sex Med 5:1590–1606

Mos J, Van Logten J, Bloetjes K, Olivier B (1991) The effects of idazoxan and 8-OH-DPAT on sexual behaviour and associated ultrasonic vocalisations in the rat. Neurosci Biobehav Rev 15:505–515

Olivier B, van Oorschot R, Waldinger MD (1998) Serotonin, serotonergic receptors, selective serotonin reuptake inhibitors and sexual behaviour. Int Clin Pychopharmacol 13(suppl 6):S9–S14

Olivier B, Chan JS, Pattij T, de Jong TR, Oosting RS, Veening JG, Waldinger MD (2006) Psychopharmacology of male rat sexual behavior: modeling human sexual dysfunctions? Int J Impot Res 18(Suppl 1):S14–S23

Pattij T, de Jong T, Uitterdijk A, Waldinger MD, Veening JG, van der Graaf PH, Olivier B (2005a) Individual differences in male rat ejaculatory behavior: searching for models to study ejaculation disorders. Eur J Neurosci 22:724–734

Pattij T, Olivier B, Waldinger MD (2005b) Animal models of ejaculatory behaviour. Curr Pharm Des 11:4069–4077

Pfaus JG (2009) Pathways of sexual desire. J Sex Med 6:1506–1533

Schapiro BP (1932) Kombiniertes Epiphysen-Präparat gegen Reizzustände am Genitale und Hypererotismus. Chemische Fabrik Promonta G.m.b.H, Hamburg

Serretti A, Chiesa A (2009) Treatment-emergent sexual dysfunction related to antidepressants: a meta-analysis. J Clin Psychopharmacol 29(3):259–266

Shifren JL, Monz BU, Russo PA et al (2008) Sexual problems and distress in United States women: prevalence and correlates. Obstet Gynecol 112: 970–978

Sinyor M, Schaffer A, Levitt A (2010) The sequenced treatment alternatives to relieve depression (STAR*D) trial: a review. Can J Psychiatry 55(3): 126–135

Spector NL, Blackwell KL (2009) Understanding the mechanisms behind Trastuzumab therapy for human epidermal growth factor receptor 2–positive breast cancer. J Clin Oncol 34:5838–5847

Tuiten A, Michiels F, Böcker KBE, Höhle D, van Honk J, de Lange RPJ, van Rooij K, Kessels R, Bloemers J, Gerritsen J, Janssen P, de Leede L, Meyer J-J, Everaerd W, Frijlink HW, Koppeschaar HPF, Olivier B, Pfaus JG (2018) Genotype scores predict drug efficacy in subtypes of female sexual interest/arousal disorder: a double-blind, randomized, placebo-controlled cross-over trial. Womens Health (Lond) 14:1745506518788970

Virag R (1982) Intracavernous injection of papaverine for erection failure. Letter to the editor. Lancet 2:938

Waldinger MD (2002) The neurobiological approach to premature ejaculation (review). J Urol 168:2359–2367

Waldinger MD (2003) Towards evidence-based drug treatment research on premature ejaculation: a critical evaluation of methodology. J Impot Res 15:309–313

Waldinger MD (2006) The need for a revival of psycho-analytic investigations into premature ejaculation. J Mens Health Gend 3:390–396

Waldinger MD (2007) Premature ejaculation: state of the art. Urol Clin N Am 34:591–599

Waldinger MD, Olivier B (1998) Selective serotonin reuptake inhibitor-induced sexual dysfunction: clinical and research considerations. Int Clin Psychopharmacol 13(suppl 6):S14–S33

Waldinger MD, Schweitzer DH (2006a) Changing paradigms from an historical DSM-III and DSM-IV view towards an evidence based definition of premature ejaculation. Part I: validity of DSM-IV-TR. J Sex Med 682-692(41):3

Waldinger MD, Schweitzer DH (2006b) Changing paradigms from an historical DSM-III and DSM-IV view towards an evidence based definition of premature ejaculation. Part II: proposals for DSM-V and ICD-11. J Sex Med 3:693–705

Waldinger MD, Schweitzer DH (2008) The use of old and recent DSM definitions of premature ejaculation in observational studies: a contribution to the present debate for a new classification of PE in the DSM-V. J Sex Med 5:1079–1087

Waldinger MD, Hengeveld MW, Zwinderman AH (1994) Paroxetine treatment of premature ejaculation: a double-blind, randomized, placebo-controlled study. Am J Psychiatry 151:1377–1379

Waldinger MD, Berendsen HHG, Blok BFM, Olivier B, Holstege G (1998a) Premature ejaculation and serotonergic antidepressants-induced delayed ejaculation: the involvement of the serotonergic system. Behav Brain Res 92:111–118

Waldinger MD, Hengeveld MW, Zwinderman AH, Olivier B (1998b) A double-blind, random-ized, placebo-controlled study with fluoxetine, fluvoxamine, paroxetine and sertraline. J Clin Psychopharmacol 18:274–281

Waldinger MD, Hengeveld MW, Zwinderman AH, Olivier B (1998c) An empirical operationalization study of DSM-IV diagnostic criteria for premature ejaculation. Int J Psychiatry Clin Pract 2:287–293

Waldinger MD, Rietschel M, Nothen MM, Hengeveld MW, Olivier B (1998d) Familial occurrence of primary premature ejaculation. Psychiatr Genet 8:37–40

Waldinger MD, van de Plas A, Pattij T, van Oorschot R, Coolen LM, Veening JG, Olivier B (2002) The selective serotonin re-uptake inhibitors fluvoxamine and paroxetine differ in sexual inhibitory effects after chronic treatment. Psychopharmacology 160:283–289

Waldinger MD, Zwinderman AH, Schweitzer DH, Olivier B (2004) Relevance of methodological design for the interpretation of efficacy of drug treatment of premature ejaculation: a systematic review and meta-analysis. Int J Impot Res 16:369–381

Waldinger MD, Quinn P, Dilleen M, Mundayat R, Schweitzer DH, Boolell MA (2005a) Multinational population survey of intravaginal ejaculation latency time. J Sex Med 2:492–497

Waldinger MD, Zwinderman AH, Olivier B, Schweitzer DH (2005b) Proposal for a definition of lifelong premature ejaculation based on epidemiological stopwatch data. J Sex Med 2:498–507

Waldinger MD, Zwinderman AH, Olivier B, Schweitzer DH (2007) The majority of men with lifelong premature ejaculation prefer daily drug treatment: an observational study in a consecutive group of Dutch men. J Sex Med 4:1028–1037

Waldinger MD, McIntosh J, Schweitzer DH (2009) A five-nation survey to assess the distribution of the intravaginal ejaculatory latency time among the general male population. J Sex Med 6:2888–2895

West SL, D'Aloisio AA, Agans RP et al (2008) Prevalence of low sexual desire and hypoactive sexual desire disorder in a nationally representative sample of US women. Arch Intern Med 168:1441–1449

Yang J et al (2011) Genome partitioning of genetic variation for complex traits using common SNPs. Nat Genet 43:519–525

13

A Paradigm Shift from DSM-5 to Research Domain Criteria: Application to Translational CNS Drug Development

William Potter and Bruce Cuthbert

Contents

14.1   Introduction – 212

14.2   The Research Domain Criteria Project – 213

14.3   RDoC in Pharmaceutical Development – 215

14.4   Examples of Translational Studies Using Brain
       Measures and Domains – 216

14.5   Background for Case Study 2:
       NIMH FAST-Psychosis Spectrum (PS) – 221

14.6   Summary – 223

       References – 225

The development of new therapeutics for mental disorders has lagged for a decade or more, associated with a marked decrease in the number of large companies working in this space. This lag reflects the difficulty of identifying new targets that have a likelihood of leading to marketable treatments, especially for broadly defined syndromes that are characterized by extensive heterogeneity and co-morbidity. Further, disorders are defined by presenting symptoms and signs rather than specific biological or behavioral tests. To address these obstacles, the US National Institute of Mental Health (NIMH) instituted two related programs. The first is the Research Domain Criteria (RDoC) initiative. RDoC is a research framework intended to focus on fundamental behavioral dimensions such as reward-related activity or cognitive control, studied in tandem with their implementing neural circuitry. The second is the implementation of precision-medicine approaches that employ fast-fail, target-engagement trial designs; these focus on demonstrations that compounds engage their hypothesized targets and result in predicted, specific change in CNS activity and/or behavior, thus optimizing translation from animal models of function. This chapter summarizes these two initiatives and provides two case-study examples that illustrate the issues involved with these new strategies. Many specific elements need to be further developed, such as more informative biomarkers and outcome measures. However, the focus on functions that are more directly related to the relevant neural circuitry and cut across current disorders offers many advantages for the next generation of translational research and drug development.

🕮 Learning Objectives

The reader should be able to respond to the following questions after reading this chapter:

- Discuss the reasons why development of new therapeutics for psychiatric disorders has stalled over the last 10–15 years, with particular emphasis on the characteristics of current diagnostic nosologies

- Show familiarity with the Research Domain Criteria (RDoC) initiated by the US National Institute of Mental Health (NIMH) and what role RDoC concepts address in drug development
- Discuss proposed new strategies and procedures for approaching all phases of new drug development and outline the rationale for the "fast-fail" approach to evaluating new compounds
- Discuss examples of recent or current clinical studies and trials that illustrate new approaches to drug development, including demonstration of target engagement and optimal translation of animal studies to human trials

14.1 Introduction

Pharmaceutical agents have been a mainstay for the treatment of mental disorders since their first systematic use in the 1950s. The appearance of blockbuster new classes such as selective serotonin reuptake inhibitors (SSRIs) and second-generation antipsychotics in succeeding decades led to a further burgeoning of the market for psychotropic drugs. Beginning around the turn of the century, however, development of new therapeutic drugs began to drop and pharmaceutical companies began to withdraw from the CNS drug space (Abbott 2010; Miller 2010). There are multiple reasons for this relatively rapid decline. First, the cost of large phase III clinical trials has become much more expensive, with costly failures. Second, extensive pharmacological research indicated that the molecular receptors involved in first-generation psychotropic drugs, discovered almost entirely by serendipity, were not particularly compelling as mechanisms for disease biology (Fibiger 2012). In fact, a major reason cited by pharmaceutical companies in exiting CNS disorder research was the lack of promising targets for neuropsychiatric disorders, particularly as compared to other areas of medicine such as cancer (Wong et al. 2010).

These difficulties with respect to disease biology played a major role in efforts to revisit the fundamental nature of psychiatric disorders and the ways in which they are diagnosed. Inconclusive studies in the early twentieth century that attempted to relate the structure and function of the nervous system to clinical observations resulted in a continued reliance upon presenting symptoms and signs for assigning a diagnosis, with subsequent tracking of course and outcome as confirmation. The promulgation of the Diagnostic and Statistical Manual – III (DSM-III) in 1980 rendered the previously subjective process into a systematic and quasi-quantitative manual, which greatly enhanced the reliability of diagnosis and played a major role in ushering in the modern era of psychiatric research. Unfortunately, an unanticipated adverse effect was the reification of the disorder categories, which became perceived as specific disease entities in spite of the manual's disclaimers to the contrary (Hyman 2010).

Definition

The Diagnostic and Statistical Manual of Mental Disorders (DSM) is one of the two major manuals for listing and defining psychiatric disorders, published by the American Psychiatric Association. The other is the Mental, Behavioural, and Neurodevelopmental Disorders chapter of the International Classification of Diseases (ICD), published by the World Health Organization. The two manuals contain nearly identical lists of disorders, but differ in how the disorders are characterized and formally defined.

The assumption that putative biological diseases could be identified by symptom reports resulted in a number of problems over time. First, it was presumed that biomarkers could be identified through appropriate clinical studies that would distinguish diagnosed patients versus controls. Second, clinical trials designs were structured and powered in terms of overall main effects for active treatment arms versus control conditions, under the supposition that the large majority of patients would respond to treatment in similar ways.

The results of such diagnostic hegemony have become clear over the subsequent decades. New technologies (such as genetics, neuroimaging, and sophisticated behavioral science) have contributed to an emerging Zeitgeist that mental disorders are broad and heterogeneous syndromes rather than specific disease entities (Hyman 2010); further, disorders can be regarded as dimensional phenomena as opposed to a binary, "sick-healthy" infectious disease model (Patel 2019). These factors are now regarded as a major reason that virtually no diagnostic tests are available for mental illnesses (Kapur et al. 2012). There are scant quantitative data to support the decision to assign a particular diagnosis on the basis of presenting dysfunction and/or distress, and after a diagnosis is established there are minimal empirical data to guide clinicians in deciding upon a modality (drugs, devices, cognitive/behavior therapies) or a specific treatment within a modality. Further, similarly sparse data are available to provide treatment guidelines at increasing levels of disorder severity. Overall, while effective treatments are available, they are generally efficacious in only around 50% of patients (Wong et al. 2010).

14.2 The Research Domain Criteria Project

A key factor in all of these problems is that the entire clinical research system became fixed around the DSM system (now in its fifth edition, having changed in details but not in essence), such that alternative conceptions of psychopathology or diagnosis were excluded from consideration – notably including grant review committees considering research applications. In an attempt to alleviate this problem, NIMH initiated in 2009 a new project to provide an alternative framework for reviewing grant applications about psychopathology. The basic strategy was to incorporate the considerable research literatures that had accumulated over the past decades from behavioral neuroscience, explicating neural and behavioral aspects of such functions as fear (Davis 2006), reward (Berridge and Robinson 2003), and behavioral regulation (Cole et al. 2014).

Definition

The Research Domain Criteria (RDoC) project is a program initiated by NIMH in 2009 to provide alternative approaches to the DSM for reviewing grant applications in psychopathology. Rather than the DSM's symptom-based syndromes, the criteria for selecting participants in RDoC studies involve functional dimensions of behavior or cognition (e.g., working memory, fear, reward learning) that may cut across multiple diagnostic groups. The goal is to build a literature based on understanding psychopathology from the perspective of dysregulation in normal functional systems, which can lead to precision diagnosis, treatment, and prevention for mental illness.

Clinical studies had already begun to include such measures, but conducted from the traditional view that aberrations would prove to be accurate biomarkers for virtually all patients in a particular diagnostic class (e.g., amygdala dysfunction in Post-Traumatic Stress Disorder (PTSD) or anterior cingulate abnormalities in ADHD, or, similarly, that genetic markers could be found for schizophrenia). However, the lack of success with such investigations was key in prompting the concerns that diagnostic categories did not align well with genetic, biological, or behavioral data. From the current perspective, several aspects were responsible for such failures. First, it is clear that current disorders are heterogeneous syndromes, so only a subset of patients in any category are likely to share any given dysfunction. Second, it is now apparent that any particular abnormality is likely to be found in patients across several different disorder classes, so that any given biomarker is not likely to be specific even to a subset of one disorder. Further, co-morbidity of disorders has proven to be quite common in patient populations, further muddying the search for specific diagnostic tests. Finally, there is now general agreement that psychopathology is dimensional in nature, so that the use of biomarkers must rest upon a quantitative basis for making treatment decisions (as found in hypertension, cholesterol levels, and other areas of medicine).

The NIMH project represented an attempt to translate burgeoning basic neuroscience data to patient samples in a direct way, as specified in Goal 1.4 of the 2008 NIMH Strategic Plan (see Cuthbert and Insel 2013). There were several marked departures from typical clinical studies. First, the emphasis was on the study of fundamental functional dimensions rather than clinical categories. Second, an important goal was to determine the full range of functional variation from normal to abnormal, so as to better explicate the transitions from normal-range functioning to various degrees of dysfunction.

Third, an important aim was to pursue multi-measure studies from a perspective of empirically examining the nature of relationships among various biological, behavioral, and self-report measures to seek an integrative understanding – as opposed to the typical reductionistic practice. This aim was directly intended to address the mind-body problems that have chronically plagued mental disorders research. The crux of the issue has been that data and concepts are typically derived separately for mental states (whether disorder classes or specific symptoms such as hallucinations) and for biological variables, with an expectation (most often disappointed) that a tight correspondence should be found with the other domain. The new project addressed this problem by calling for the compilation of a set of *functional constructs* that require two simultaneous requirements in order to be included: data for a valid psychological or behavioral construct (such as working memory or reactions to threat) *and* evidence for a neural circuit or system that plays a major role in implementing the function.

The new project was instantiated by a series of workshops in which experts from various fields evaluated the evidence for potential constructs according to the two criteria above. The constructs were grouped into superordinate *domains* based on the current state of the animal and human behavioral neuroscience literature. At the current time, there are six such domains (each containing three to six constructs): negative valence, positive valence, cognitive systems, systems for social processes, arousal/regulatory systems, and sensorimotor systems (NIMH 2019). It is important to emphasize that the constructs and domains so established are considered as useful exemplars for the general principles of the RDoC approach – not as a delimited set – and new or modified constructs/domains are strongly encouraged on a continual basis as new data emerge. The new project was named the Research Domain Criteria project to reflect the framework's organization and emphasize on empirically derived domains and constructs.

In implementing RDoC studies, the usual clinical design was inverted: rather than recruiting patients based on presenting symptoms and seeking correlative biomarkers, the focus was upon functional dimensions as the independent variables. Patients could be recruited from one or multiple diagnostic classes that seemed generally relevant to the function of interest (e.g., threat reactivity for fear/anxiety disorders, working memory for psychotic disorders), and the experimental design was then oriented around the gradations in functioning as studied across multiple measures that might include circuit activity, behavioral/cognitive functioning, and symptom reports.

> The RDoC paradigm was developed to enable research that could provide direct translation of knowledge about normal behavioral-brain operations to an understanding of the varying degrees of dysregulation in these functions that comprise mental disorders, thus supporting novel ideas about the nature and etiology of psychopathology that can lead to advances in diagnosis and interventions.

14.3 RDoC in Pharmaceutical Development

RDoC was originally conceived as an experimental framework for psychopathology research that was needed to transcend the exclusive focus on DSM disorders in clinical and translational research that had evolved as de facto requirements in NIH/NIMH study sections. However, as noted above, the pharmaceutical industry began to withdraw from CNS drug development shortly before RDoC was developed, and a major reason cited for this departure was the lack of promising drug targets, given the increasing recognition that DSM and ICD disorders were heterogeneous syndromes rather than specific disease entities (Wong et al. 2010). As a result, attention turned to RDoC as one way to conduct research into specific functions and mechanisms that could offer new target possibilities.

These developments also coincided with the move by NIMH toward a fast-fail, experimental medicine approach to clinical trials (Insel 2012; Insel and Gogtay 2014). The key to this approach was to emphasize early trials that involved proof of mechanism in acknowledgment of the fact that the large majority of new compounds will fail, and so the goal is to "fail fast" in order to minimize lost resources and move on to the next candidates (Paul et al. 2010). The aim of the new trials was to establish a path from engaging a biological target to demonstrating a decrease in symptoms and/or an increase in performance with respect to a particular function associated with the target. The ideal sequence (which admittedly could not be followed in all cases) for such a development program was to pass the following steps one at a time:

1. Demonstrate target engagement of the compound, ideally with a PET (positron emission tomography) imaging ligand demonstrating occupancy of the receptors of interest
2. Demonstrate that successful target engagement resulted in some type of relevant physiological activity, such as changes in EEG or resting-state MRI (magnetic resonance imaging)
3. Show that drug administration resulted in some change in functioning in the hypothesized direction, whether on some type of behavioral task or relevant circuit activity during the task (or both)
4. Demonstrate that clinical improvement was related to the changes in functioning in step 3

If a failure occurred at any step, then the effort was deemed a failure and the project was shelved. The idea was to avoid false conclusions along the way; for example, a promising clinical outcome in the absence of demonstrating target engagement might be due to other mechanisms or to spurious influences on the results, thus eventuating in a subsequent, more expensive and time-consuming failure. A success, of course, provided a more solid basis from which to invest in further development.

> **Definition**
>
> Endophenotype in psychiatry has come to mean any biomarker or functional measure that is associated with a clinical diagnosis but is quantitative and thought to reflect some discrete biological system with the additional property of being reasonably heritable. Endophenotypes have been identified and studied as likely more reflective of some specific, core pathophysiology than the range of clinical symptoms that generate DSM diagnoses.

The RDoC concept aligns with the experimental medicine approach to study promising targets based on specific aspects of nervous system activity or behavior. Further, the emphasis on focused constructs or endophenotypes, along with the normal-to-abnormal dimensional approach, interfaces much more directly with animal models in terms of relevant concepts of abnormal functioning (Young et al. 2017). Central to such translation is the idea that, while far from perfect, the homology from animals to humans can be evaluated much more precisely with this kind of approach (Anderzhanova et al. 2017).

A number of projects reflect this new type of translational paradigm in moving toward new targets or compounds. For instance, collaborative efforts in the EU such as the IMI-NEWMEDS project (Artigas et al. 2017) have been developed to relate circuit activity to drugs and behaviors across species, entailing an iterative bidirectional translational effort. Such large collaborative efforts require choos-

ing standard ways of measuring, for instance, behavioral/cognitive functioning and circuit activity across laboratories and clinical research sites. Similarly, to further advance the RDoC approach, efforts are underway to provide a set of common tools for each of the originally articulated domains with a path to refining these (NIMH 2016). In parallel to various initiatives to identify a set of common measures to be used in future clinical and translational research, there have been a number of focused efforts to include evidence of common brain effects in animals and humans as an essential component of validating novel molecular targets. Some examples of these follow as exemplifying what is entailed by applying RDoC principles to translational approaches for moving from the bench to bedside.

> The emphasis on specific aspects of functioning (e.g., working memory) in RDoC is well aligned with new drug development paradigms that emphasize experimental-medicine and fast-fail concepts, as shown by the use of endophenotypes and the greatly enhanced potential for homologous measures between animals and humans that can speed the search for potential therapeutic effects of novel agents.

14.4 Examples of Translational Studies Using Brain Measures and Domains

It is now generally appreciated that since animals cannot provide verbal reports of what they are experiencing, they cannot be used as true models of those syndromal conditions which are diagnosed, at least in part, on the basis of verbal responses. Moreover, in the absence of knowing a disease's pathophysiology, it is impossible to have a valid disease model. To the extent, however, that brain processes measurable across species have a conserved relationship to a particular function, it is reasonable to test the predictive potential of observing drug effects on intact animal brains for what will be observed in human brains. One can then link our growing knowledge of how brain processes relate to a range of quantifiable human behaviors and functions, taking the domains described earlier as examples.

Over the last two decades, both academic and industry investigators have begun to pursue the possibilities of approaching novel drug development not from the perspective of, for example, a "model of depression" in rodents such the Porsolt forced swim test (Porsolt et al. 1978) but from the perspective of having an effect on more domain-based constructs such as cognitive performance. The fact that many agents have been advanced to the clinic, at least in part dependent on some preclinical evidence for enhanced cognitive function, and failed to show effects in humans highlights a major limitation of relying on behaviors as an accurate reflection of, for instance, encoding or retrieval of memories. This problem in translation has been noted for at least a decade (Shineman et al. 2011) and has given rise to many suggestions on how to improve translatability to humans, most notably coupling behavioral and functional measures of neural substrates that are affected by compounds (Elmer et al. 2016). If one could better link effects of drugs on specific brain processes in the animal which have been established to be core to something like encoding of memories to the behavioral readouts of memory function, then predictions to humans might be increased since once could establish similar brain effects in both animals and humans.

A nearer term approach to, for instance, detection of drug effects on cognition has been to use a broader range of domain type measures in the context of a clinical trial to see if subgroups of individuals characterized by a domain rather than diagnosis per se might show differential responses in terms of effects on different domains of cognition (Dawson 2015). Indeed, in one elegant study comparing otherwise healthy individuals in terms of high vs low schizotypy, low-dose amisulpiride improved N-back performance (a memory-related measure) in the former while impairing it in the latter group (Koychev et al. 2012).

Despite such examples of using more extensive and refined ways of subgrouping clinical populations and their cognitive responses dating back more than a decade, we have yet to be able to select a group of subjects in whom a drug can be shown to enhance a specific cognitive domain. Furthermore, even if clinical domain measures prove of high utility, there is no reasonable back-translation pathway whereby one would divide, for instance, animals into high- and low-schizotypy populations. It is with this background that we focus on whether incorporating the demonstration of brain effects as well as performance measures will facilitate translation from the laboratory to the clinic.

> Since clinical symptoms of syndromal psychiatric diseases cannot be reliably modeled in animals, focus is shifting to analogous relationships of brain processes with performance measures across species in order to better translate drug effects in animals to humans. This now allows for the bidirectional translation of the effects of drugs on domains between humans and animals.

The biggest effort to date to put all of the elements together to further the translation from the lab to the clinic of novel agents with potential in major psychiatric disorders has been the **IMI-NEWMEDS** initiative which was launched in 2009 (Artigas et al. 2017). Two of the major areas of focus were utilization of electrophysiological approaches from animals to humans and "back translation" of functional fMRI from humans into animals, both with the aim of having at least analogous, if not homologous, readouts of drug action in brain across species. Electrophysiology provides the simplest and clearest case of at least an aspect of brain function that can be linked to level of consciousness (as applied in anesthesiology), levels of wakefulness (hypnotics and agents used to counteract sleepiness), and aberrant firing of clusters of cells (epilepsies). Linking normal waking or resting-state EEG to circuit function that subserves, for instance, cognitive and emotional responses has to date proven elusive. The extent to which EEG measures translate well enough across species to serve such goals remains to be determined.

In contrast, the back translation of fMRI BOLD has already proven its value, showing that similar patterns of activation can be observed in both rats and humans following acute administration of psychotomimetic agents such as ketamine (De Simoni et al. 2013) and phencyclidine (Gozzi et al. 2008). These findings provided the basis for a series of studies both by industry and more recently by the NIMH to assess whether novel agents targeted to metabotropic glutamate receptors were indeed having their desired brain effects on humans. This example constitutes one of the case studies that will be presented later.

Linking circuit-based measures of functional connectivity and their pharmacological modulation to specific behavioral paradigms and validating a circuit-based approach for evaluating established and novel compounds remains a work-in progress. The most robust preclinical examples have depended on paradigms such as reversal of phencyclidine effects on brain circuits by clozapine or other antipsychotic drugs (Santana et al. 2011; Lladó-Pelfort et al. 2016). To date, however, there is no example of which we are aware in which novel agents have been evaluated with respect to their ability to reverse disrupted oscillatory activity in rodents followed by testing in humans with either drug-induced or disease-associated disrupted oscillatory activity. As described below, much work is required to establish whether an induced abnormal state in humans can be reversed with a compound shown to reverse the same induced state in animals.

An ambitious new project with a significant translational component, PRISM (Psychiatric Ratings Using Intermediate Stratified Markers), has been developed subsequent to the IMI-NEWMEDS initiative and starts with a clinical symptom, social withdrawal, treated as a domain of interest which cuts across diagnoses. Schizophrenia and Alzheimer's dementia are highlighted as two very different diagnostic groups in which social withdrawal can be a major symptom (▶ https://prism-project.eu). The PRISM project seeks to provide new classification tools "based primarily on quantitative biological parameters, focusing on this psychiatric domain common

to these two disorders" (Kas et al. 2019). As further outlined in an editorial describing the PRISM project that introduces papers describing many of the specific approaches, the plan is to characterize newly recruited subjects with measures of "social withdrawal, attention, sensory processing, and working memory utilizing digital, brain imaging, EEG and epigenetic biomarkers.... Furthermore, a preclinical platform will be implemented to allow backtranslation from human findings into rodents." (Kas et al. 2019).

For PRISM, as described by Danjou et al. (2019), measures of brain function include a range of electrophysiological paradigms including resting-state EEG, auditory mismatch negativity, auditory and visual-based oddball paradigms, facial emotion processing ERPs (event-related potentials), and auditory steady-state response. Given that the relationship between measures of social withdrawal and these electrophysiological parameters has not previously been systematically explored, this constitutes a discovery effort. As discussed in another paper among the series on PRISM, there is considerable complementarity between the approach to social withdrawal and the proposed RDoC domain of social communication (Cuthbert 2019) which is recognized as relatively unexplored in terms of brain circuitry and other measures. Notably, the PRISM project starts with a clinically defined functional impairment to see if this is associated with quantifiable aspects of brain function, whereas in a translational case study described below, one starts with a clinical domain for which there is reasonable evidence linking it to quantifiable circuit function.

Interestingly, in the PRISM project, there is also a focus on attention and working memory, which may indeed be important for social communication but are more generally identified as representing components of cognition. The process and pros and cons of various measures are described in detail by Gilmour et al. (2019), who emphasize the challenges involved in dissecting cognition in specific domains. We highlight this point since cognitive domains in and of themselves are seen as potentially separable in terms of therapeutic interventions that might be developed. It has long been known that attentional performance can be enhanced by drugs such as amphetamine and that so-called psycho-stimulants are effective treatments in ADHD, reducing, not stimulating, hyperactive behavior. Given this serendipitous success story, it seems likely that other aspects of cognition might also be improved with a pharmacologic intervention. Thus, the already referenced efforts to generate analogous tasks in animals and humans to assess the impact of drugs on one or another component of cognitive function informed the selection of measures being implemented by PRISM. Many of the investigators in PRISM had previously been involved in designing and carrying out "experimental medicine" studies in humans with a focus on a measure analogous to one in animals rather than a traditional DSM disorder-related clinical severity measure (Dawson 2015), laying some of the groundwork for the current project.

One recently funded NIMH study to test the approach being pursued in the IMI-NEWMEDS effort builds on preclinical findings that D1 partial agonists can modulate a brain circuit that may be important in the control of working memory. Earlier studies looking for a drug effect on cognition in psychiatric patients independent of clinical state or severity have utilized a range of cognitive measures that do not have precise analogues from rodent species (e.g., novel object recognition) to humans. Human studies utilize such measures as the Digit Symbol Substitution Test, as was done with the serotonergic antidepressant vortioxetine (Dawson 2015), with the question still open as to whether effects of vortioxetine can be linked to any specific cognitive domain or molecular attribute of the drug (Bennabi et al. 2019). In the current approach, to more precisely translate findings in non-human primates that D1 agonism can improve working memory (Goldman-Rakich et al. 2004; Arnsten et al. 2017), the primary measure becomes a BOLD signal in a specific brain region in response to a specially designed spatial working memory task in humans (Cho et al. 2018). This is more than a subtle difference in terms of what one prioritizes as a readout of drug action since it is initially agnostic to the relationship between the magnitude of the response one measures in the brain and how well one performs on a particular trial. In other words, for the first step, one does not worry about whether there is an established relationship between the magnitude of brain response and performance. One seeks to establish first that a brain measure can be altered and only subsequently explores the relationship of that effect to behavior, reported feelings, or performance of a task.

> Large consortia efforts have been initiated and take a range of approaches to see if effects of a drug on some brain function are similar in animals and humans, in order to lay the groundwork for relating brain function to some clinically important domain such as cognition across species. Results from such approaches will become available over the 4 to 5 years from early 2020, the date of this writing.

The clinical need and large potential market for compounds that substantially improve cognitive performance have understandably stimulated many studies to this end and continue to do so. Other examples of translational studies utilizing functional brain measures and/or domains of human behavior include resolving whether or not a compound produces either a targeted "resting state" brain effect or a brain effect coupled to a functional domain. The two examples which follow have been selected as translational case studies emerging out of a National Institute of Mental Health (NIMH) initiative to use a contract mechanism, coined "fast-fail," to perform early-stage pharmacodynamic (PD) trials of drugs with novel mechanisms. The drugs selected were ones that industry had advanced into humans but at least temporarily shelved for want of definitive dosage data and evidence of brain effects that might support future clinical trials. The two contracts were Fast-Fail Trials in Mood and Anxiety Spectrum Disorders (FAST-MAS), awarded to Duke University (principal investigator: Andrew Krystal), and Fast-Fail Trials in Psychotic Spectrum Disorders (FAST-PS), awarded to Research Foundation for Mental Hygiene (principal investigator: Jeffrey Lieberman).

FAST-MAS

The study conducted under the Fast-Fail Trials in Mood and Anxiety Spectrum Disorders (FAST-MAS) contract provides the best exemplar to date of putting together all the components of what were selected to be the criteria for a translational target validation approach from animals to humans. These criteria, derived from the ideal sequence of studies outlined in an earlier section, included (1) a specific and testable hypothesis, (2) PET-imaging quantification of receptor occupancy, (3) brain functional target engagement measures, (4) a target-selective, CNS-penetrant, Investigational New Drug (IND)-ready compound, and (5) functional measure based on RDoC.

Thus, the FAST-MAS study (ClinicalTrials. gov Identifier: NCT02218736) focused on a core symptom domain within the broad spectrum of mood and anxiety – anhedonia, the inability to experience pleasure – and introduced measures of reward circuitry as the "outcome." From the perspective of taking RDoC principles into account, anhedonia falls under a range of reward-related RDoC constructs ("Reward Responsiveness," "Reward Learning," and "Reward Valuation"; NIMH 2018). Preclinical studies had shown that κ-opioid receptor (KOR) antagonism increases the activation of reward-related brain circuitry (the ventral striatum) to improve reward-associated function and/or reverse anhedonic behaviors (Carlezon et al. 2006;, Bruinjnzeel 2009; Ebner et al. 2010, Rorick-Kehn et al. 2013). A KOR antagonist, JNJ-67953964 (formerly known as LY2456302 and CERC-501), met the other criteria for target selection based on existing PET target engagement data showing close to full saturation of the receptor using the specific KOR PET tracer [^{11}C]PKAB (LY2879788 (Zheng et al. 2013), evidence of human safety, and compound availability (Lowe et al. 2014).

The monetary incentive delay task (MID) was selected as a tool for exploring whether κ-antagonism in humans altered reward circuitry function, as this task reliably demonstrates ventral striatal activation in anticipation of reward (Oldham et al. 2018).

> The monetary incentive delay task (MID) consists of a task which involves three possible outcomes – monetary reward, monetary loss, or neither – that can be performed while undergoing fMRI visualization of brain activity. Versions of this task have been widely employed for two decades to explore reward processing in human brain.

Although not explicitly built on how the role of κ-receptors was demonstrated in animal studies (e.g., cocaine and not money as a reward), the motivational aspects were deemed comparable. Given the many components of this paradigm allowing for post hoc ways of finding evidence of drug effect, the contract required specification of an a priori outcome measure. Mean fMRI ventral striatal activation in the MID task in anticipation of gain, testing JNJ-67953964 compared with placebo, was selected (Krystal et al. 2018). Interestingly, another measure from the paradigm would have yielded an even larger drug effect size than the prespecified one.

Specifically, JNJ-67953964 compared with placebo enhanced the mean and maximum fMRI ventral striatal activation in the MID task in anticipation of gain. The degree of enhancement, however, was almost double using a "secondary" outcome measure, ventral striatal activation to anticipation of loss (as opposed to gain) (Krystal et al. 2020). In retrospect, one might argue that responses to anticipation of loss might be as much or more informative regarding how the brain processes the possibility of reward which is offered in the context of possible loss as opposed to situations in which

14

there is only the possibility of reward without risk of loss. Alternatively, recent data increasingly indicate that the striatum is part of a general motivational system that plays an important role in anticipating either reward or loss, as opposed to the traditional view confined to reward (Oldham et al. 2018). More generally, finding the most robust translational functional measures will continue to be an iterative process for the foreseeable future, given the complexities involved within as well as across species in the constructs being explored.

There is evidence that such translational studies as the one summarized above on KOR antagonism are having an effect on investment decision. Subsequent to the FAST-MAS study, we understand that there has been a rise in companies pursuing development of treatments for anhedonia and exploring anhedonia or apathy as a potential indication. At least one company, Takeda, followed the precedent set by FAST-MAS by embarking upon a development path which includes a small phase 2a proof of mechanism study using the MID test. The study seeks to determine if motivation/reward deficits observed in schizophrenia can be attenuated with the G-protein-coupled receptor 139 (GPR139) agonist, TAK-041 (ClinicalTrials.gov Identifier: NCT03319953). Whether the CNS effect in the MID by itself, without some evidence of positive effects on a clinical measure such as the Snaith-Hamilton Pleasure Scale (SHAPS; Snaith et al. 1995), would stimulate continued investment is an open question. In the FAST-MAS studies, modest positive effects on the SHAPS were observed after KOR antagonism, although not nearly as robust as the effects on the BOLD signal. Put another way, would knowing that one had an effect on a brain functional circuit take precedence over lack of a finding on a clinical measure at an early stage of development?

14.5 Background for Case Study 2: NIMH FAST-Psychosis Spectrum (PS)

The other NIMH contracted study to be discussed here was focused exclusively on a translational approach to deciding whether the doses of a compound that failed in expensive phase 3 studies (Downing et al. 2014; Adams et al. 2014) were high enough to interpret the study as ruling out the molecular mechanism of the drug in the treatment of schizophrenia. The field had been excited when an mGluR2/3 agonist, pomaglumetad (POMA/LY2140023), was reported to show efficacy in a phase 2 proof of concept study in schizophrenia (Patil et al. 2007). It was not known at the time of the clinical trials whether the phase 3 doses had the hypothesized effect on brain function. Dose selection had been based solely on theoretical extrapolations as to brain effects if similar cerebrospinal fluid (CSF) concentrations could be achieved that equaled those in rat CSF at pharmacologically active doses (Lowe et al. 2016).

To date, there is still no PET ligand that would allow assessment of mGluR2/3 receptor occupancy following either an agonist or antagonist in humans. And for agonists, even if a PET ligand is available for the receptor in question, one might still achieve functional effects at levels of occupancy too low to detect through displacement of a radiolabeled ligand. In such instances, the field has looked for translatable functional readouts such as spectral power using EEG or, in the case of mGluR2/3 agonism, an fMRI pharmacoBOLD signal paradigm.

The rationale for pursuing POMA in schizophrenia was based on rodent findings that induction of glutamate release in the brain by psychotomimetic compounds such as phencyclidine (PCP) and ketamine could be blocked by prior treatment with an mGluR2/3 agonist (Moghaddam and Adams 1998; Lorrain et al. 2003). The possibility was therefore explored that a translatable functional readout could be developed if the same dose of a compound associated with glutamate release in rats produced a robust increase in the BOLD signal in the same species. Such

PCP-induced increases were demonstrated in rats and then shown to be blocked by prior treatment with an mGlu2/3 agonist (as well as clozapine), providing a readout that could reasonably be inferred to represent an effect on glutamate release (Gozzi et al. 2008). The fact that clozapine also blocked the BOLD response to ketamine points to the potential of other mechanisms that may not involve reduction of glutamate release to block some of ketamine's effects. The preclinical characterization of the mGluR2/3 agonist, however, appears to rule out any direct non-glutamatergic actions with only indirect actions on, for instance, measures of dopamine turnover in CSF (Lowe et al. 2016).

Ideally one would have a more direct measure of glutamate release that was translatable but none have been developed to date. It is possible to measure analytes that reflect total glutamate in brain with MRS, but the contribution of released glutamate to these measures remains to be determined. Nonetheless, as noted below, it was recently of interest to compare fMRI and MRS responses in humans. When the paradigm employed in rats focusing only on pharmacoBOLD was translated into an acute human healthy volunteer study, only the highest dose of POMA (160 mg) was able to attenuate the BOLD signal at the group level, with a high degree of variability whereby some subjects showed total blockade and others none at all, with no observed relationship to blood concentrations (Mehta et al. 2018). The maximum dose of POMA used in the negative phase 3 studies was 80 mg bid, so given the modest acute effect of 160 mg, it seemed unlikely that the clinical trial dose would have been sufficient to block the ketamine BOLD response in most subjects.

Case Study 2

FAST-PS Study

Given this background, a two-part FAST-PS multi-site study designed to compare 40 mg (the phase 2 positive study dose) and 160 mg (twice the phase 3 dose that failed) BID for 10 days was planned to address the question of whether high enough doses of POMA had been given in any of the efficacy studies to achieve sustained inhibition of the BOLD increase following ketamine. Even though no preclinical studies were available to establish that MRS could be used to detect acute effects of ketamine on glutamate concentrations, the question was raised whether this might be demonstrable in humans. Therefore, prior to undertaking the study with POMA, a three-site study (ClinicalTrials.gov Identifier: NCT02134951) was designed not only to compare both BOLD and MRS signals after ketamine but also to establish that the paradigm could be implemented across sites with potential application to clinical trials in mind (Javitt et al. 2018). The paradigm for detecting ketamine-induced increases in the BOLD signal was selected to see if the earlier pharmacoBOLD findings (DeSimoni et al. 2013) could be replicated and, as a measure of glutamate, the MRS signal of glutamate + glutamine (MRS-Glx) was used. Following ketamine infusion, a robust BOLD response to ketamine was observed at all sites, replicating the earlier single site study, while MRS-Glx revealed only a small but significant increase vs placebo at one time point (15 minutes) which returned to baseline by 30 minutes (Javitt et al. 2018).

With results from the methodologic studies in hand, effects of POMA (ClinicalTrials.gov Identifier: NCT02919774; Kantrowitz et al. 2020) following administration of 40 mg BID (used in the positive phase 2 trial) and 160 mg BID were assessed. Following 10 days of POMA administration, no effect at either dose on the ketamine BOLD response was detected. Thus, to the extent that reduction of the BOLD response is a valid means of detecting functional engagement of mGluR2 receptors, the lack of efficacy of POMA in phase 3 trials cannot be used as clear evidence against the hypothesis that mGluR2 agonism has therapeutic potential in schizophrenia since the functional readout shown in the rat was not achieved in humans. Put another way, the question remains open about the therapeutic potential of the target since the doses tested to date do not, upon repeated administration, reduce the BOLD response to ketamine.

A Paradigm Shift from DSM-5 to Research Domain Criteria: Application...

223   **14**

If we accept that the ketamine BOLD signal reversal paradigm is a valid translatable functional measure of mGluR2/3 agonism, then the above findings argue for additional clinical studies in schizophrenia if a tolerable dose can be found that does block the ketamine response. To revisit a compound after it has failed in expensive phase 3 studies is not an easy decision and puts a high degree of reliance on translation of a paradigm developed in rats. Translatable readouts of brain effects of drugs, however, can have very high degrees of reliability. The field accepts that if one has a PET ligand for a receptor, then homologous receptor occupancy curves of an orthosteric antagonist provide equivalent information whether in rats, non-human primates, or humans. Similarly, for at least certain types of EEG power spectrum effects, animals and humans are not that different in terms of drug responses. And once the technical hurdles are addressed, one can translate drug-induced fMRI BOLD responses in rats into humans.

> The FAST case studies show that technologies are now sufficiently robust to allow for quantitative translation between animals and humans with regard to drug effects on domain- and/or mechanism-specific brain functions; only in the last decade has this capability become generally recognized and accepted.

There are obvious limits to translations from rats to human based on major differences in brain size and structural details which are only partly addressed when working from non-human primates to humans. But as technology advances with such undertakings as the brain initiative (Mott et al. 2018), we can expect that more and more sophisticated paradigms will emerge for translating brain effects from animals to humans. Such studies may not predict precise human therapeutic effects but can be taken as reliable indicators of whether the molecular and functional brain effects one hopes to test are indeed being testing by the doses used in clinical efficacy studies.

Another finding from this translation of a paradigm worked out in rats to humans was that not all human subjects showed a ketamine response, even though the use of an infusion and determination of ketamine concentrations showed equivalent exposure to that observed in the majority of individuals who did show a robust response. In the absence of understanding why no BOLD response was detectable in some individuals, the subsequent study involving administration of POMA included demonstration of BOLD response to ketamine as an entry criterion. We emphasize this detail as an aspect of translating pharmacodynamic measures from animals to humans since one is often left with unexplainable variability in the latter. The strategy of selecting individuals based on demonstration of a similar effect in human brain to that observed in rat brain was chosen, in the example presented here, as the best way forward to deal with the issue of interpretation of data at the individual rather than group level. The important more general methodologic lesson learned is that for functional measures involving responses, it is important to be sure that subjects show the expected response before trying to modify it.

14.6 Summary

The intent of this chapter has been to provide the reader with an overview of recent precision medicine approaches to drug development for psychiatric disorders and to illustrate the kinds of theoretical and practical issues that are encountered through a consideration of some current studies. The need for new methods of drug development has developed jointly from the realization that current psychiatric syndromes represent poor targets for compounds with novel mechanisms and from the resultant large-scale withdrawal of pharmaceutical companies from psychiatric projects.

One recent concept with particular relevance for mental disorders has been the elaboration of "fast-fail" approaches that acknowledge at the outset the low success rates of bringing new compounds to market. Rather than moving rapidly to expensive phase III trials on the basis of relatively modest phase II efficacy data, the idea is to start with the demonstration of target engagement and move successively through a series of steps that (if successful) culminate in a solid chain of causation for the compound. On the other hand, failure at any one of the steps minimizes costs and lost time by discontinuing the program at that point.

The heterogeneity and co-morbidity of current diagnostic categories compound the difficulty of implementing fast-fail efforts in psychiatric disorders, as only a modest percentage of patients with a given diagnosis may manifest a dysfunction in the mechanism that is targeted by a particular compound. (This fact, combined with variability in samples, seems likely to be one major reason why development programs for so many drugs show inconsistent results from one trial to the next.) The NIMH RDoC program was developed to enable novel psychopathology research by encouraging studies focusing on functional dimensions of behavior for which the relevant neural circuits can be specified. Such a shift has the potential for better alignment of a particular clinical problem (e.g., anhedonia) with a more homogeneous patient sample and also provides more direct translation from animal to human studies.

The two case studies that were presented provide examples of the issues that can arise in the course of a translational study that is built on a strong foundation of animal research. The FAST-MAS project benefited on the pharmacological side from the ready availability of a compound and an appropriate PET ligand to confirm dosage and target engagement, and on the functional side, from a specific hypothesis (regarding anhedonia) and tasks that were appropriate to the goals of the study. The study outcome was also positive, in that the main outcome measure was significant and a second measure showed a further promising result. While it would have been yet more encouraging if some clinical significance had been established, the study was not powered to detect this eventual goal. The development program remains viable.

In contrast, the FAST-PS study was designed to address the setbacks that resulted from an unsuccessful phase III trial for which a suitable PET ligand was not available and the dosing level was estimated based on extrapolation from rodent models. To follow-up whether the phase III doses were sufficient, it was necessary to adopt a target engagement measure that relied on a demonstration that the experimental compound could block the release of glutamate caused by a psychotomimetic drug (ketamine). In turn, this required a preliminary experiment to confirm that pharmacoBOLD and MRS measures could accurately measure glutamate release. When the experimental compound was finally tested, neither a low nor a high dose showed any effect of blocking the glutamate released by ketamine. As discussed above, this indeterminate outcome poses a potentially expensive dilemma for a sponsor. Glutamate release was not blocked by a dose that was twice as high as that used in the phase III trial, so it remains possible that a yet higher dose might be successful; on the other hand, the failure to demonstrate any effects raises the question of whether an effective dose would be too high to be tolerated.

These two case studies indicate that paradigms and methods translated from model paradigms in animals can be auspicious, but not always smooth sailing. However, the field is only at the beginning of new approaches for pharmaceutical development. Projects such as PRISM (Kas et al. 2019) include assessments of how directly one can translate back and forth between animals and humans, taking into account that test batteries have been developed over the past decades that provide tasks as nearly identical as possible (e.g., Robbins et al. 1998). Moreover, tests based on cognitive neuroscience studies in humans are also appearing (e.g., Smucny et al. 2019). Further, computational psychiatry has been growing at a rapid pace, providing tasks that model precise relationships between brain operations and behavioral functions (Paulus et al. 2016). Thus, the transition to outcome measures based on the tight translation from animals to humans, and to model-based tests that offer data-driven assessments of brain-behavior relationships, are advancing rapidly.

At the same time, fast-fail type projects are increasing in both academia and industry as investigators become more familiar with the distinct nature of such projects. Further, the NIH BRAIN project is rapidly developing innovative technologies for measuring brain activity with unprecedented scope in both animals and humans, including more precise MRI techniques (Mott et al. 2018). The use of such advanced measures to serve as biomarkers, combined with the growth of data-driven computational techniques to identify precision phenotypes, appear poised to pinpoint more homogeneous (and potentially trans-diagnostic) groups of patients for clinical trials in the spirit of the RDoC framework. Given the expertise and resources required to implement studies of brain function in the context of translational pharmacology, team science is already imperative and will be more and more required. We believe this will prove a worthwhile effort in the pursuit of novel pharmacological treatments for neuropsychiatric disorders.

Conclusion

In the twentieth century, syndromal, symptom-based diagnostic classes exerted almost total hegemony over all aspects of psychiatric research. Drug development was thus dependent upon so-called animal models of diagnostic categories which were not linked to underlying biology. Major developments in transcending the limits of traditional psychiatric diagnoses and the maturation of methods for exploring living human brain function have led to a new twenty-first-century approach. We can now focus on domains of function such as those specified by RDoC and relate them to brain processes that can be studied in both animals and humans. This allows for selection of novel drugs that we can rule in or out as having brain effects in humans that link to

functions which we know to be important and are variably impaired in many DSM diagnosed conditions. It is predicted that this approach will provide a much more efficient and productive means of identifying and developing drugs for abnormalities observed in psychiatric conditions.

Competing Interests   The authors report no biomedical financial interests or potential conflicts of interest.

References

Abbott A (2010) The drug deadlock. Nature 468: 158–159

Adams DH, Zhang L, Millen BA et al (2014) Pomaglumetad Methionil (LY2140023 Monohydrate) and Aripiprazole in Patients with Schizophrenia: a Phase 3, Multicenter, Double-Blind Comparison. Schizophr Res Treatment 2014:758212. https://doi.org/10.1155/2014/758212. PMCID: PMC3977437

Anderzhanova E, Kirmeier T, Wotjak C (2017) Animal models in psychiatric research: the RDoC system as a new framework for endophenotype-oriented translational neuroscience. Neurobiol Stress 7:47–56

Arnsten AF, Girgis RR, Gray DL et al (2017) Novel dopamine therapeutics for cognitive deficits in schizophrenia. Biol Psychiatry 81:67–77

Artigas F, Schenker E, Celada P et al (2017) Defining the brain circuits involved in psychiatric disorders: IMI-NEWMEDS. Nat Rev Drug Discov 15:1–2

Bennabi D, Haffen E, Van Waes V (2019) Vortioxetine for cognitive enhancement in major depression: from animal models to clinical research. Front Psych 10:1–14

Berridge KC, Robinson TE (2003) Parsing reward. Trends Neurosci 26:507–513

Bruijnzeel AW (2009) Kappa-opioid receptor signaling and brain reward function. Brain Research Review 62:127–146

Carlezon WA Jr, Béguin C, DiNieri JA, Baumann MH, Richards MR, Todtenkopf MS, Rothman RB, Ma Z, Lee DY, Cohen BM (2006) Depressive-like effects of the kappa-opioid receptor agonist salvinorin A on behavior and neurochemistry in rats. J Pharmacol Exp Ther 316:440–447

Cho YT, Lam NH, Starc M et al (2018) Effects of reward on spatial working memory in schizophrenia. J Abnorm Psychol 127:695–709

Cole MW, Repovš G, Anticevic A (2014) The frontoparietal control system: a central role in mental health. Neuroscientist 20:652–664

Cuthbert BN, Insel TR (2013) Toward the future of psychiatric diagnosis: the seven pillars of RDoC. BMC Med 11:127

Cuthbert BN (2019) The PRISM project: social withdrawal from an RDoC perspective. Neurosci Biobehav Rev 97:34–37

Danjou P, Viardot G, Maurice D et al (2019) Electrophysiological assessment methodology of sensory processing dysfunction in schizophrenia and dementia of the Alzheimer type. Neurosci Biobehav Rev 97:70–84

Dawson GR (2015) Experimental medicine in psychiatry: new approaches in schizophrenia, depression and cognition. In: Robbins TW, Sahakian BJ (eds) Translational neuropsychopharmacology. Current topics in behavioral neurosciences, vol 28. Springer, Cham, pp 475–497

Davis M (2006) Neural systems involved in fear and anxiety measured with fear-potentiated startle. Am Psychol 61:741–756

DeSimoni S, Schwarz AJ, O'Daly OG et al (2013) Test-retest reliability of the BOLD pharmacological MRI response to ketamine in healthy volunteers. Neuroimage 64:75–90

Downing AM, Kinon BJ, Millen BA et al (2014) A double-blind, placebo-controlled comparator study of LY2140023 monohydrate in patients with schizophrenia. BMC Psychiatry 14:351–353

Ebner SR, Roitman MF, Potter DN et al (2010) Depressive-like effects of the kappa opioid receptor agonist salvinorin A are associated with decreased phasic dopamine release in the nucleus accumbens. Psychopharmacology (Berlin) 210:241–252

Elmer GI, Brown PL, Shepard PD (2016) Engaging research domain criteria (RDoC): neurocircuitry in search of meaning. Schizophr Bull 42:1090–1095

Fibiger HC (2012) Psychiatry, the pharmaceutical industry, and the road to better therapeutics. Schizophr Bull 38:649–650

Gilmour G, Porcelli S, Bertaina-Anglade V et al (2019) Relating constructs of attention and working memory to social withdrawal in Alzheimer's disease and schizophrenia: issues regarding paradigm selection. Neurosci Biobehav Rev 97:47–69

Goldman-Rakic PS, Castner SA, Svensson TH et al (2004) Targeting the dopamine D1 receptor in schizophrenia: insights for cognitive dysfunction. Psychopharmacology 174:3–16

Gozzi A, Large C, Schwarz A et al (2008) Differential effects of antipsychotic and glutamatergic agents on the fMRI response to phencyclidine. Neuropsychopharmacology 33:1690–1703

Hyman SE (2010) The diagnosis of mental disorders: the problem of reification. Annu Rev Clin Psychol 6:155–179

Insel TR (2012) Next-generation treatments for mental disorders. Sci Transl Med 4:155

Insel TR, Gogtay N (2014) National Institute of Mental Health clinical trials: new opportunities, new expectations. JAMA Psychiat 71:745–746

Javitt DC, Carter CS, Krystal JH et al (2018) Utility of imaging-based biomarkers for glutamate-targeted drug development in psychotic disorders: a randomized clinical trial. JAMA Psychiat 75:11–19

Kantrowitz JT, Grinband J, Goff DC et al (2020) Proof of mechanism and target engagement of glutamatergic drugs for the treatment of schizophrenia: RCTs of pomaglumetad and TS-134 on ketamine-induced psychotic symptoms and pharmacoBOLD in healthy volunteers. Neuropsychopharmacology 45(11):1842–1850. https://doi.org/10.1038/s41386-020-0706-z

Kapur S, Phillips AG, Insel TR (2012) Why has it taken so long for biological psychiatry to develop clinical tests and what to do about it? Mol Psychiatry 17:1174–1179

Kas MJ, Serretti A, Marston H (2019) Quantitative neurosymptomatics: linking quantitative biology to neuropsychiatry. Neurosci Biobehav Rev 97:1–2

Koychev I, McMullen K, Lees J et al (2012) A validation of cognitive biomarkers for the early identification of cognitive enhancing agents in schizotypy: a three-center double-blind placebo-controlled study. Eur Neuropsychopharmacol 22:469–481

Krystal AD, Pizzagalli DA, Mathew SJ et al (2018) The first implementation of the NIMH FAST-FAIL approach to psychiatric drug development. Nat Rev Drug Discov 18:82–84

Krystal AD, Pizzagalli DA, Smoski M et al (2020) A randomized proof-of-mechanism trial applying the 'fast-fail' approach to evaluating κ-opioid antagonism as a treatment for anhedonia. Nat Med 26(5):760–768. https://doi.org/10.1038/s41591-020-0806-7

Lladó-Pelfort L, Troyano-Rodriguez E, van den Munkhof HE et al (2016) Phencyclidine-induced disruption of oscillatory activity in prefrontal cortex: effects of antipsychotic drugs and receptor ligands. Eur Neuropsychopharmacol 26:614–625

Lorrain DS, Baccei CS, Bristow LJ et al (2003) Effects of ketamine and N-methyl-D-aspartate on glutamate and dopamine release in the rat prefrontal cortex: modulation by a group II selective metabotropic glutamate receptor agonist LY379268. Neuroscience 117:697–706

Lowe S, Dean R, Ackermann B et al (2016) Effects of a novel mGlu2/3 receptor agonist prodrug, LY2140023 monohydrate, on central monoamine turnover as determined in human and rat cerebrospinal fluid. Psychopharmacology (Berlin) 219:959–970

Lowe SL, Wong CJ, Witcher J et al (2014) Safety, tolerability, and pharmacokinetic evaluation of single- and multiple-ascending doses of a novel kappa opioid receptor antagonist LY2456302 and drug interaction with ethanol in healthy subjects. J Clin Pharmacol 54:968–978

Mehta MA, Schmechtig A, Kotoula V et al (2018) Group II metabotropic glutamate receptor agonist prodrugs LY2979165 and LY2140023 attenuate the functional imaging response to ketamine in healthy subjects. Psychopharmacology (Berlin) 235:1875–1886

Miller G (2010) Is pharma running out of brainy ideas? Science 329:502–504

Moghaddam B, Adams BW (1998) Reversal of phencyclidine effects by a group II metabotropic glutamate receptor agonist in rats. Science 281:1349–1352

Mott MC, Gordon JA, Koroshetz WJ (2018) The NIH BRAIN initiative: Advancing neurotechnologies, integrating disciplines. PLoS Biology 16(11):e3000066. https://doi.org/10.1371/journal/pbio.3000066

National Institute of Mental Health (2016) Behavioral assessment methods for RDoC constructs. Report by the National Advisory Mental Health Council Workgroup on Tasks and Measures for Research Domain Criteria. https://www.nimh.nih.gov/about/advisory-boards-and-groups/namhc/reports/behavioral-assessment-methods-for-rdoc-constructs.shtml. Accessed 3 Jan 2020

National Institute of Mental Health (2018) RDoC changes to the matrix CMAT Workgroup update: proposed positive valence domain revisions. https://www.nimh.nih.gov/about/advisory-boards-and-groups/namhc/reports/rdoc-changes-to-the-matrix-cmat-workgroup-update-proposed-positive-valence-domain-revisions.shtml. Accessed 3 Jan 2020

National Institute of Mental Health (2019) Definitions of the RDoC domains and constructs. https://www.nimh.nih.gov/research/research-funded-by-nimh/rdoc/definitions-of-the-rdoc-domains-and-constructs.shtml. Accessed 3 Jan 2020

Oldham S, Murawski C, Fornito A et al (2018) The anticipation and outcome phases of reward and loss processing: a neuroimaging meta-analysis of the monetary incentive delay task. Hum Brain Mapp 39:3398–3418

Patel V (2019) Reimagining outcomes requires reimagining mental health conditions. World Psychiatry 18:286–287

Patil ST, Zhang L, Martenyi F et al (2007) Activation of mGlu2/3 receptors as a new approach to treat schizophrenia: a randomized phase 2 clinical trial. Nat Med 13:1102–1107

Paul SM, Mytelka DS, Dunwiddie CT et al (2010) How to improve R & D productivity: the pharmaceutical industry's grand challenge. Nat Rev Drug Discov 9:203–214

Paulus MP, Huys QJM, Maia TV (2016) A roadmap for the development of applied computational psychiatry. Biol Psychiatry Cogn Neurosci Neuroimaging 1:386–392

Porsolt RD, Bertin A, Blavet N et al (1978) Behavioural despair in rats: a new model sensitive to antidepressant treatments. Eur J Pharmacol 47:379–391

Robbins TW, James M, Owen AM, Sahakian BJ et al (1998) A study of performance of tests from the CANTAB battery sensitive to frontal lobe dysfunction in a large sample of normal volunteers: implications for theories of executive functioning and cognitive aging. Cambridge neuropsychological test automated battery. J Int Neuropsychol Soc 4:474–490

Rorick-Kehn LM, Witkin JM, Statnick MA et al (2013) LY2456302 is a novel, potent, orally-bioavailable small molecule kappa-selective antagonist with activity in animal models predictive of efficacy in mood and addictive disorders. Neuropharmacology 77C:131–144

Santana N, Troyano-Rodriguez E, Mengod G et al (2011) Activation of thalamocortical networks by the N-methyl-d-aspartate receptor antagonist phencyclidine: reversal by clozapine. Biol Psychiatry 69:918–927

Shineman DW, Basi GS, Bizon JL et al (2011) Accelerating drug discovery for Alzheimer's disease: best practices for preclinical animal studies. Alzheimers Res Therapy 3:28. https://doi.org/10.1186/alzrt90

Smucny J, Barch DM, Gold JM et al (2019) Cross-diagnostic analysis of cognitive control in mental illness: insights from the CNTRACS consortium. Schizophr Res 208:377–383

Snaith RP, Hamilton M, Morley S et al (1995) A scale for the assessment of hedonic tone: the Snaith–Hamilton pleasure scale. Br J Psychiatry 167:99–103

Wong EF, Yocca F, Smith MA et al (2010) Challenges and opportunities for drug discovery in psychiatric disorders: the drug hunters' perspective. Int J Neuropsychopharmacol 13:1269–1284

Young JW, Winstanley CA, Brady AM et al (2017) Research domain criteria versus DSM-V: how does this debate affect attempts to model corticostriatal dysfunction in animals? Neurosci Biobehav Rev 76:301–316

Zheng MQ, Nabulsi N, Kim SJ et al (2013) Synthesis and evaluation of 11C-LY2795050 as a κ-opioid receptor antagonist radiotracer for PET imaging. J Nucl Med 54:455–463

Clinical Development and Regulatory Approval

Contents

Chapter 15   The Special Challenges of Developing CNS
              Drugs – 231
              Wim Riedel

Chapter 16   Phase 1 Clinical Trials in
              Psychopharmacology – 235
              Eef Theunissen

Chapter 17   Early Development of Erenumab for
              Migraine Prophylaxis – 245
              Gabriel Vargas

Chapter 18   Microdosing Psychedelics as a Promising
              New Pharmacotherapeutic – 257
              Kim P. C. Kuypers

Chapter 19   Partnering with the FDA – 275
              Katie McCarthy and Niki Gallo

The Special Challenges of Developing CNS Drugs

Wim Riedel

Contents

15.1   Clinical Development Challenges: Applied
           Translational Neuroscience – 232

15.2   Clinical Development Challenges: A Special Case
           of Repurposing Illicit Drugs? – 232

15.3   Clinical Development Challenges: Some General
           Methodological and Regulatory Issues – 233

           References – 234

The section in this book on Clinical Development is a mix of achieving goals by means of answering questions through scientific experimentation and following regulatory agreed guidelines. This is reflected by chapters describing the paths and guidelines for phase 1 (▶ Chap. 16), answering bottom-up questions by utilising biomarkers for peripheral target engagement to determine dose selection for clinical phase 2 studies (▶ Chap. 17). Subsequently, the chapter on microdosing of psychedelics uses the top-down perspective to find underpinning of a mechanism for data pointing to antidepressant effects of drugs taken out of the medical- or drug development context (▶ Chap. 18). Finally, the chapter on interactions with FDA binds it altogether defining the best match for new data to medical needs and how to manage expectations in developers and regulators mutually and how to keep track of it (▶ Chap. 19).

> 💬 **Learning Objectives**
> - Know typical challenges of neuroscience drug development
> - Understand the difference between top-down and bottom-up approach
> - Know that illicit drugs can be converted into investigational new drugs

15.1 Clinical Development Challenges: Applied Translational Neuroscience

The example provided by Vargas is typical for Drug Development underpinned by scientific method in this case exemplified by passing the three pillars of survival to answer the main questions: does the drug find its way to the target (PK/PD), does it bind and does it have functional effects (but see ▶ Chap. 17 for details). This is a very good example of applied translational neuroscience by a Pharma company in terms of hitting all the goalposts and achieving registration.

15.2 Clinical Development Challenges: A Special Case of Repurposing Illicit Drugs?

Similar to the repurposing of a drug that is also used by many as an illicit drug, i.e. ketamine for treatment-resistant depression, the example provided by Kuypers Chap. 18 illustrates the seeking of a new application by microdosing of psychedelic drugs, within the broader but also relatively new framework of clinical and academic interest in psychedelic drugs for treating depression. Here clearly, the first and most important pillar of survival has yet to be passed: there seems to be no biomarker of depression and this will be the biggest hurdle on the way to success in this area where the last two hurdles seem to be already cleared before development began from a perspective of an illicit drug – psilocybin is the active ingredient in magic mushrooms – moving its way from the black to the white market. It is already known that it has an effect which is why many users claim to be taking it and there are already several imaging studies showing evidence of on target effects. Here it is however not important whether macro- or microdoses of psychedelics will work, it is the methods followed in clinical drug development using placebo conditions that are important to illustrate. Several small companies of various signatures have picked this up: the non-profit organisation MAPS (= Multidisciplinary Association for Psychedelic Studies; ▶ https://maps.org) has advocated the use of MDMA, ever since its scheduling in 1985, as MDMA-assisted psychotherapy for PTSD and potential other therapeutic targets (Greer and Tolbert 1998). Currently, they have phase 2 studies running in Europe as well as phase 3 studies in the USA. Compass Pathways (▶ https://compasspathways.com) is a commercial biotech company that prepares psilocybin for phase 3 development in treatment-resistant depression. In addition MindMed (▶ https://www.mindmed.co) and Usona (▶ https://www.usonainstitute.org/) are profit and non-profit companies, respectively, that have focused on clinical development of microdoses of psilocybin for depression and associated indications (see also: Ona and Bouso 2019).

15.3 Clinical Development Challenges: Some General Methodological and Regulatory Issues

A number of general methodological and regulatory issues in clinical development especially in neurosciences deserve to be mentioned. Besides the challenges mentioned in the chapter by Vargas (▶ Chap. 17), i.e.: Unclear pathophysiology, Difficulties in translating models, Inaccessibility of the target organ, Subjective endpoints and Lack of biomarkers & diagnostics, I would like to add the following:

— Proactive or receptive safety assessment
In phase 1 what we typically do is receptive safety assessment, i.e. the focus is on registering what is spontaneously reported by healthy volunteer subjects. The method of always systematic probing of the same questions for side effects always, would be the more favorable method. It is even recommendable that subjects should be trained into drug connoisseurs first (Riedel 2019).

— Mechanistic study – must-do or nice to have?
Mechanistic studies which used to be 'nice-to-have' are 'must-do', provided the feasibility of biomarkers that can be discovered or a priori known (candidate-)biomarkers, yet these are scarce in Neuroscience (▶ Chap. 17).

— A bottleneck is the speed at which doctors can enroll patients.
Time is of the essence in clinical development studies; therefore, the more strict the in- and exclusion criteria are, the longer the clinical trials will last.

— Maintaining adherence to patient inclusion criteria

Another consequence of strict in- and exclusion criteria in clinical development studies is non-adherence by those who select the patients. All too often, criteria are stretched as to allow patients to participate in clinical trials.

— Placebo effects
Placebo effects occur both in patients as well as in observers, although in the latter they are mostly referred to as experimenter expectancy effects. For this reason, use of a truly nondistinguishable placebo, thereby maintaining double-blind status, is essential (Rosnow and Rosenthal 1997).

— The bottom-up versus top-down perspective
Rush et al. (2001) focused on how drug-naïve human volunteers experience or feel drug effects. Healthy volunteers received different doses of amphetamine, methylphenidate and placebo. These subjects were asked to rate if the drug that they had taken was a sedative/downer, a stimulant/upper, or placebo. The most striking result is that 75% of healthy volunteers rated d-amphetamine and methylphenidate as sedative/downer and 25% as stimulant/upper; none thought it was placebo. Placebo was identified as a 'placebo/blank' (63%) or 'sedative/downer' (37%); none thought it was a stimulant. Drug effects are most commonly misinterpreted in drug-naïve young men. There is an age effect: Experiencing drug effects is a learning process. There is also a gender effect: Females are better at interpreting drug effects, probably because they are better at interpreting their own bodily signals. In brief, this is about the question whether a drug effect is purely determined by the biology of the drug (bottom-up) or also influenced by the brain (top-down): in terms of what a subject expects (Riedel 2019). This ultimately means that the Information for Volunteers on the Informed Consent Form co-determines therapeutic- and side effects of CNS drugs. Studies in phase 1 often recruit healthy young male volunteers who are often

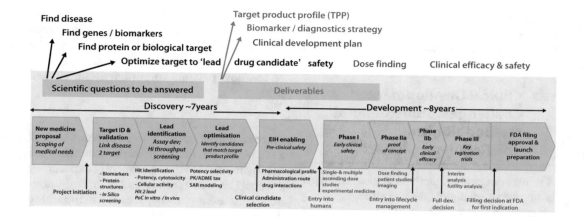

drug-naive. It might be beneficial to recruit subjects with substantial experience in psychoactive drug use, medicinal and/or non-medicinal ('drug-connoisseurs'), although this obviously raises ethical issues.

Conclusion

Drug development for central nervous system (CNS) indications remains highly challenging. Many investors have pulled out, especially when it concerns developing new drugs for psychiatric indications (rather than neurological indications). While the latter is more an area where drug development can lean on the translational neuroscience model, i.e. with defined neurochemical targets with associated biomarkers, psychiatric indications lack biomarkers, perhaps because they even lack biological substrates. Depression might not be a biological disease after all. However, this does not exclude the possibility that mood can be changed with biological manipulations such as is seen in illicit drug use. For depression, it means that re-purposing of 'illicit' drugs into meaningful treatments for mood disorders may be an obtainable goal. Finally, Van Os (2016) states that schizophrenia is not a brain disorder, which in essence means there is no biological substrate either. Likewise this does

not exclude the possibility of influencing the status of 'psychosis susceptibility syndrome' patients as Van Os (2016) calls it, with pharmacological agents, but we may need to reconsider the feasibility of the translational neuroscience model in these indications and pay much more attention to the top-down approach in which there is an important place for the interaction between bottom-up drug effects and top-down expectancy effects.

References

Greer GR, Tolbert R (1998) A method of conducting therapeutic sessions with MDMA. J Psychoactive Drugs 30(4):371–379

Ona G, Bouso JC (2019) Can psychedelics be the treatment for the crisis in psychopharmacology? Preprints, 2019010249.https://doi.org/10.20944/preprints201901.0249.v1

Riedel WJ (2019) Drugs or brains: who obeys? Maastricht University. https://doi.org/10.26481/spe.20190215wr

Rosnow RL, Rosenthal R (eds) (1997) People studying people. Artifact and ethics in behavioral research. W. H. Freeman & Co, New York

Rush CR et al (2001) Reinforcing and subject-rated effects of methylphenidate and d-amphetamine in non-drug-abusing humans. J Clin Psychopharmacol 21(3):273–286

Van Os J (2016) "Schizophrenia" does not exist. BMJ 352:i375

15

Phase 1 Clinical Trials in Psychopharmacology

Eef Theunissen

Contents

16.1   Introduction – 236

16.2   Regulations and Guidelines – 237

16.3   Ethical Approval – 238
16.3.1   Objectives – 239
16.3.2   Study Designs – 239
16.3.3   Methodology – 240

References – 243

Phase 1 trials are an important step in drug development and are usually the first that involve human participants. Their main objective is most often to determine a safe and effective dose range for further clinical development of the drug. Specific guidelines and regulations apply to these type of trials and have been laid down by health authorities. These guidelines are aimed at minimizing the risks and maximizing the benefits and concern the trial's objective, design, methodology, and organization. To obtain approval for a clinical trial, a Clinical Trial Application, outlining how these guidelines are implemented, has to be reviewed and approved by competent authorities and an ethical committee.

In this chapter, an overview is given of the unique features of phase 1 trials, and the specific guidelines that apply to setting up, submitting, and conducting phase 1 trials in the European Union.

🔁 Learning Objectives

- To learn what phase 1 trials are in the process of drug development.
- To get familiar with the specific guidelines for submitting a phase 1 clinical trial application.
- To learn which guidelines and rules of conduct apply to performing a phase 1 study.

16.1 Introduction

Clinical development, or drug development, refers to the whole process of bringing a new drug (or device) onto the market. That is, before a drug or device is suitable for patients, it has to go through an extensive process of research and development. The sponsor, usually the pharmaceutical company, will have to determine the safety and efficacy of the product, to receive a marketing authorization ("license") from the relevant regulatory health authorities (EMA or FDA, see below). First, preclinical research conducted in animals will answer elementary questions about the drug's safety. Next, *clinical research* in human participants has the purpose of testing whether the

drug is effective and safe in humans. Typically, these trials aim to test new drugs treating or preventing a disease or a condition. In order to bring a new compound to the market, clinical trials follow a typical series of studies starting from small-scale, phase 1 studies to post-marketing phase 4 studies (Friedman et al. 2010; European Medicines Agency 1998b; Derhaschnig and Jilma 2010).

> **Definition**
>
> The term "drug" used in this chapter, is considered synonymous with "investigational medicinal product."
>
> *Investigational medicinal product or IMP* is "a pharmaceutical form of an active substance or placebo being tested or used as a reference in a clinical trial, including products already having a marketing authorization but used or assembled (formulated or packaged) in a way different from the authorized form, or when used for an unauthorized indication, or when used to gain further information about the authorized form." (European Parliament and Council of the European Union 2001).

Phase 1 trials are the first stage of testing an investigational medicinal product (IMP) in human participants. This phase involves testing multiple doses in a relatively small sample (10–200 participants) of healthy volunteers. The primary aims are to determine effective dose range, safety and tolerability, pharmacokinetics, and pharmacodynamics of the drug.

Phase 2 trials aim at evaluating the efficacy of the IMP in the target patient group. This is typically done in a much larger sample (100–300 participants). This level of trial is usually randomized, double-blinded, and placebo-controlled and is intended to identify further side effects. Approximately 30% of phase 2 trials progress to the next phase (Thomas et al. 2016), which is in fact the lowest transition success rate of all four stages.

Phase 3 trials are subsequently set up to confirm effectiveness shown in the previous phase. In this stage, an even larger sample is used to augment statistical reliability and to identify less common side effects. In these trials, the IMP will often be compared to a known effective treatment (i.e., active comparator). Phase 3 trials are usually the longest and most expensive to conduct (Thomas et al. 2016).

Phase 4 trials take place after a drug has been marketed for sale. These post-marketing studies concentrate on long-term safety and effectiveness following use of the drug in the real world. These studies provide additional information on the drug's risks and benefits, and can identify any side effects resulting from long-term use or interactions with other drugs. Thousands of participants are typically included in phase 4 studies.

The European Medicines Agency (*EMA*) and the local Ethical Committee (EC) (or the Food and Drug Administration (*FDA*) and Investigational Review Boards (IRB) in the USA) ensure that drug experiments in humans are done in accordance with internationally agreed ethical standards and rules of conduct. Phase 1 trials are an important milestone in the development of new medicines, as this is when they are administered to humans for the first time (hence also called a *first-in-humans* trial, previously known as "first-in-men" trial).

Specific guidelines that apply to setting up and conducting phase 1 trials will be discussed in this chapter. This chapter will provide a starting point to familiarize researchers with some of the most prominent aspects that have to be taken into account when setting up, submitting, and conducting a phase 1 study in the European Union. For complete and up-to-date guidelines and instructions, consult the EMA or contact your local EC.

16.2 Regulations and Guidelines

A trial needs to comply with the appropriate national legislation and regulations of its country. Next to that, all clinical trials performed in the EU have to comply with the guidelines of the Clinical Trials Directive (*Directive 2001/20/EC*). This concerns the implementation of the *Good Clinical Practice* (GCP). The purpose of the GCP guideline is to protect the rights of human participants taking part in clinical trials and to warrant the scientific validity and reliability of the collected data. The rights, safety, and well-being of the trial participant, as formulated by the principles of the Declaration of Helsinki (World Medical Association 2001), are the most important considerations of this guideline and should prevail over the interests of science and society (Dixon 1999).

> **Definition**
>
> The *Clinical Trials Directive 2001/20/EC* is a European legal act, aimed at simplifying and harmonizing the administrative requirements for clinical trials in Europe. It harmonizes the rules for the approval of a clinical trial and sets rules on safety reporting in the context of a clinical trial. Volume 10 of Eudralex, the collection of rules and regulations governing medicinal products in the European Union, contains the guidelines and recommendations on the application of the Clinical Trials Directive (European Parliament and Council of the European Union 2001).
>
> The Clinical Trials Directive will be replaced by the Clinical Trials Regulation in 2020 (Official Journal of the European Union 2014).

It is also essential that the quality and safety of medicines and drugs for human use is assured. Therefore, clinical trials also have to comply with the guidelines of *Good Manufacturing Practice* (GMP), which have been laid down by the *Commission Directive 2003/94/EC*. These guidelines govern the production, distribution, and supply of a drug to make sure that the medicinal product is of consistent and high quality, is appropriate for the intended use, and meets the requirements of the marketing authorization or clinical trial organization (Gouveia et al. 2015; European Commission 2003).

> **Declaration of Helsinki**
>
> The World Medical Association (WMA) developed a set of ethical principles for experiments with human participants. The Declaration was originally adopted in June 1964 in Helsinki, and has undergone 7 major revisions, the most recent in 2013 (Fortaleza). It was an answer in trying to find a balance between the need for knowledge concerning the efficacy and safety of interventions on the one hand, and the risk of harming human participants with untested drugs on the other hand (World Medical Association 2001).

16.3 Ethical Approval

To obtain approval to conduct a clinical trial, a *Clinical Trial Application* (CTA) has to be submitted to a competent authority (CA) and ethics committee (EC). The EC, also known as Institutional Review Board (IRB), reviews whether the clinical trial will be performed according to Good Clinical Practice (GCP) and the appropriate legal requirements. Besides submitting the application to the EC, a valid request for authorization also has to be submitted to the CA of the country in which the trial is planned. A trial may start only if the ethics committee and the competent authority conclude that the anticipated therapeutic benefits justify the risks. Compliance with this prerequisite should be monitored permanently. The requirements for the CTA are described in Directive 2001/20/EC, and additional documents may be required depending on the individual member state (European Parliament and Council of the European Union 2001). The main documents to be submitted are the trial protocol, the Investigator's Brochure (IB), the Investigational Medicinal Product Dossier (IMPD), and the EudraCT application form.

Study protocol: A trial's **objective, design, methodology**, statistical analyses, and **organization** are described in a study protocol.

Investigator's Brochure: The IB summarizes the information of an investigational medicinal product. It contains background information on the properties and history of the IMP, information on the sponsor's name, and the identity of the product (e.g., trade and generic names), and a summary of relevant non-clinical and clinical data (European Parliament and Council of the European Union 2001).

Investigational Medicinal Product Dossier (IMPD): The IMPD contains available information about the IMP and includes summaries of information related to the quality (chemistry, manufacturing, and controls), data from non-clinical studies and clinical trials and experience, as well as an overall risk and benefit assessment (Wood 2009).

The preclinical and clinical information thus can be supplied in the IB but also in the IMPD. By cross referencing between the two documents, overlap can be avoided.

EudraCT form is the application form used for Clinical Trial Applications in the EU. Information on the clinical trial, the investigational medicinal product, the study design, and the different parties involved (e.g., sponsor, applicant, investigators). For most clinical trials, the EudraCT form is uploaded to the EU clinical trials register, making certain information on the trial available to the general public (European Commission n.d.).

Definition

Ethics committee (EC): an independent body in a Member State, consisting of healthcare professionals and nonmedical members, whose responsibility it is to protect the rights, safety, and wellbeing of human subjects involved in a trial and to provide public assurance of that protection, by, among other things, expressing an opinion on the trial protocol, the suitability of the investigators and the adequacy of facilities, and on the methods and documents to be used to inform trial subjects and obtain their informed consent (European Parliament and Council of the European Union 2001).

<table>
<tr><td>

Definition

Competent Authority (CA): The competent authority of each member state is responsible for the regulation of human medicines and the authorization of clinical trials. The authorization and oversight of a clinical trial is the responsibility of the competent authority of the country where the trial is taking place. Each trial must be approved by the competent authority country where the protocol was submitted (European Medicines Agency n.d.).

</td></tr>
</table>

16.3.1 Objectives

The primary objectives of any study should be clear and explicitly stated. Phase 1 trials typically involve one or a combination of the following objectives:

■ ■ Safety and Tolerability

The safety of a medical product concerns the medical risk to the participant, usually assessed by laboratory tests (e.g., clinical chemistry and hematology), vital signs, adverse events, and other specific safety tests (e.g., ECGs). The tolerability of the medical product is the degree to which evident adverse effects are acceptable for the participant. Safety and tolerability are studied for the expected therapeutic dose range and are typically done by using both single and multiple dose administration (European Medicines Agency 1998a).

■ ■ Pharmacokinetics (PK)

A drug's pharmacokinetics (i.e., the drug's absorption, distribution, metabolism, and excretion) provide a lot of information even when the study sample is still relatively small. These analyses provide valuable information to determine the appropriate dose and dose intervals, which is essential for the next phase in the drug's development. In addition, pharmacokinetics are important to be able to

assess the clearance of the drug and to predict possible accumulation of the drug and potential drug-drug interactions (Dunnington et al. 2018).

■ ■ Pharmacodynamics (PD)

How the drug affects the body is studied in pharmacodynamics, providing information on the relationship between the concentration of the drug and biological and physiological effect. This knowledge on the activity, efficacy, and potential side effects of the drug can subsequently be used in deciding on the therapeutic dosage and dose regimen (Pacey et al. 2011).

■ ■ Therapeutic Effect

In phase 1 studies, the therapeutic benefit, i.e., is the intended beneficial effect of the drug, is not the primary objective but can occasionally be studied as a secondary objective. Whether or not it is possible to study the therapeutic effect is dependent of the population and the drug target being studied (European Medicines Agency 1998b).

To achieve their objectives, researchers should first of all ask the relevant questions, and subsequently try to answer these by designing, conducting, and analyzing their clinical trial according to sound scientific principles.

> Phase 1 clinical trials typically aim to determine safety and tolerability, pharmacokinetics, and pharmacodynamics of the expected therapeutic dose range of the drug, and occasionally also study the intended therapeutic effect.

16.3.2 Study Designs

The general rule for any study is that the trial is designed in such a way that it will provide the requested data, using as little participants as needed and ensuring the safety of these participants. This also means that all available information on PK, PD, level of risk, and the number of doses of the IMP to be studied is taken into account in determining the appropriate study design (The Association of the British Pharmaceutical Industry 2018).

Some typical study designs in phase 1 trials are as follows:

- *Sentinel dosing*: To reduce the risk of exposing all participants in a cohort simultaneously, an important advice for phase 1 studies is that the administration of the first dose is given to a limited number of participants (often only one with the active compound) at a time (European Medicines Agency 2017).
- *Single ascending dose trials*: These are trials in which a single dose of the drug is given to a limited number of participants (usually three), in a sequential way, so one by one. After drug administration, participants are monitored and tested for a certain amount of time in order to investigate tolerability, safety, pharmacokinetics, and sometimes pharmacodynamics. If there are no side effects and the outcome data are as predicted, a new batch of participants is subsequently given an increased dose.
- *Multiple Ascending Dose Studies*: This type of study is usually performed after successful completion of single ascending dose trials. In these studies, each subject receives multiple doses of the study drug. The goal of these studies is to determine the pharmacokinetics and its metabolites at steady-state level, to identify if drug accumulation occurs, and to determine maximum tolerated dose (Norfleet and Cox Gad 2010).

It has become also more common to combine the single-dose and multiple-dose trials of an IMP, and even add a trial on the effect of food or age, in order to reduce the cost and time of drug development (European Medicines Agency 2017).

> Phase 1 clinical trials typically use as little participants as needed. Therefore, they often apply sentinel dosing and choose a single or multiple ascending dose set-up.

16.3.3 Methodology

16.3.3.1 Participants

The *sample size*, or number of participants, should be determined and justified in the protocol. Whereas studies in later phases of drug development base their sample size on formal hypotheses and sample size calculations, the guidelines for phase 1 studies are less clear. As the focus of phase 1 studies lies on safety, the sample size should limit unneeded exposure to possible harmful treatments. Sample size will be dependent on PK and PD of the IMP, the study objectives as well as previous comparable studies (Shen et al. 2019; European Medicines Agency 2017).

The *type of participants* should also be determined beforehand. Most phase 1 studies are conducted in healthy participants. This has the advantage that participants are easy to recruit, and it avoids difficulties interpreting the results due to concomitant medication or interfering diseases. However, certain research questions can only be answered in a patient population, or it is more appropriate to include patients, due to the risk-benefit balance, e.g., with highly experimental drugs for life-threatening diseases or when an invasive administration route is essential (Shen et al. 2019; European Medicines Agency 2017; The Association of the British Pharmaceutical Industry 2018).

16.3.3.2 Determining the Dose

An important aspect in a phase 1 trial is deciding on the starting dose. Typically, the dose is determined by using all available preclinical data on pharmacology, toxicology, PK and PD. Data from clinical studies with a drug that has a similar mechanism of action can also be taken into account. For the starting dose, it is again important to find a balance between minimizing the risks (e.g., toxicity) and maximizing the benefits (eliciting a pharmacological response). For determining the dose range studied in a trial, the protocol needs to justify and outline the dose range that will be used, the dose escalation steps, and the maximum exposure. Available data from previous doses (e.g., from previous cohorts) should be carefully taken into account before deciding on increasing a dose and the magnitude of the increase (Shen et al. 2019; European Medicines Agency 2017; The Association of the British Pharmaceutical Industry 2018).

16.3.3.3 Organization

■ **Safety**

Whether the study is done in healthy participants or in patients, safety of the participants has to be guaranteed (European Parliament and Council of the European Union 2001). To limit the risks and to minimize the variability in participants, clear *in- and exclusion criteria* have to be specified in the protocol and should be checked carefully when participants enter the study. In addition, the protocol should describe and define how and at what time points the participant's health is assessed before, during, and after participation in the trial.

Before study inclusion, the investigator checks the eligibility of a participant by performing a *screening*. This includes checking medical history (including use of medication and drugs of abuse), a physical examination (including an electrocardiogram), safety tests in blood and urine samples (hematology, blood chemistry, and urinalysis), and including drug and pregnancy screening (The Association of the British Pharmaceutical Industry 2018).

During a trial, the participant's health and safety is *monitored* via routine assessments, such as checking for adverse events, physical exam, vital signs, safety tests of blood and urine, ECG, and saturation measurements, but it can also include trial-specific tests (European Medicines Agency 2017; The Association of the British Pharmaceutical Industry 2018).

Participants who complete the trial, but also those who drop out prematurely, are required to have a *follow-up*, which usually includes the same assessments as before and during the study trial. In case some parameters are not within the normal ranges as defined in the protocol, the follow-up should be extended until these values return to normal.

■ **Risk Assessment and Management**

Each phase I trial should provide for a strategy to minimize any risk during the study. Again, this needs to be described clearly in the protocol.

■■ **Equipment**

Phase 1 trials should take place in an appropriate facility with access to relevant medical equipment, including specific antidotes (if they exist). The facility should enable close supervision of the participants during and after administration of the IMP.

■■ **Staff**

All investigators should be trained regarding the relevant procedures of the study. Medical supervision should be done by a medical doctor who is familiar with the specifics of the IMP and phase 1 trials. All staff should be GCP trained (European Medicines Agency 2017).

■■ **Reporting Adverse Events**

Adverse events (AEs) are defined as any undesirable experience occurring to a participant during the study, whether or not they are considered related to the IMP or the experimental intervention. All AEs reported spontaneously by the subject or observed by researchers have to be recorded. With regard to serious adverse events (SAEs) and suspected unexpected serious adverse reactions (SUSARs), the protocol should define a plan for a rapid communication between the investigators, IRB, competent authority, and participants (European Medicines Agency 2017).

Safety procedures should be in place so that all researchers involved know how to respond in case of adverse events. The medical doctor will determine the grade (severity) of the adverse event and on what action should be taken. An emergency procedure should be in place, describing how transfer of the participant to the nearest hospital is arranged, in case of life-threatening adverse events (European Medicines Agency 2017).

■■ **Data and Safety Monitoring Board (DSMB)**

A DSMB is not always required for a phase 1 study, but is relevant in trials where interim data analysis is needed to ensure the safety of research participants. A DSMB consists of independent experts who review the clinical study data for incidental events, and clinical study performance at predefined intervals. The DSMB subsequently advices the researchers whether the trial can continue, should be adjusted, or should be terminated (European Medicines Agency 2005).

■■ Stop Rules

Clear stop rules, indicating under which circumstances the trial or the administration of the IMP should be stopped prematurely, have to be defined in the protocol. These rules are set in place to maximize the benefit and minimize the harm for the participants, and are based on the principles of safety (e.g., unexpected severe adverse event), benefit (interim analyses prove the hypothesis early), and futility (successful termination of the study does not seem possible) (European Medicines Agency 2017).

■■ Emergency Unblinding

For double-blind studies, the protocol should foresee a clear procedure for rapid unblinding of a participant's treatment allocation in case of an emergency. All researchers involved should be able to perform this unbinding, in order to avoid any time delay (European Medicines Agency 2017).

> Phase 1 trials' methodology and organization should apply the guidelines which are aimed at minimizing the risks and maximizing the benefits. This applies to selecting the right number and type of participants, determining a safe dose range, monitoring the participants safety, and assessing and managing risks during the trial.

Case Study

Novel Psychoactive Substances in Phase 1 Trials

Around 2010, massive amounts of novel psychoactive substances (NPS) flooded the recreational drug market. NPS consist of earlier developed pharmaceutical compounds and newly synthesized substances which mimic the effects of traditional drugs, such as cocaine, XTC, or cannabis. Yet, they are not controlled by the United Nations drug conventions, and when individual substances are banned, the clandestine producers modify the chemical structures to circumvent this regulation (European Monitoring Centre for Drugs and Drug Addiction and Eurojust 2016).

Usually, little is known about the PK, PD, or adverse effects of these NPS. Most of what we know comes from users' self-reports, hospital reports, and survey studies (Logan et al. 2017; Wood and Dargan 2012). Systematic toxicological and pharmacological studies with NPS in humans are virtually missing, but absolutely needed to define the health risk profiles.

In 2017, our group successfully completed the first controlled experimental study with a popular NPS (a synthetic cannabinoid) (Theunissen et al. 2018). Although the drug in question was not under investigation for drug development, and there was no therapeutic benefit to be expected, the study was technically a first-in-human study. Even though millions on the street had used the drug before, no controlled experiment had been performed in humans. Therefore, the study was set up and conducted according to the phase 1 guidelines.

16

❱ Conclusion

Phase 1 trials are an important step in clinical drug development as they are the first introduction of the new drug to humans. They are focused on the safety aspects of the drug rather than the efficacy in treating a disease. This chapter provides an overview of the most important rules and regulations set by the European Union, in order to minimize the risks for participants taking part in the trial and maximizes the benefits, i.e., the resulting knowledge that is needed for further development of the drug.

References

Derhaschnig U, Jilma B (2010) Phase-I studies and first-in-human trials. In: Müller M (ed) Clinical pharmacology: current topics and case studies. Springer, Vienna, pp 89–99

Dixon JR (1999) The international conference on harmonization good clinical practice guideline. Qual Assur 6(2):65–74

Dunnington K, Benrimoh N, Brandquist C, Cardillo-Marricco N, Di Spirito M, Grenier J (2018) Application of pharmacokinetics in early drug development. In: Pharmacokinetics and adverse effects of drugs-mechanisms and risks factors. IntechOpen

European Commission (2003) Commission Directive 2003/94/EC laying down the principles and guidelines of good manufacturing practice in respect of medicinal products for human use and investigational medicinal products for human use. https://eur-lex.europa.eu/LexUriServ/LexUriServ.do?uri=OJ:L:2003:262:0022:0026:en:PDF. Accessed 16 April 2020

European Commission (n.d.) European Union Drug Regulating Authorities Clinical Trials Database. https://eudract.ema.europa.eu/. Accessed 6 April 2020

European Medicines Agency (1998a) ICH E9 statistical principles for clinical trials (CPMP/ICH/363/96). http://www.ema.europa.eu/docs/en_GB/document_library/Scientific_guideline/2009/09/WC500002928.pdf. Accessed 6 April 2020

European Medicines Agency (1998b) ICH Topic E 8 General Considerations for Clinical Trials. Note for guidance on general considerations for clinical trials (CPMP/ICH/291/95). https://www.ema.europa.eu/en/documents/scientific-guideline/ich-e-8-general-considerations-clinical-trials-step-5_en.pdf. Accessed 1 May 2020

European Medicines Agency (2005) Guideline on data monitoring committees EMEA/CHMP/EWP/5872/03 Corr. https://www.ema.europa.eu/en/documents/scientific-guideline/guideline-data-monitoring-committees_en.pdf. Accessed 5 May 2020

European Medicines Agency (2017) Guideline on strategies to identify and mitigate risks for first-in-human and early clinical trials with investigational medicinal products (EMEA/CHMP/SWP/28367/07 Rev. 1). https://www.ema.europa.eu/en/documents/scientific-guideline/guideline-strategies-identify-mitigate-risks-first-human-early-clinical-trials-investigational_en.pdf. Accessed 15 April 2020

European Medicines Agency (n.d.) National competent authorities (human). https://www.ema.europa.eu/en/partners-networks/eu-partners/eu-member-states/national-competent-authorities-human. Accessed 4 May 2020

European Monitoring Centre for Drugs and Drug Addiction and Eurojust (2016) New psychoactive substances in Europe: legislation and prosecution — current challenges and solutions, EMCDDA–Eurojust joint publication. Publications Office of the European Union, Luxembourg

European Parliament and Council of the European Union (2001) Directive 2001/20/EC of the European parliament and of the council of 4 April 2001 on the approximation of the laws, regulations and administrative provisions of the Member States relating to the implementation of good clinical practice in the conduct of clinical trials on medicinal products for human use. https://ec.europa.eu/health/sites/health/files/files/eudralex/vol-1/dir_2001_20/dir_2001_20_en.pdf. Accessed 4 May 2020

Friedman LM, Furberg C, DeMets DL, Reboussin DM, Granger CB (2010) Fundamentals of clinical trials, vol 4. Springer, New York

Gouveia BG, Rijo P, Gonçalo TS, Reis CP (2015) Good manufacturing practices for medicinal products for human use. J Pharm Bioallied Sci 7(2):87. https://doi.org/10.4103/0975-7406.154424

Logan BK, Mohr ALA, Friscia M, Krotulski AJ, Papsun DM, Kacinko SL, Ropero-Miller JD, Huestis MA (2017) Reports of adverse events associated with use of novel psychoactive substances, 2013–2016: a review. J Anal Toxicol 41(7):573–610. https://doi.org/10.1093/jat/bkx031

Norfleet, E., & Cox Gad, S. (2010). Phase I clinical trials. Pharmaceutical Sciences Encyclopedia: Drug Discovery, Development, and Manufacturing, 1–10

Official Journal of the European Union (2014) Regulation (EU) No 536/2014 of the European Parliament and of the Council of 16 April 2014 on clinical trials in medicinal products for human use and repealing Directive 2001/20/EC. https://ec.europa.eu/health/sites/health/files/files/eudralex/vol-1/reg_2014_536/reg_2014_536_en.pdf. Accessed 6 April 2020

Pacey S, Workman P, Sarker D (2011) Pharmacokinetics and pharmacodynamics in drug development. In: Encyclopedia of cancer. Springer, Berlin Heidelberg, pp 2845–2848

Shen J, Swift B, Mamelok R, Pine S, Sinclair J, Attar M (2019) Design and conduct considerations for first-in-human trials. Clin Transl Sci 12(1):6–19. https://doi.org/10.1111/cts.12582

The Association of the British Pharmaceutical Industry (2018) Guidelines for Phase I clinical trials. 2018 edition. https://www.abpi.org.uk/media/4992/guidelines-for-phase-i-clinical-trials-2018-edition-20180626.pdf. Accessed 13 March 2020

Theunissen EL, Hutten NR, Mason NL, Toennes SW, Kuypers KP, Sousa Fernandes Perna EB, Ramaekers JG (2018) Neurocognition and subjective experience following acute doses of the synthetic cannabinoid JWH-018: a phase 1, placebo-controlled, pilot study. Br J Pharmacol 175(1):18–28

Thomas DW, Burns J, Audette J, Carroll A, Dow-Hygelund C, Hay M (2016) Clinical development success rates 2006–2015. https://www.bio.org/sites/default/files/legacy/bioorg/docs/Clinical%20Development%20Success%20Rates%202006-2015%20-%20BIO,%20Biomedtracker,%20Amplion%202016.pdf. Accessed 6 May 2020

Wood LF (2009) Investigational medicinal products dossier. In: Targeted regulatory writing techniques: clinical documents for drugs and biologics. Springer, Basel, pp 121–124

Wood DM, Dargan PI (2012) Novel psychoactive substances: how to understand the acute toxicity associated with the use of these substances. Ther Drug Monit 34(4):363–367. https://doi.org/10.1097/FTD.0b013e31825b954b

World Medical Association (2001) World medical association declaration of Helsinki. Ethical principles for medical research involving human subjects. Bull World Health Organ 79(4):373

Early Development of Erenumab for Migraine Prophylaxis

Gabriel Vargas

Contents

17.1 Introduction – 246

17.2 Challenges in Neuroscience Drug Development – 246
17.2.1 How Biomarkers May Be Used in Early Development – 247
17.2.2 Early Development – 248

17.3 Phase 2 Dose Selection – 252

References – 254

Modern neuroscience drug development is challenging, success rates are low, and costs are high. A rigorous early development program, which utilizes biomarkers to answer fundamental questions before advancement to later development, will allow us to progress and develop neuroscience medications with higher success rates. Fundamental questions to be answered in early development include the following: does the drug get to the target, does it bind, and does it have an effect? This chapter illustrates some of these concepts by highlighting the development of erenumab, a monoclonal antibody targeting the Calcitonin-Gene related protein (CGRP) receptor for migraine prophylaxis, as a case study of how a peripheral target engagement biomarker was utilized to determine dose selection for clinical phase 2 studies.

> **Learning Objectives**
> - Understand some of the challenges of neuroscience drug development.
> - Learn about the use of target engagement biomarkers.
> - Learn about translational approaches in neuroscience.

17.1 Introduction

Modern drug development is challenging, and success rates are very low. This is particularly true for neuroscience where the success rate has been estimated to be below 3%. This figure is based on data from >28,000 research and development (R&D) projects (i.e., patents filed) between 1990 and 2004 from the Pharmaceutical Industry Database. This database is maintained by the Institutions, Markets, Technologies (IMT) center in Lucca, Italy, and comprises data sets regarding R&D activity and collaborations (Pammolli et al. 2011). The recent failures with beta-site amyloid precursor protein cleaving enzyme (BACE) inhibitors for Alzheimer's disease from various companies are just the latest examples of the inherent difficulties in developing new medications targeting neuropsychiatric disorders (Imbimbo and Watling 2019).

However, not only are success rates low, but the timing of the failure is also problematic. The later the drug fails in the drug development process, the more costly that failure becomes. Unfortunately, these costly late-phase program failures are rising (Pammolli et al. 2011). The costs associated with late stage failure include most prominently, the financial costs of conducting large, expensive phase 3 trials, as well as the opportunity cost of not spending money and resources on other programs which may have been more likely to succeed.

A recent analysis of the economics of drug development demonstrated that the cost of developing a new drug roughly doubles every 9 years. This phenomenon was termed Eroom's law, a play on the well-known dictum from computer science, Moore's law, which is that overall processing power for computers will double every 2 years (Scannell et al. 2012). Taken together, these trends of low success rates and increasing costs spell trouble ahead for the drug development enterprise. It is essential to improve the way we develop drugs if we are to have an impact on the low success rates and increasing costs.

17.2 Challenges in Neuroscience Drug Development

One of the reasons for the low success rates in drug development is our inability to predict efficacy in early development. Despite advances in the preclinical stage of drug development where through rigorous and efficient toxicology assessments we have improved our ability to select safer compounds, we still have a big problem with efficacy. Currently programs most commonly fail due to lack of efficacy, not safety (Arrowsmith 2011).

Efficacy failure reflects our lack of understanding of the human disease and particularly its heterogeneity (Arrowsmith 2011). This is particularly true in neuroscience where our understanding of these disorders is limited. There are many reasons for this limited understanding, including, among others:

- Unclear pathophysiology
- Difficulties in translating models
- Inaccessibility of the target organ
- Subjective endpoints
- Lack of biomarkers and diagnostics

17.2.1 How Biomarkers May Be Used in Early Development

One way forward to meet the challenge in neuroscience drug development is to develop and employ biomarkers which can be used in early development to determine whether a drug has the right properties to progresses further to late development.

> **Definition**
>
> A biomarker can be defined as follows: A characteristic that is objectively measured and evaluated as an indicator of normal biologic or pathogenic processes or pharmacological responses to a therapeutic intervention (NIH Biomarkers Definitions Working Group 2001).

Biomarkers are commonly used to aid in prognosis, patient stratification, and diagnosis (Vargas 2009; Nikolcheva et al. 2011). However, in early development, the type of biomarkers most used are those that provide an indication of the drug's impact on biology. This type of biomarker is called a target engagement biomarker. Target engagement refers to the action of a molecule on its target. For example, for glycine transporter inhibitors which were being developed for the treatment of the negative symptoms of schizophrenia, increases in the cerebrospinal fluid levels of glycine (Hofmann et al. 2016) were used to determine the level of biological effect of the drug. This measurement follows directly from the mechanism of action of the drug and provides information that the drug has reached its site of action, bound the target, and had a biologically relevant effect.

Target engagement biomarkers are essential in neuroscience where it is difficult to rely on clinical measures in early development. These measures are often subjective, and there are issues with the uniformity of rater assessments and the small sample sizes of early studies. A target engagement biomarker, which elucidates the core features of the mechanism of action of a drug, will enable the advancement of only those molecules that have the right pharmacokinetic properties to 1) reach their target receptor, 2) to bind to the receptor (2), and 3) to have a biological effect which can be measured. Scientists at Pfizer have labeled this approach *the three pillars of survival*.

These pillars (See ❏ Table 17.1) capture some of the fundamental questions in early development. The benefits of addressing and answering these questions were described in an article from Pfizer (Morgan et al. 2012). In the publication they reviewed Pfizer trials to determine success rates based on whether these trials had successfully addressed the three pillars. Their conclusion was that successful assessment of the three pillars correlated with higher success rates. The review covered 44 Phase 2 programs that were run between 2005 and 2009. Out of 14 trials in which all three pillars were addressed, 8 programs went on to phase 3 development. In marked contrast, from 12 trials in which none of these pillars were met, not a single program went on to further development.

> ❯ The impact of target engagement biomarkers cannot be overstated. They provide critical data which enables decision-making on whether a drug has reached its target and has had an effect. Many companies now require evidence of the type of biological impact, which these biomarkers are able to provide before proceeding further in clinical development.

The usefulness of target engagement biomarkers and the determination of the three pillars of survival are demonstrated in the development of the migraine prophylactic erenumab (trade name Aimovig™ ▶ Chap. 11). For the

☐ **Table 17.1** These fundamental determinants defined as the *three pillars of survival* all determine the likelihood of candidate survival in Phase 2 trials. A positive assessment of each of these increases the probability of success for the programs

Three pillars of survival	Questions
An integrated understanding of the fundamental pharmacokinetic/pharmacodynamic principles of exposure at the site of action	*Does the drug get to the target (in CNS this often is dependent on the drug crossing the blood brain barrier)?*
Target binding	*Does it bind?*
Expression of functional pharmacological activity	*Does it have an effect?*

Adapted from Morgan et al. 2012

rest of this chapter, the early development of erenumab will be used as an example of the use of target engagement biomarkers in neuroscience.

17.2.2 Early Development

Early development refers to the period from the time a new molecular entity (NME)

Definition

A new molecular entity (NME) is a drug that contains an active moiety that has never been approved by regulatory agencies anywhere in the world.

---

is about to enter humans for the first time until after it has completed its initial phase 2 study. Prior to entry into humans the project team developing the NME must prepare enabling studies for generating data for regulatory agency submission. The data which must be provided to regulatory agencies regarding the NME include the following: (1) Toxicology data to provide an understanding of safety in preclinical species and how this would translate to human doses. (2) Chemistry and Manufacturing Controls (CMC) is the term used to describe the data for the manufacture and testing of a medicinal prod-

uct. CMC data provide assurance that the NME is being made and measured correctly. (3) A clinical development plan which has a strong rationale for disease indication, dosing to be used and makes note of any potential safety issues and does not dose outside of safety margins calculated from the toxicology data.

Once the investigational new drug (IND) submission is accepted by the regulatory agency, the single ascending dose (SAD) part of the first-in-human (FIH) study begins. The initial dose of an FIH study is a very important decision and takes into account data from animals where either an established model of disease or a target engagement model can be used to then translate the dosing to the human study.

Once the drug is shown to be tolerated without any safety findings, the dose is increased until either a maximum tolerated dose is found or the dose approaches a level that is considered sufficient to meet full target engagement.

Following the SAD, the multiple ascending dose (MAD) begins which follows the same process but consists of giving a dose usually daily for 1 to 2 weeks.

It is during the FIH study that target engagement biomarkers are used to determine early indications of efficacy and whether the three pillars are met.

Erenumab Early Drug Development

Migraine is a common debilitating disorder, which affects over 35 million individuals in the USA. A migraine is more than just a severe headache. It is a neurovascular disorder in which patients experience moderate to severe pain, which can be debilitating at times, and is associated with throbbing and aggravation on movement. Thet episodes typically last anywhere from 4 to 72 hours. Patients also frequently experience nausea or vomiting and photophobia or phonophobia. Around 20% of patients also are affected by an aura, which are neurological disturbances in vision (Goadsby et al. 2002; Charles 2017). Employers in the USA lose more than $13 billion each year as a result of 113 million lost workdays due to migraine.

People with migraine use about twice the medical resources – including prescription medications and office and emergency room visits – as non-sufferers. It is the most common neurological condition in the world. When occurring frequently, it is highly debilitating and leads to a poor quality of life affecting school, work, and social relationships.

Migraine can be divided into episodic and chronic, which are defined based on frequency of attacks. Patients with more than 15 occurrences of migraine a month are considered chronic, those with fewer than 15 occurrences a month are episodic. If patients experience more than 3–4 migraines per month, it is recommended that prophylactic treatment be considered.

One problem with prophylactic treatment, however, is that until recently the prophylactic medications available for migraine were poorly tolerated or had poor efficacy. This changed with the advent of drugs targeting the Calcitonin-Gene related protein (CGRP) pathway. CGRP is a peptide which was first identified in the early 1980s with the discovery that the calcitonin gene encodes CGRP and that the peptide is expressed in both the central and peripheral nervous systems (Rosenfeld et al. 1983; Russell et al. 2014).

Migraine pathophysiology is thought to be due to a neurovascular response in which patients are overly sensitive to certain stimuli (triggers), which can lead to release of CGRP that causes pain and vasodilation. CGRP is expressed in areas of the brain involved in migraine including the trigeminal ganglion and was found to be elevated in patients experiencing migraines (Goadsby et al. 1990; Goadsby and Edvinsson 1993; Edvinsson 2008). Interestingly, it was also shown to produce migraine-like attacks in patients susceptible to migraines but only headaches in subjects without a migraine history (Lassen et al. 2002; Hansen et al. 2010).

The development of antibodies against CGRP was initiated based on the clinical trials conducted by Boehringer Ingelheim on BIBN 4096 BS and by Merck on MK-0974. It was demonstrated that targeting this pathway was efficacious in the treatment of acute migraine attacks (Olesen et al. 2004; Ho et al. 2008).

The first drug approved with a mechanism of action targeting the CGRP pathway was erenumab (trade name Aimovig™). Erenumab is a human monoclonal antibody that binds to the CGRP receptor and interferes with the CRGP molecule binding to its receptor (de Hoon et al. 2018). Other migraine prophylactic medications targeting this pathway have developed antibodies which bind to the CGRP ligand instead of the receptor but have the same goal that is to interrupt the CGRP signaling pathway (Bigal et al. 2015; Charles et al. 2019).

The early development of erenumab involved an approach, which leveraged the use of biomarkers to translate the preclinical data findings to humans. Since there are no well-established models of migraine in preclinical species, it was necessary to rely on a peripheral target engagement model. As previously mentioned, use of target engagement biomarkers is an extremely valuable approach to demonstrate that the drug is having an effect. This is particularly true for neuroscience drugs where the effect is difficult to capture in the typically small early development trials. Instead of relying on a subjective clinical rating scale – which is usually the way to determine efficacy in later development – the target engagement approach allows one to use an alternative strategy. We rely on a characteristic which is dependent on the mechanism of action of the drug to determine that the molecule is reaching its target, binding, and having an effect. In other words, it enables the early development program to address the three pillars of survival.

The target engagement biomarker model used for erenumab, and indeed for all the antibodies targeting the CGRP pathway, is capsaicin-induced dermal blood flow (CIDBF) (See ◘ Figs. 17.1 and 17.2). Initially pioneered by Merck (Hershey et al. 2005; Li et al. 2015), CIDBF takes advantage of the peripheral effects of CGRP. In the skin, the CGRP receptor is expressed in blood vessels where it functions to promote vasodilation and hence increased blood flow. The stimulation of CGRP-induced vasodilation can be triggered by the application of capsaicin to the skin where it binds and stimulates the TRPV1 receptor. This in turn leads to the release of CGRP (See ◘ Fig. 17.1) (Sinclair et al. 2010), which causes the blood vessel to dilate and this is detected by changes in blood velocity using a laser doppler imager (See ◘ Fig. 17.2).

The assessment of CGRP blockade is conducted by measuring capsaicin-induced dermal blood flow in the presence or absence of erenumab (or other anti-CGRP antibodies) and measuring whether the drug inhibits the increased blood flow. This biomarker was initially used in

non-human primates and then it was shown that the results translated well to humans (Shi et al. 2016). This allowed testing of several different antibodies preclinically before choosing the best candidate to move into clinical development. The CIDBF biomarker used in cynomolgus monkeys enabled the demonstration of in vivo dose-dependent target coverage for erenumab and helped to select the doses for the FIH studies.

In FIH studies, this target engagement biomarker was used to demonstrate that erenumab dose-dependently prevented CIDBF (See ◘ Fig. 17.3). The response is very steep which means that small changes in erenumab concentration around the inflection point can lead to major changes in CIDBF blockade. Once past the steep part of the dose response curve, increasing concentrations of erenumab lead to a longer blockade of the CIDBF response but not a greater extent of blockade.

One of the benefits of establishing a target engagement biomarker such as CIDBF is that the three pillars of survival can be addressed. And if all three pillars are in place, then the probability of success is expected to be much higher for a compound.

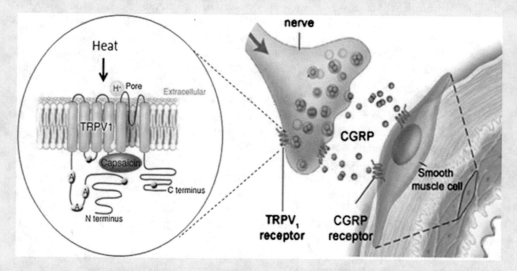

◘ **Fig. 17.1** Capsaicin-induced CGRP release. In the skin, the CGRP receptor is expressed in blood vessels on smooth muscle cells where it functions to promote vasodilation leading to increased blood flow. The stimulation of CGRP-induced vasodilation can be triggered by the application of capsaicin to the skin. Capsaicin applied to the skin binds to and stimulates the TRPV1 receptor, which in turn leads to the release of CGRP. This released CGRP causes the blood vessel to dilate, which can be detected by changes in blood velocity using a laser doppler imager. (Adapted from Buntinx et al. 2015)

17

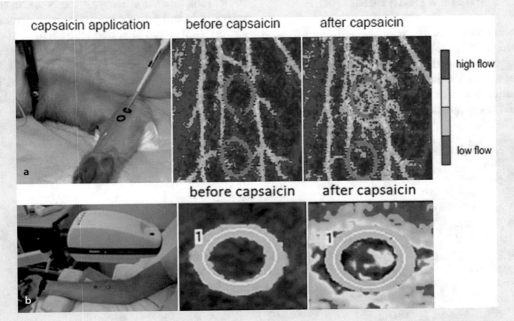

○ **Fig. 17.2** Capsaicin-induced dermal blood flow model. This biomarker model can be used in primates **a** and the data translated to humans **b**. Capsaicin application to the skin is done within plastic O-rings. In A the top ring has been administered capsaicin, while the bottom ring has been administered saline. As can be seen in the third panel of **a**, the capsaicin clearly induces an increase in dermal blood flow, which is measured with laser Doppler imaging. In **b**, the volunteer also has two O-rings and the O-ring to the right is being measured with a laser doppler machine. The third panel of **b** shows the increased blood flow caused by administration with capsaicin. (Adapted from Buntinx et al. 2015)

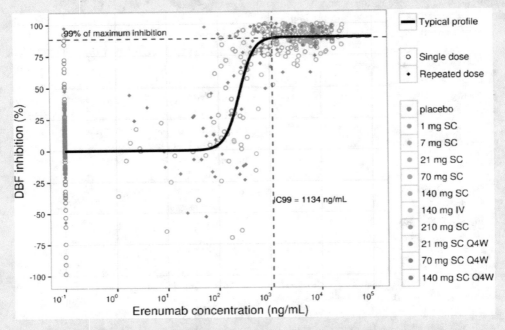

○ **Fig. 17.3** Relationship between erenumab concentrations and dermal blood flow (DBF) inhibition. The symbols represent individual observations. IV, intravenous; Q4W, every 4 weeks; SC subcutaneous. (From Vu et al. 2017)

17.3 Phase 2 Dose Selection

A fundamental piece of data which must be provided by the FIH studies is what dose to use in the initial phase 2 study. Phase 2 studies are larger than FIH studies and only enroll patients unlike the FIH studies where mostly healthy subjects are recruited. Their purpose is to demonstrate the efficacy of the NME in the relevant patient population and to also demonstrate that it is well tolerated without any safety issues which might preclude further development. The dose is selected based on any efficacy measures which might have been used in FIH studies or more likely in the case of neuroscience, a target engagement biomarker.

Case Study

Application of the CIDBF model to establish the relationship between erenumab concentration and dermal blood flow inhibition enabled us to predict the erenumab exposures, magnitude, and duration of peripheral target engagement, and their theoretical relationship to efficacy in the migraine patient population. This approach was used to guide dose selection in phase 2 studies. We used the erenumab concentration-CIDBF model to select doses for a phase 2 study with the assumption that maximum CIDBF inhibition is required for migraine efficacy. Three doses, 7, 21, and 70 mg administered subcutaneously (SC) monthly were selected for the phase 2 study.

Based on CIDBF simulations, our prediction was that both 21 mg and 70 mg SC monthly doses would be effective in the episodic migraine population to be tested in the phase 2 program. Our expectation was that, based on CIDBF exposure response modeling, a 70 mg SC monthly regimen would have better efficacy. The 7 mg SC dose was added to demonstrate a noneffective dose.

Since these predictions of migraine efficacy were based on a *peripheral* target engagement model whose relationship to *clinical efficacy* was unknown, we chose the highest dose for the phase 2 study to cover an exposure that is predicted to be well over the IC99. Indeed, this 70 mg dose exceeded IC99 by sevenfold after the first administration and tenfold after the third administration of antibody (Vu et al. 2017).

The phase 2 study was a 12-week long multicenter, randomized, double-blind, placebo-controlled trial with 483 patients aged 18–60 years with 4 to 14 migraine days per month. Results demonstrated a separation from placebo for only the 70 mg SC monthly dose and not for the lower doses of 7 mg and 21 mg. The mean change in monthly migraine days at week 12 was −3·4 days with erenumab 70 mg versus −2·3 days with placebo (difference −1·1 days, $p = 0·021$) (See ◻ Fig. 17.4). The mean reductions in monthly migraine days with the 7 mg (−2·2) and the 21 mg (−2·4) doses were not significantly different from that with placebo (Sun et al. 2016). These results are important in terms of how we can use target engagement biomarkers in neuroscience drug development. The CIDBF model, despite being a well-validated biomarker of peripheral target engagement, did not precisely predict the exposures needed to obtain antimigraine efficacy (Sun et al. 2016). It did, however, enable the estimation of doses that were likely to be efficacious, and it did predict that the 7 mg SC dose, which by the CIDBF model was not efficacious, was also found to be without clinical efficacy.

One possibility for the discrepancy between the CIDBF model prediction of doses and clinical efficacy is that accessing the migraine-relevant neurovascular compartments is harder to accomplish than accessing peripheral sites. Therefore, larger doses than those producing maximal peripheral CGRP receptor inhibition may be necessary for anti-migraine efficacy. The take-home message is that these models of target engagement are very useful and necessary but likely allow us to set the lower boundary of the drug exposure which is needed. To achieve clinical efficacy, we may need to administer medications at exposures substantially above what is predicted from solely a peripheral target engagement model. Indeed, studies conducted with higher doses of erenumab have demonstrated that a dose of 140 mg monthly SQ leads to a greater reduction of migraine days per month (3.2 in the 70-mg erenumab group and 3.7 in the 140-mg erenumab group, as compared with 1.8 days in the placebo group ($P < 0.001$ for each dose vs. placebo) (Goadsby et al. 2017).

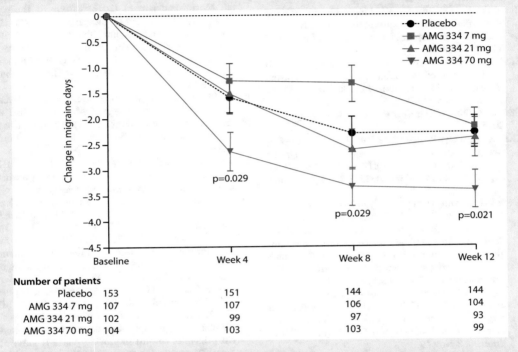

Number of patients

	Baseline	Week 4	Week 8	Week 12
Placebo	153	151	144	144
AMG 334 7 mg	107	107	106	104
AMG 334 21 mg	102	99	97	93
AMG 334 70 mg	104	103	103	99

▫ **Fig. 17.4** Change from baseline in mean monthly migraine days. Data are least squares means (SE). p value is for placebo group versus 70 mg dose group. AMG 334 = erenumab. (Adapted from Sun et al. 2016)

❯ Conclusion

The success rates in neuroscience drug development are low and the costs are high. To improve our probability of success, an early development neuroscience program must include target engagement biomarkers that allow assessment of the three pillars of survival. A positive assessment of these three pillars improves the chances of the new chemical entity progressing to late development. In the case of erenumab the translational biomarker approach used took advantage of CIDBF, a peripheral target engagement biomarker. Use of this biomarker demonstrated that erenumab maximally inhibited the CGRP receptor in the periphery with high potency and reproducibility, with favorable PK characteristics that translate to a prolonged CGRP receptor blockade.

Capsaicin-induced dermal blood flow was essential for demonstrating target engagement and to select phase 2 dosing. However, the phase 2 data demonstrates that CIDBF overestimated the potential efficacy but did allow for the selection of doses that demonstrated efficacy. The FDA approved erenumab for migraine prophylaxis in 2018 at dose levels of 70 mg and 140 mg SC per month.

References

Arrowsmith J (2011) Trial watch phase II failures: 2008–2010. Nat Rev Drug Discov 10:328–329

Bigal ME, Walter S, Rapoport AM (2015) Therapeutic antibodies against CGRP or its receptor. Br J Clin Pharmacol 79:886–895

Biomarkers Definition Working Group (2001) Biomarkers and surrogate endpoints: preferred definitions and conceptual framework. Clin Pharmacol Therap 69:89–95

Buntinx L, Vermeersch L, de Hoon J (2015) Development of anti-migraine therapeutics using the capsaicin-induced dermal blood flow model. Br J Clin Pharmacol 80(5):26

Charles A (2017) Migraine. N Engl J Med. August 10, 2017

Charles A, Pozo-Rosich P (2019) Targeting calcitonin gene-related peptide: a new era in migraine therapy. Lancet 394(10210):1765–1774. https://doi.org/10.1016/S0140-6736(19)32504-8. Epub 2019 Oct 23

de Hoon J, Van Hecken A, Vandermeulen C, Yan L, Smith B, Chen JS, Bautista E, Hamilton L, Waksman J, Vu T, Vargas G (2018) Phase I, randomized, double-blind, placebo-controlled, single-dose, and multiple-dose studies of Erenumab in healthy subjects and patients with migraine. Clin Pharmacol Ther 103(5):815–825. https://doi.org/10.1002/cpt.799. Epub 2017 Oct 24

Edvinsson L (2008) CGRP-receptor antagonism in migraine treatment. Lancet 372(9656):2089–2090. https://doi.org/10.1016/S0140-6736(08)61710-9. Epub 2008 Nov 25

Goadsby PJ, Edvinsson L (1993) The trigeminovascular system and migraine: studies characterizing cerebrovascular and neuropeptide changes seen in humans and cats. Ann Neurol 33:48–56

Goadsby PJ, Edvinsson L, Ekman R (1990) Vasoactive peptide release in the extracerebral circulation of humans during migraine headache. Ann Neurol 28:183–187

Goadsby PJ, Lipton RB, Ferrari MD (2002) Migraine – current understanding and treatment. N Engl J Med 346:257–270

Goadsby PJ, Reuter U, Hallström Y, Broessner G, Bonner JH, Zhang F, Sapra S, Picard H, Mikol DD, Lenz RA (2017) A controlled trial of Erenumab for episodic migraine. N Engl J Med 377(22):2123–2132. https://doi.org/10.1056/NEJMoa1705848

Hansen JM, Hauge AW, Olesen J, Ashina M (2010) Calcitonin gene-related peptide triggers migraine-like attacks in patients with migraine with aura. Cephalalgia 30:1179–1186

Hershey JC, Corcoran HA, Baskin EP, Salvatore CA, Mosser S, Williams TM, Koblan KS, Hargreaves RJ, Kane SA (2005) Investigation of the species selectivity of a nonpeptide CGRP receptor antagonist using a novel pharmacodynamic assay. Regul Pept 127(1–3):71–77

Ho TW, Mannix LK, Fan X, Assaid C, Furtek C, Jones CJ, Lines CR, Rapoport AM (2008) Randomized controlled trial of an oral CGRP receptor antagonist, MK-0974, in acute treatment of migraine. Neurology 70:1304–1312

Hofmann C, Pizzagalli F, Boetsch C, Alberati D, Ereshefsky L, Jhee S, Patat A, Boutouyrie-Dumont B, Martin-Facklam M (2016) Effects of the glycine reuptake inhibitors bitopertin and RG7118 on glycine in cerebrospinal fluid: results of two proofs of mechanism studies in healthy volunteers. Psychopharmacology (Berl) 233(13):2429–2439. https://doi.org/10.1007/s00213-016-4317-7. Epub 2016 May 14

Imbimbo BP, Watling M (2019) Investigational BACE inhibitors for the treatment of Alzheimer's disease. Expert Opin Investig Drugs 28(11):967–975. https://doi.org/10.1080/13543784.2019.1683160. Epub 2019 Oct 29

Lassen LH, Haderslev PA, Jacobsen VB, Iversen HK, Sperling B, Olesen J (2002) CGRP may play a causative role in migraine. Cephalalgia 22(1):54–61

Li CC, Vermeersch S, Denney WS, Kennedy WP, Palcza J, Gipson A, Han TH, Blanchard R, De Lepeleire I, Depre M, Murphy MG, Van Dyck K, de Hoon JN (2015) Characterizing the PK/PD relationship for inhibition of capsaicin-induced dermal vasodilation by MK-3207, an oral CGRP receptor antagonist. Br J Clin Pharmacol 79:831–837

Morgan P, Van Der Graaf PH, Arrowsmith J, Feltner DE, Drummond KS, Wegner CD, Street SD (2012) Can the flow of medicines be improved? Fundamental pharmacokinetic and pharmacological principles toward improving phase II survival. Drug Discov Today 17(9–10):419–424. https://doi.org/10.1016/j.drudis.2011.12.020. Epub 2011 Dec 29

Nikolcheva T, Jäger S, Bush TA, Vargas G (2011) Challenges in the development of companion diagnostics for neuropsychiatric disorders. Expert Rev Mol Diagn 11(8):829–837. https://doi.org/10.1586/erm.11.67

Olesen J, Diener HC, Husstedt IW, Goadsby PJ, Hall D, Meier U, Pollentier S, Lesko LM (2004) Calcitonin gene-related peptide receptor antagonist BIBN 4096 BS for the acute treatment of migraine. N Engl J Med 350:1104–1110

Pammolli F, Magazzini L, Riccaboni M (2011) The productivity crisis in pharmaceutical R&D. Nat Rev Drug Discov 10(6):428–438. https://doi.org/10.1038/nrd3405

Rosenfeld MG, Mermod JJ, Amara SG et al (1983) Production of a novel neuropeptide encoded by the calcitonin gene via tissue-specific RNA processing. Nature 304:129–135

Russell FA, King R, Smillie SJ, Kodji X, Brain SD (2014) Calcitonin gene-related peptide: physiology and pathophysiology. Physiol Rev 94:1099–1142

Scannell JW, Blanckley A, Boldon H, Warrington B (2012) Diagnosing the decline in pharmaceutical R&D efficiency. Nat Rev Drug Discov 11(3):191–200. https://doi.org/10.1038/nrd3681

Shi L, Lehto SG, Zhu DX, Sun H, Zhang J, Smith BP, Immke DC, Wild KD, Xu C (2016) Pharmacologic characterization of AMG 334, a potent and selective human monoclonal antibody against the calcitonin gene-related peptide receptor. J Pharmacol Exp Ther 356(1):223–231. https://doi.org/10.1124/jpet.115.227793. Epub 2015 Nov 11

Sinclair SR, Kane SA, Van der Schueren BJ, Xiao A, Willson KJ, Boyle J, de Lepeleire I, Xu Y, Hickey L, Denney WS, Li CC, Palcza J, Vanmolkot FH, Depré M, Van Hecken A, Murphy MG, Ho TW, de Hoon JN (2010) Inhibition of capsaicin-induced increase in dermal blood flow by the oral CGRP receptor antagonist, telcagepant (MK-0974). Br J Clin Pharmacol 69(1):15–22. https://doi.org/10.1111/j.1365-2125.2009.03543.x

Sun H, Dodick DW, Silberstein S, Goadsby PJ, Reuter U, Ashina M, Saper J, Cady R, Chon Y, Dietrich J, Lenz R (2016) Safety and efficacy of AMG 334 for prevention of episodic migraine: a randomised, double-blind, placebo-controlled, phase 2 trial. Lancet Neurol 15(4):382–390. https://doi.org/10.1016/S1474-4422(16)00019-3. Epub 2016 Feb 12

Vargas GA (2009) Personalized healthcare: how to improve outcomes by increasing benefit and decreasing risk through the use of biomarkers. Biomark Med 3(6):701–709. https://doi.org/10.2217/bmm.09.74.

Vu T, Ma P, Chen JS, de Hoon J, Van Hecken A, Yan L, Wu LS, Hamilton L, Vargas G (2017) Pharmacokinetic-Pharmacodynamic relationship of Erenumab (AMG 334) and capsaicin-induced dermal blood flow in healthy and migraine subjects. Pharm Res 34(9):1784–1795. https://doi.org/10.1007/s11095-017-2183-6. Epub 2017 Jun 7

Microdosing Psychedelics as a Promising New Pharmacotherapeutic

Kim P. C. Kuypers

Contents

18.1 **Background** – 258

18.2 **Clinical Research with Low Psilocybin Doses** – 259

18.3 **Clinical Research with Low LSD Doses** – 262
18.3.1 Older Research with LSD – 263
18.3.2 Recent Research with LSD – 268

18.4 **Discussion** – 269
18.4.1 LSD and Psilocybin: What Is the Difference – 270
18.4.2 Full Psychedelic Doses Show Therapeutic Value: Do We Need Microdoses? – 270
18.4.3 Is there an "Ideal" Dosing Schedule? – 270
18.4.4 Unwanted Effects – 271
18.4.5 Does Expectation Play a Role? – 271
18.4.6 The Importance of User Claims and Survey Studies in Drug Development – 272

References – 272

© The Author(s), under exclusive license to Springer Nature Switzerland AG 2021
R. Schreiber (ed.), *Modern CNS Drug Discovery*, https://doi.org/10.1007/978-3-030-62351-7_18

Microdosing psychedelics, the repeated use of small doses of substances such as psilocybin and lysergic acid diethylamide, has gained popular and scientific attention in recent years. While some users claim microdosing psychedelics has therapeutic value, to date only a handful of (placebo-controlled) experimental studies in human volunteers have been conducted testing the effects of low doses on physiological, subjective state, and performance measures. This chapter aims to answer, based on the scientific knowledge we have so far, whether microdosing psychedelics has therapeutic potential. Reviewed studies demonstrated that low doses were in general well tolerated. Single doses produced subtle, beneficial effects on selective performance measures and subjective states. The fact that most studies were conducted in small samples of healthy (young) volunteers hampers generalization to other populations. However, the observed cognitive and affective effects might be of help in some psychiatric disorders such as attention-deficit hyperactivity disorder or depression. Future placebo-controlled studies in patient populations are needed to conclude about the (therapeutic) potential of microdosing psychedelics.

🔁 Learning Objectives

- To learn what psychedelics are, and more specifically LSD and psilocybin
- To learn what microdosing psychedelics is
- To take note of scientific evidence of microdosing psychedelics and to understand the limitations, but also to understand its therapeutic potential
- To take note of the questions that yet have to be answered

18.1 Background

Sixty years after pharmaceutical agents made their way to the psychiatric ward, it is clear that they are not a panacea (Penn and Tracy 2012). A considerable proportion of patients never remits, and there is notable despair among patients due to this unmet need (Penn and Tracy 2012). Recently psychedelics have regained attention

as therapeutic aid after a scientific radio silence of a few decades (Carhart-Harris and Goodwin 2017). Psychedelics, the collective name for psychoactive substances characterized by their hallucinogenic effect, include lysergic acid diethylamide (LSD) and psilocybin as prototypical examples (Osmond 1957). They cause their effects on behavior and cognition primarily via agonism of the serotonin (5-HT) 2A receptor in the brain (Nichols 2016) and are considered to be a relatively safe class of substances that generally do not cause addiction but rather show potential in the treatment of addiction (Nichols 2016; Sakloth et al. 2019).

> **Definition**
>
> In the first half of the twentieth century, until the scheduling of psychedelics as Schedule I drugs in the 1970s, a lot of research with these substances took place (Chi and Gold 2020). Thereafter there was a relative standstill in this research field, until a few years ago, which now leads us to talk of a "*second wave of psychedelic research.*"

> **Definition**
>
> *Lysergic acid diethylamide (LSD)* was isolated from ergot preparations in 1938 by Albert Hofmann who also discovered its hallucinogenic properties (Lee 2010). After oral ingestion of 100–200 mcg LSD (base), the effects peak around 2.5 hour and last approximately for 8–12 hours (Dolder et al. 2017).

> **Definition**
>
> *Psilocybin* (4-phosphoryloxy-N,N-dimethyltryptamine) is found globally in certain types of mushrooms and truffles. In 1957 Albert Hofmann isolated psilocybin from Central American mushrooms (Passie 2019). Once ingested, psilocybin converts to the psychoactive psilocin. Its subjective effects peak between 60 and 90 minutes after oral ingestion of 15–25 mg psilocybin and last approximately for 4–6 hours (Tylš et al. 2014).

Preliminary data from patient studies suggest that already one or two regular doses of a psychedelic in combination with some kind of psychological assistance are effective in the combatting of severe, persistent psychiatric disorders (dos Santos et al. 2016; Fuentes et al. 2020; Johnson and Griffiths 2017). Anecdotal evidence suggests that even very low doses of these substances without therapeutic care are effective in the treatment of certain psychopathologies (Johnstad 2018). The repeated use of these substances in small, sub-hallucinogenic, or sub-perceptual doses, which is usually one-tenth of a recreational dose, is also referred to as microdosing (Fadiman 2011). Users claim this to be linked with positive effects on creativity, productivity, mood and relationships with others, and their environment, and general well-being (Anderson et al. 2019b; Fadiman and Korb 2019; Hutten et al. 2019a; Lea et al. 2020; Polito and Stevenson 2019). Motives vary from personal development to cognitive improvement, and also mental health improvement (Fadiman and Korb 2019; Hutten et al. 2019a; Lea et al. 2020; Polito and Stevenson 2019; Johnstad 2018; Webb et al. 2019).

Case Study

Microdosing in the Context of Psychedelic Research and of Drug Development

Oral doses producing hallucinogenic effects typically used in human research are 100 to 200 mcg LSD, usually given as fixed dose, and 15 mg of psilocybin on average, which is usually dosed per body weight (Liechti 2017; Studerus et al. 2011). In general, a microdose is considered one-tenth of a full psychedelic dose, which would be 10–20 micrograms (mcg) of LSD and/or 0.3–0.5 g of psilocybin-containing mushrooms (Hutten et al. 2019a; Kuypers et al. 2019). The included reviewed experimental studies demonstrated beneficial effects of LSD (base) in doses ranging from 10 to 20 mcg of LSD and for psilocybin after administration of <1–3 mg psilocybin.

Of note, the definition of microdosing in the context of psychedelic research differs from that used in drug development. There, a microdose is one percent of the pharmacologically active dose, up to a maximum of 100 µg. It is considered to be sub-pharmacological and is used in phase 0 studies to collect early pharmacokinetic parameter information, before proceeding with "higher" doses (Garner and Lappin 2006; Bertino Jr. et al. 2007). In that context, a microdose would be 1–2 mcg LSD and 3–5 mg of psilocybin-containing mushrooms.

Decades earlier, Albert Hofmann, the "discoverer" of LSD and its hallucinogenic effects, mentioned that "very small doses, perhaps 25 micrograms", could be useful as an antidepressant (Ghose 2015; Horowitz 1976) or as a substitute for Ritalin (Fadiman 2017; Horowitz 1976). Most of the evidence about its therapeutic effectivity comes from survey studies in which users retrospectively report on their effects (Hutten et al. 2019b). Nonetheless, a handful of clinical studies have been conducted. This chapter will give an overview of the scientific knowledge about microdosing psychedelics, based on these clinical studies, and try to answer whether there is ground to believe there is therapeutic potential in microdosing psychedelics. Additionally, it will point out which questions remain unanswered and will give some food for thought for drug developers.

18.2 Clinical Research with Low Psilocybin Doses

To date, six studies have investigated the effects of (low) doses of psilocybin on physiological measures, subjective state, and cognitive performance in healthy volunteers (Griffiths et al. 2011; Hasler et al. 2004; Madsen et al. 2019; Prochazkova et al. 2018) and in patients with OCD (Moreno et al. 2006), and anxiety and/or depressive symptoms in cancer patients (Griffiths et al. 2016) (◘ Table 18.1). One of those studies was an uncontrolled naturalistic study in a group of people who self-administered psilocybin-containing truffles in a social setting (Prochazkova et al. 2018). One-and-a-half hour after ingesting 0.35–0.41 g of truffles, divergent and convergent thinking was increased Compared to a non-drug (Prochazkova et al.

■ Table 18.1 Methodological details of the included experimental studies with low doses of psilocybin

Author	Aim, to test…	Design (number of conditions)	Intervention (Dosage, Route of Administration)	Sample (N)	Average age	Setting	Physiological Measures [Findings[a]]	Subjective State Measures [Findings[a]]	Performance Measures [Findings[a]]
Griffiths et al. (2011)	Or to compare the ascending and descending sequences of drug–dose exposure	Within-group design (5 drug conditions)	Psilocybin (0, 5, **10, 20,** 30 mg/70 kg)	Healthy volunteers (18)	46	Aesthetic living-room-like environment with two monitors present; couch, eye mask, headphones for music	Blood pressure [+], heart rate [+]	Questionnaire: Drug effects [+]	
Griffiths et al. (2016)	The effects of psilocybin on symptoms of anxiety and depression in cancer patients	Within-group design (2 drug conditions)	Psilocybin (1 or 3 mg/70 kg and **22** or **30** mcg/70 kg); *half received the low dose first, half the high dose*	Cancer patients with anxiety and/ or depressive symptoms (N (1 mg) = 38; N (3 mg) = 12)	56.3	Aesthetic living-room-like environment with two monitors present	Blood pressure [+], heart rate [+]	Questionnaire: Affective effects [+]	
Hasler et al. (2004)	Or to explore the potential dose–response relationship of psilocybin on various parameters	Within-group design (5 drug conditions)	Psilocybin (mcg/ kg bodyweight, 0, 45, **115, 215, 315**)	Healthy volunteers (8)	29.5	Psychiatrists were present during the session	24-hr ECG [−]; hormones [−], standard clinical chemical parameters [−]	Questionnaire: Drug effects [+], affective effects [+]	Paper-and-pencil test: Concentration [−]

Study	Aim	Design	Drug/dose	Population (n)	Age	Setting	Positron emission tomography	Questionnaire	Paper-and-pencil test
Madsen et al. (2019)	The relationship between the subjective psychedelic experience, plasma psilocin levels, and 5-HT2AR occupancy in the human brain	Within-group design (repeated measures of participant receiving a specific dose)	Psilocybin (3, 6, **12, 15, 18, 24, 30 mg**)	Healthy volunteers (8)	33	Two psychologists providing interpersonal support were present; during PET scans music was played	Positron emission tomography: 5-HT2A receptor binding [+]	Questionnaire: Drug [+]	
Moreno et al. (2006)	The safety, tolerability, and clinical effects of psilocybin in patients with OCD	Within-group design (4 drug conditions)	Psilocybin (25, **100, 200, 300** mcg/kg, baseline)	OCD patients (9)	40.9	Participants were asked to wear eyeshades, listen to music and minimize the interaction during the session; trained sitters were present		Questionnaire: OCD symptoms [+]	
Prochazkova et al. (2018)	The cognitive-enhancing potential of microdosing psychedelics	Within-group design, naturalistic study	Psilocybin truffles: average dose of 37 g, baseline[b]	Healthy volunteers (38)	31.1	Non-laboratory environment (microdosing event); tasks were conducted in a group setting free from outside distraction			Paper-and-pencil test: Intelligence [-], convergent [-], divergent [+] thinking

a[+] = presence, [–] = absence of low-dose psilocybin effect relative to placebo or another control condition; doses in bold are not considered microdose and are not considered in the last column where the effects are shown relative to the reference condition

b In this naturalistic study, dried psilocybin-containing mushrooms were taken by participants

2018). Because of the uncontrolled nature of this study, placebo-controlled experimental studies are needed to conclude with certainty that the observed effects are due to the intervention and not due to learning effects, expectation, or the social context (Prochazkova et al. 2018). While they calculated the dose people would have to take to have one-tenth of a recreational dose, and they had the truffles analyzed afterward, it appeared that 1 g contained 1.6 mcg psilocybin, so the psilocybin dose was on average 0.6 mcg of psilocybin, which can be regarded as a very low dose.

In the three placebo-controlled experimental studies in healthy volunteers, subtle effects on physiological measures (Hasler et al. 2004) and subjective state were found (Griffiths et al. 2011; Madsen et al. 2019) compared to placebo. Heart rate decreased 6 hours after ingestion of 2.3 mg psilocybin/70 kg person (45 mcg psilocybin/kg bodyweight) together with a slight increase in drowsiness, and heightened sensitivity and intensification of preexisting mood states that were most prominent effects at this dosage level. A slightly higher dose (3 mg, fixed dose) in the study by Madsen et al. (2020) led to a psychedelic experience of average (40%) intensity (Madsen et al. 2019). With a 5 mg/70 kg bodyweight dose -which is double compared to Hasler et al (2004)-, Griffiths and colleagues (2011) demonstrated mild psychedelic effects compared to placebo (Griffiths et al. 2011). Madsen et al. (2019) demonstrated that psilocybin (3–30 mg) binds to the 5-HT2A receptor, with a positive relation between the degree of receptor binding (%) and the intensity of the psychedelic experience (Madsen et al. 2019). The 5-HT2A binding after the 3 mg dose was 43%; the highest psilocybin dose (30 mg) led to a psychedelic experience of maximum intensity (100%) and a binding rate of 65% (Madsen et al. 2019). Hasler et al. (2004) who also assessed several hormone levels and performance by means of a paper-and-pencil test did not show effects compared to placebo (Hasler et al. 2004).

Moreno et al. (2006) reported findings of their small-scaled study in which they administered on separate occasions a range of psilocybin doses to patients ($N = 9$) with obsessive compulsive disorder. Besides a low dose of psilocybin (25 mcg/kg body-weight = 1.75 mg/70 kg person), three higher doses were included (100, 200, 300 mcg/70 kg person). Relative to baseline, the low psilocybin dose induced symptom reduction (Moreno et al. 2006), suggesting its potential to positively change cognitive habitual and control processes, an effect that might also be relevant in other psychopathologies like depression (Vanderhasselt et al. 2014). Nonetheless, future studies in (large) patient samples have to confirm this. Griffiths et al. (2016) tested the effects of low (1 or 3 mg) and high doses (22 or 30 mg) of psilocybin in patients with life-threatening cancer (Griffiths et al. 2016). While the high dose produced the most changes in subjective state compared to baseline, and more pronounced changes compared to the low dose, the low dose decreased anxiety and depression ratings compared to baseline. With respect to adverse effects, both physical and psychological effects were noticed, that were more frequent in the high-dose condition, compared to the low-dose condition. Physical discomfort (any type) occurred in 8% of the low-dose sessions compared to 21% in the high-dose sessions; psychological discomfort (any type) occurred in 12% of the low-dose sessions compared to 32% in the high-dose sessions. An episode of anxiety occurred in 15% of the low-dose sessions and in 26% of the high-dose sessions (Griffiths et al. 2016).

> Only a handful of experimental studies have investigated the effects of low doses of psilocybin on human behavior and performance. While subtle effects on physiological and psychological parameters were shown, no firm conclusions can be drawn seen the small amount of data.

18.3 Clinical Research with Low LSD Doses

In the first half of the twentieth century, a number of LSD studies in healthy volunteers and patients took place. These aimed to investigate among others the effects of a broad range of LSD doses on physiological and performance measures, some also including low LSD doses (Abramson et al. 1955; Greiner et al. 1958; Isbell et al. 1956; McGlothlin et al. 1967) (◻ Table 18.2). More recently, five placebo-controlled experimental studies aiming to assess the effects of low LSD doses on cognition, subjective experience, mood, and brain activity took place (Bershad et al. 2020; Bershad et al. 2019; Family et al. 2019; Gasser et al. 2015; Yanakieva et al. 2019). Four of these studies were conducted in healthy volunteers, one in people suffering from anxiety related to life-threatening diseases (◻ Table 18.3).

Case Study

LSD Base or Tartrate

An important addition when talking about LSD doses is to specify whether LSD base or LSD ("salt") tartrate was used, especially in this microdosing range. As Liechti (2017) clarified this in a commentary to a microdosing paper, "a dose of 100 μg LSD base corresponds to 123 μg LSD tartrate" (Kuypers et al. 2019). Apparently, the older research used LSD tartrate, whereas modern research, with full psychedelic doses, uses LSD base (Liechti 2017).

18.3.1 Older Research with LSD

Abramson et al. (1955) who conducted a range of experiments with LSD in the 1950s combined the data of 141 experimental sessions in 31 participants; eight of them received a dose between 1 and 25 mcg LSD. Because observers did not use a standardized questionnaire in the different experimental sessions, data was clustered into five symptom classes – euphoria, dysphoria, changes in perception, neurotic behavior, and psychotic symptoms. The effects of the low doses of LSD on the selected parameters were demonstrated to be very mild or placebo-like with only a slight increase in euphoria and an absence of effects on psychotic behavior, distortions in perception, and dysphoria which were increased by higher doses of LSD, and neurotic behavior, which was not affected by any LSD dose (Abramson

et al. 1955). Of note, these findings should be, as the authors also stated, regarded as descriptive due to the confounding factors of different measurements, settings, doses, and small sample size for some doses. In another study, Greiner et al. (1958) administered five different doses of LSD and placebo to 14 healthy male volunteers after which effects were measured up to 4 hours after treatment on subjective state, physiological measures, performance. While they mentioned it was a double-blind design, they did not describe how many doses participants received, which was more than one as they stated that three participants received placebo, two 4 mcg, six 7 mcg, two 12 mcg, six 20 mcg, and five 40 mcg LSD, where the latter dose does not classify as "microdose" (Greiner et al. 1958). Participants noticed effects on mood starting from 7 mcg, but they did not experience the changes in mood states that we observed by the experimenter (Greiner et al. 1958). Interesting was the cycling pattern of depressed and euphoric mood states. Linked to that, the authors expressed their concern about mood changes potentially negatively affecting higher-order cognitive processes like planning and motivation (Greiner et al. 1958). Of note, no statistical analyses were performed, and seen the sample size per dose, and the way effects were reported, i.e., it was marked as a change when it was seen in more than 50% of the group, this paper should merely be seen as qualitative, descriptive research.

In two of the older studies included in this chapter, LSD was given repeatedly and effects were assessed on a range of physiological and subjective state measurements (Isbell et al. 1956; McGlothlin et al. 1967). Isbell et al. (1956) published on the findings of six experiments in which a range of LSD doses (0.25–2 mcg/kg or 10–180 mcg) was administered in several regimens to test a number of questions. Only the findings of the three studies administering low doses of LSD are described here (Isbell et al. 1956). These all included subjective state and physiological measures. LSD (0.25 mcg/kg = 17.5 mcg) did not produce differentiating effects from placebo. Interestingly, repeated administration of low doses (10–30 mcg), twice daily for three days, produced transient tolerance to the mental

■ Table 18.2 Methodological details of the included older experimental studies with low doses of LSD

Author	Aim, to test...	Design (number of conditions)	Intervention (dosage, route of administration)	Sample (N)	Average age	Physiological measures [findings]	Subjective state measures [findings]	Performance measures [findings]
Abramson et al. (1955)	The mental effects of a range of LSD doses	Not stated (6 dosing categories)	LSD (0, 1–25, 26–50, 51–75, 76–100, 101+ mcg, p.o.)	Non-psychotic, adult volunteers (31); *Some participants received multiple doses, though it was not clear who; 8 received a low dose*	Not stated		Questions related to 5 mental state 'themes': Euphoria [±], dysphoria [−], perception and psychotic behavior [−], neurotic behavior [−]	
Greiner et al. (1958)	The dose-response effects	Double-blind, not clear how many participants received more than one dose	LSD (0, 4, 7, 12, 20, 40 mcg, p.o.)	Healthy male volunteers (14); 0 mcg (3); 4 mcg (2); 7 mcg (6); 12 mcg (2); 20 mcg (6); 40 mcg (5)	Not stated	Galvanic skin conductance [+, 7 mcg], pupil diameter [+, 12–40 mcg]	Mood [+; 7–40 mcg] and perception [body, alertness, emotion thought] [+, 20–40 mcg]; psychiatric information: Changes in mood [cycles of depression, euphoria] and behavior [+, 7–40 mcg]	Psychiatric information: Changes behavior [+, 7–40 mcg]

18

Isbell et al. (1956)	I. The dose-response effects; II. the reproducibility of the LSD effect using a specific measure; III. the tolerance to LSD after (long term) daily intake (4 sub-experiments; only 2 with low doses)	I. Within subject; II. within subject; III. within subject	I. LSD (0, 0.25, **0.5**, **0.75**, **1**, **1.5**, **2** mcg/kg, p.o., intervals of 1 week); II. LSD (**60 mcg**, p.o.); IIIa. LSD twice a day for three days (10 mcg, 20 mcg, **30 mcg**, or placebo, p.o.), day 4: LSD (**75 mcg**, p.o.), 3 days placebo (or LSD, p.o.), day followed by LSD (**75 mcg**, p.o.); IIIb. LSD, once daily for 7 days (20–**75 mcg**, p.o.)	Former morphine addicts, all male, abstinent for at least 3 months; I. N = 8; II. N = 11; IIIa. N = 11; IIIb. N = 12	Not stated	Pupil size, blood pressure [−, I.], knee jerk [−, I.]	Mental effects questionnaire and observation [−, I.; +, IIIa-b. reversible tolerance to LSD effects after dosing for three days in a row; tolerance disappeared after 3 days of abstinence]
McGlothlin et al. (1967)	The mental effects of a high dose of LSD	Mixed within-between group, within: 5 test days (baseline, 3 experimental sessions, 2 follow-ups at 2 weeks and 6 months); between: treatment (3 drug conditions)	LSD (25, **200 mcg**), amphetamine (20 mg; 5 immediate +15 sustained release)	Healthy volunteers (N = 24/group)	Not stated	Galvanic skin response (anxiety) *[because of the nature of statistics that were performed, no findings are presented]*	Questionnaire: Anxiety, attitude, value, aesthetic sensitivity, creativity *[because of the nature of statistics that were performed, no findings are presented]*

a[+] = presence, [−] = absence of low dose LSD effect relative to placebo or another control condition; doses in bold are not considered a microdose and are not considered in the last column where the effects are shown relative to the reference condition

■ Table 18.3 Methodological details of the more recent experimental studies with low doses of LSD included

Author	Aim, to test…	Design (number of conditions)	Intervention (dosage, route of administration)	Sample (N)	Average age	Physiological measures [findings]	Subjective state measures [findings]	Performance measures [findings]
Bershad et al. (2019)	The effect of low LSD doses on subjective experience and cognitive behavior	Within-group design (4 drug conditions)	LSD tartrate (0, 6.5, 13, 26 mcg, p.o.)	Healthy volunteers (20)	25	Blood pressure [+]; heart rate [−], basal temperature [−]	Questionnaires: Drug [+] and affective [+] effects	Computer tests: working memory [−], cognitive performance [−], emotion recognition [−], social inclusion [−], convergent thinking [−]
Bershad et al. (2020)	The neural effects of a low dose of LSD on resting-state cerebral blood flow (CBF) and connectivity using functional magnetic resonance imaging	Within-group design (2 drug conditions)	LSD tartrate (0, 13 mcg)	Healthy volunteers (20)	25	Blood pressure [+], fMRI: Connectivity analysis [+]	Questionnaires: Drug [+] and affective effects [+]	
Family et al. (2019)	The safety, tolerability, pharmacokinetics, and pharmacodynamics of repeated low-dose LSD administration	Mixed within-between group design (4 drug groups); repeated dosing (dose each 3 days; 21 days in total)	LSD tartrate (0, 5, 10, 20 mcg)	Healthy volunteers (12/group)	62.9	Blood pressure [−], pulse rate [−], clinical laboratory evaluations (i.e., hematology, blood chemistry, urinalysis) [−], electrocardiogram parameters [−], physical examination [−]	Questionnaires: Drug [+]	Computer tests: reaction time [−], visual memory and learning [−], visual attention [−], spatial working memory [−], balance and proprioception [−]

18

Gasser et al. (2014)	The safety and efficacy of LSD-assisted psychotherapy	Double-blind (2 drug conditions)	LSD free base (20, **200** mcg, p.o.)	Patients with anxiety associated with life-threatening diseases (11)	51.7	Heart rate [−], blood pressure [−]	Adverse events [+, more participants reported anger, anxiety, and abnormal thinking compared to after 200 mcg dose], self-report questionnaire for state and trait anxiety [+, state and trait anxiety increased and decreased after the 200 mcg crossover]	
Yanaki-eva et al. (2019)	The effects of repeated low doses of LSD on time perception	Mixed within-between group design (4 groups); repeated dosing (dose each 3 days; 21 days in total)	LSD tartrate (0, 5, 10, 20 mcg)	Healthy volunteers (12/group)	62.9		Questionnaires: Drug [+]	Computer test: time reproduction [+]

a[+] = presence, [−] = absence of low-dose LSD effect relative to placebo or another control condition; doses in bold are not considered a microdose and are not considered in the last column where the effects are shown relative to the reference condition

effects of a subsequent higher dose of LSD (75 mcg) (Isbell et al. 1956). Based on these findings, it can be suggested that daily microdosing would not be an efficient practice and also that an abstinence period of three days would be long enough to reinstate the mental response to a higher dose (75 mcg) of LSD.

McGlothlin et al. (1967) aimed to test the long-lasting effects of three subsequent administrations of a high dose of LSD (200 mcg) on measures of subjective state in healthy volunteers. Next to the high-dose group, there were two control groups respectively receiving single doses of amphetamine (20 mg) or LSD (25 mcg), also on three separate occasions (McGlothlin et al. 1967). Volunteers were tested at baseline, after administration of the treatment and at 2 weeks and 6 months posttreatment. Unfortunately for the aim of this chapter, the data of the two control groups were combined to conduct statistical analyses as the findings of both groups allegedly did not differ systematically. Nonetheless, from this it can be inferred that LSD (25 mcg) has a similar effect pattern as the stimulant amphetamine (20 mg) in the mentioned doses. Interestingly, throughout their paper, the authors give percentages of people in the three groups that have experienced subjective state effects acutely, at 2 weeks, and at 6 months of follow-up. Whereas LSD 25 mcg was labeled as "pleasant" by the majority (78%) and without lasting effects (65%), LSD 200 mcg was labeled as "dramatic and intense" (71%) with some lasting effects (42%). Of note, the 25 mcg LSD group was included as a control group in the hope they would experience enough visual or auditory hallucinations and therefore realize they received LSD which would be a good control for prior expectations. The same proportion of people in the LSD 25 mcg and amphetamine group thought they received LSD on one or more sessions (McGlothlin et al. 1967).

> In general, the older LSD studies are atypical in the light of current methodological standards which are placebo-controlled, double-blind experimental studies with validated questionnaires and objective performance measures. Nonetheless, these studies produced some interesting findings, also due to the fact they made use of qualitative data, something that is valued again today.

18.3.2 Recent Research with LSD

In two placebo-controlled experimental studies in young healthy volunteers, the acute effects of single low doses of LSD (6.5, 13, and 26 mcg) were assessed on subjective state, performance measures, and physiological measures in a within-subject design (Bershad et al. 2019). LSD led to a statistically significant, though clinically irrelevant, increase in systolic blood pressure (13 and 26 mcg) and diastolic blood pressure (26 mcg) 2 hours after LSD administration, compared to placebo. Participants felt "under the influence" after taking 13 and 26 micrograms LSD and drug liking increased together with drug disliking. They felt better, friendlier, and also more anxious compared to placebo (Bershad et al. 2019). LSD (13 mcg) induced changed brain connectivity in the limbic ("emotion") system which was linked to the changes in positive mood (Bershad et al. 2020). No acute LSD effects were found on heart rate and basal body temperature, other mood states (vigor, depression, anger, confusion or fatigue), cognitive skills, or social behavior compared to placebo (Bershad et al. 2019). Performance was assessed with tasks sensitive to the effects of full psychedelic doses of a psychedelic (Pokorny et al. 2017; Pokorny et al. 2019). With regard to potential persisting effects, participants were asked to fill out a mood questionnaire the days following the administration of LSD and placebo. While no mood effects were shown, it has to be remarked that the questionnaire was only completed by 55% of the participants (Bershad et al. 2019).

In another placebo-controlled study, in older healthy volunteers, LSD was given repeatedly each 3 days, for 21 days in total, in a between-subjects design (Yanakieva et al. 2019; Family et al. 2019). LSD doses were 0, 5, 10, and 20 mcg, and cognitive performance was tested at different times and with different measures alongside physiological and pharmacokinetic measures. The only statistically significant effect was an overestimation of time intervals of 2000 ms (and longer) in a time perception task after the fourth dose. These effects were most pronounced for the 10 mcg dose (Yanakieva et al. 2019). Blood was collected to determine LSD concentrations after doses 1 and 6. LSD plasma concentrations were detectable after the 10 and 20 mcg dose, with an observed peak approximately half an hour after administration. The total blood concentration after dose 1 and dose 6 did not differ, which demonstrates that this parameter is not affected by repeated doses. The average half-life over all data points was 8.25 ± 7.5 hours, which is comparable to that of a full psychedelic dose (200 mcg), i.e., 8.9 ± 5.9 hours (Dolder et al. 2015). There were not significantly more adverse events in the LSD groups compared to the placebo group, though mild-to-moderate headache was more often reported in the LSD group (Family et al. 2019).

Gasser et al. (2014) investigated the safety and efficacy of LSD in combination with psychotherapy in patients with anxiety related to life-threatening diseases. One group received 200 mcg of LSD twice and the other (control) group 20 mcg of LSD, also twice, followed by an open-label crossover to 200 mcg once the treatment blind was broken. The low dose was thought to produce short-lived, mild LSD effects that would not substantially facilitate the therapeutic process. Two regular psychotherapy sessions followed each LSD session. Self-rated anxiety (trait and state) decreased after two high dose sessions with LSD, an effect that was sustained up to 12 months after treatment. In the low-dose group however, anxiety increased after two sessions with LSD 20 mcg and decreased after the crossover to LSD 200 mcg, an effect which was also demonstrated at 12 months of follow-up. The number, frequency, and intensity of drug-related adverse events were higher in the high-dose condition compared to the low-dose condition, though anger, anxiety, and abnormal thinking were more frequent in the low-dose condition. LSD did not affect physiological parameters. Of note, the total sample size was very small, and the low dose was only given to three participants. The authors correctly remarked that the (fluctuating) medical conditions of the participants could have influenced the psychological state, and hence the self-rated anxiety (Gasser et al. 2014). While the data seems to suggest that a low dose of LSD does not support the therapeutic process, this study did not aim to test this hypothesis, and future studies in larger samples should corroborate this.

> Similar to psilocybin, only a handful of recent experimental studies have investigated the effects of low doses of LSD on human behavior and performance. The difference with psilocybin studies however is that most of the LSD studies were set up to investigate specifically the effects of low doses rather than including them as an active control. In addition to subtle effects on physiological and psychological parameters, an effect on brain connectivity was shown. Nonetheless no solid conclusions can be drawn because of the small amount of data.

18.4 Discussion

The aim of this chapter was to review the scientific knowledge on microdosing psychedelics in order to understand its potential as future pharmacotherapeutic agent in the treatment of psychiatric conditions. The reviewed (placebo-) controlled studies reveal subtle effects of low doses of LSD (10–20 mcg, base) and psilocybin (<1–3 mg) on selective performance measures (time perception, divergent/convergent thinking), subjective states (cycling mood patterns, anxiety, positive mood), and physiological parameters (blood pressure, pupil size, brain connectivity). The slowed down time perception, the decrease in OCD symptoms and the enhancement of convergent and divergent thinking (Bershad et al. 2019; Moreno et al. 2006; Yanakieva et al. 2019), and the changes in brain connectivity related to positive mood (Bershad et al. 2020) suggest the potential of a low dose of a psychedelic to respectively enhance a state of being "in the present," the balance between controlled and habitual cognitive processes, and the positive mood states, something which could be of value in a range of psychiatric conditions. With regard to its safety, it was demonstrated that low doses are well tolerated and have no-to-minimal effects on physiological parameters. Nonetheless, while low (LSD) doses were experienced as pleasant (Bershad et al. 2019; Isbell et al. 1956), it was also shown that drug disliking (Bershad et al. 2019) and anxiety increased (Gasser et al. 2014) and that a cycling pattern of depressive and euphoric mood changes can occur (Greiner et al. 1958). Apparent limitations are the relatively small number of studies, the small sample sizes of (young) healthy volunteers, the lack of performance assessments, and the single doses in most cases, without assessing persisting effects. Therefore, generalizability to other groups about beneficial or harmful effects after repeated dosing is limited and something future placebo-controlled trials will have to address.

18.4.1 LSD and Psilocybin: What Is the Difference

Interestingly, users sometimes attribute other effects to different psychedelics, in which LSD is more associated with cognitive and/or stimulant effects and psilocybin with emotional or well-being effects (Anderson et al. 2019b; Johnstad 2018). This stronger stimulant character of LSD compared to psilocybin was seen by some as an advantage while others experienced it as uncomfortable (Johnstad 2018). In this light, one older study (Abramson and Rolo 1965), -not included in this chapter-, showed that participants trained to discriminate between the effects of a range of substances sometimes confused low doses of psilocybin with LSD. This suggests that the effects of psilocybin and LSD in low doses can be similar. Of note, the reason why this paper was not included in this chapter was because the effects experienced by the participants were not described and therefore did not contribute to a better understanding of the effects of low psychedelic doses on aforementioned measures (Abramson and Rolo 1965).

Additionally, McGlothlin et al. (1967) showed that LSD (25 mcg) indeed induces

stimulant effects, as the effects were similar to those of amphetamine (20 mg) (McGlothlin et al. 1967). Notwithstanding this does not confirm that psilocybin and LSD would have dissimilar effects; it rather supports the claims by users that LSD in low doses has stimulant effects (Johnstad 2018; Anderson et al. 2019a). Therefore, in light of therapy with low doses of LSD and/or psilocybin, it is necessary to know whether they have a different, and perhaps complimentary, effect pattern that could be employed (successively) to treat different symptoms ("cognitive" or "affective") observed in one psychiatric disorder.

18.4.2 Full Psychedelic Doses Show Therapeutic Value: Do We Need Microdoses?

The effect on selective cognitive processes in the reviewed studies resembles the effects of full psychedelic doses in a milder way (Mason et al. 2019; Boardman et al. 1957) and without impairing other cognitive processes (Pokorny et al. 2019). This suggests that low doses of psychedelics could play a therapeutic role in pathologies where no psychedelic experience is necessary like developmental disorders such as ADHD, neurodegenerative disorders like Parkinson's disease, and physical disorders like cluster headache (REF). Also it might be used in a context where a psychedelic experience is potentially not wanted, thinking of cases where consent to this experience, and the capacity to put this experience into perspective is limited like in the elderly, in those who suffer from dementia, in children, or in intellectually disabled people.

18.4.3 Is there an "Ideal" Dosing Schedule?

With regard to the dosing schedule, only one recent study aimed to test the effects of repeated LSD doses on psychological and cognitive functions (Family et al. 2019; Yanakieva et al. 2019). It was shown that LSD blood concentrations were not affected after repeated dosing when leaving two dose-free days in between. Of note, previously it was demonstrated by Isbell et al. (1956) that (transient) tolerance to the effects of (a higher dose of) LSD (75 mcg) occurred after repeated dosing with low doses of LSD (10–30 mcg), given twice daily for three days in a row (Isbell et al. 1956). After three dose-less days the tolerance was reversed. This indicates that interspersing dosing days with dose-less days is more effective than daily dosing; a practice that was already suggested by Fadiman (Fadiman 2011).

18.4.4 Unwanted Effects

It is known that psychiatric treatment is not a one-size-fits-all approach (Penn and Tracy 2012) with non-response as a prominent reason one prominent reason to discontinue a therapy. Next to that, treatment adherence also hinges on the tolerability of unwanted effects caused by the pharmacological treatment (e.g., Predictable 2006; Meaux et al. 2006). When microdosing psychedelics would become the next antidepressant or replacement of Ritalin, as Albert Hoffman suggested (Horowitz 1976), it should, in my view, minimally be as efficacious as these aforementioned treatments, but moreover produce less adverse effects.

While little is reported in the media about possible negative effects (Anderson et al. 2019b), users do report negative effects when explicitly asked for this. Mostly, these effects, physical discomfort and increased feelings of anxiety, occur when under the influence, though some users state to have unpleasant "free" days (Fadiman and Korb 2019; Hutten et al. 2019a; Lea et al. 2020). In the reviewed experimental studies, a single acute dose was generally well tolerated by healthy volunteers, although cycling mood phases and anxiety were also observed (Greiner et al. 1958; Gasser et al. 2015). Repeated doses did not produce more adverse effects than placebo, although mild headache was mentioned more often after microdoses of LSD(Family et al. 2019). All in all, the negative effects linked with low doses of psychedelics seem rather mild. Nonetheless, placebo-controlled trials in patients and

pharmacovigilance data in the future will provide clarity on this.

Importantly, as anxiety might arise, future clinical trials in patients should therefore consider to not send patients, for example, home after they have been administered their microdose. An interesting note was that the presence of anxiety might signify latent emotions coming to the surface, something that could accelerate a healing process, in a therapeutic context, as these emotions can then be discussed with the therapist (Anderson et al. 2019a). This psychological support might not only have to be limited to the dosing day as previously mentioned as users can also experience less pleasant dose-less days (Lea et al. 2019).

18.4.5 Does Expectation Play a Role?

While only a small proportion of the reviewed studies used performance measures to assess the effects of low doses of psychedelics, it is nonetheless clear that the effects are not that pronounced as expected based on media coverage and user reports. Usually microdosing psychedelics is portrayed as a performance enhancer, a boost to creativity, and/or a social or mood enhancer (Anderson et al. 2019a; Hutten et al. 2019a). The question then rises whether the effects are in part due to expectancy hence a "placebo effect." To investigate this, Polito and Stevenson (2019) tested the performance of users and compared this to the expectancies a user group had about the effects of microdosing. Surprisingly, the reported effects did not match the expected effects (Polito and Stevenson 2019); in addition, survey research has shown that people who really expect certain effects sometimes stop with microdosing because the effects did not meet their expectations (Hutten et al. 2019c).

In placebo-controlled studies, the placebo correct for this expectancy effect; however, it can also mask the subtle effects caused by the active treatment by decreasing the difference between the active treatment and placebo. While in all the reviewed studies, the chance that participants would receive LSD or psilocybin was equal to, or above 50%, the effect pattern, with selective effects on specific measures stems positive, in that the demonstrated effects are "real" and due to the administered substance rather than created by expectation. Additional precautions next to the inclusion of placebo in future studies could be to not reveal beforehand the exact substance participants will receive, but instead present a list with substances they *could* receive (Abramson and Rolo 1965), or use additional "active" treatments (McGlothlin et al. 1967) to control for expectancy and placebo effects. While there are benefits to this approach, limitations are the increased study costs, the expected higher attrition rate when using a within-subject study with more conditions, and the increased complexity of statistical analyses.

18.4.6 The Importance of User Claims and Survey Studies in Drug Development

As indicated earlier, microdosing psychedelics is in its infancy and relatively few placebo-controlled experimental studies have taken place. Nonetheless, a lot of "pre-testing" has already been done by users who seek to enhance a certain aspect of their functioning, or by people who self-medicate for specific reasons. Based on this information, targeted hypotheses can be formulated which is, in my view, an opportunity from a drug development perspective. Next to self-medicating for psychiatric disorders, people also used it to treat their physical dysfunctions (like headache) (Hutten et al. 2019b) or treat their pre-menstrual disorder (Fadiman and Korb 2019). Additionally, from user reports we learn that when microdosing is used in a targeted and structured way, it is relatively safe with a low addiction potential and low undesirable effects (Andersson and Kjellgren 2019). It also became clear that not everyone benefited from microdosing psychedelics, something that might be due biological and/or psychological factors as genetic factors (5-HT2A receptor gene) or how the person interprets things that she/he experiences (Andersson and Kjellgren 2019).

Conclusion

While it is yet unclear whether psychedelic microdosing is of therapeutic value due to the dearth of studies in patient samples, the aforementioned effects on selective processes suggest that low doses of psychedelics *could* play a role in psychiatric disorders that have dysfunctional cognitive flexibility or mood for example. It has to be noted that these findings should be considered as preliminary as they are based on a relatively small number of studies, including in general healthy (young) volunteers. Nonetheless, this field shows promise, and the interesting fact for drug developers is that there is a wealth of user reports of which hypotheses can be generated. In general, it can be concluded that to date we lack evidence to confirm the statement by Albert Hofmann that low doses of a hallucinogenic could be useful as an antidepressant, or as a substitute for Ritalin, as for this placebo-controlled experimental studies in patient populations with suited measures are warranted.

- Most clinical psychedelic microdosing studies to date have been conducted in small samples of healthy (young) volunteers, administering single doses, testing the acute effect.
- Preliminary findings show beneficial effects on selective cognitive processes and mood, and connectivity between brain areas involved in affective processes.
- Repeated dosing studies in clinical samples are lacking.
- A limited number of studies have included performance measures, rather than physiological and subjective state measurements.

References

Abramson HA, Kornetsky C, Jarvik ME, Kaufman MR, Ferguson MW (1955) Lysergic acid diethylamide (Lsd-25): Xi. Content analysis of clinical reactions. J Psychol 40:53–60
Abramson HA, Rolo A (1965) Lysergic acid diethylamide (LSD-25). 38. Comparison with action

of methysergide and psilocybin on test subjects. J Asthma Res 3:81–96

Anderson T, Petranker R, Rosenbaum D, Weissman CR, Dinh-Williams LA, Hui K, Hapke E, Farb NAS (2019a) Microdosing psychedelics: personality, mental health, and creativity differences in microdosers. Psychopharmacology 236:731–740

Anderson T, Petranker R, Christopher A, Rosenbaum D, Weissman C, Dinh-Williams L-A, Hui K, Hapke E (2019b) Psychedelic microdosing benefits and challenges: an empirical codebook. Harm Reduct J 16:43

Andersson M, Kjellgren A (2019) Twenty percent better with 20 micrograms? A qualitative study of psychedelic microdosing self-rapports and discussions on YouTube. Harm Reduct J 16:63

Bershad AK, Preller KH, Lee R, Keedy S, Wren-Jarvis J, Bremmer MP, de Wit H (2020) Preliminary report on the effects of a low dose of LSD on resting state amygdalar functional connectivity. Biol Psychiatry Cogn Neurosci Neuroimaging 5(4):461–467

Bershad AK, Schepers ST, Bremmer MP, Lee R, de Wit H (2019) Acute subjective and behavioral effects of microdoses of lysergic acid diethylamide in healthy human volunteers. Biol Psychiatry 86:792–800

Bertino JS Jr, Greenberg HE, Reed MD (2007) American College of Clinical Pharmacology position statement on the use of microdosing in the drug development process. J Clin Pharmacol 47:418+

Boardman WK, Goldstone S, Lhamon WT (1957) Effects of lysergic acid diethylamide (LSD) on the time sense of Normals: a preliminary report. AMA Arch Neurol Psychiatry 78:321–324

Carhart-Harris RL, Goodwin GM (2017) The therapeutic potential of psychedelic drugs: past, present, and future. Neuropsychopharmacology 42:2105–2113

Chi T, Gold JA (2020) A review of emerging therapeutic potential of psychedelic drugs in the treatment of psychiatric illnesses. J Neurol Sci 411:116715

Dolder PC, Schmid Y, Haschke M, Rentsch KM, Liechti ME (2015) Pharmacokinetics and concentration-effect relationship of oral LSD in humans. Int J Neuropsychopharmacol 19:pyv072

Dolder PC, Schmid Y, Steuer AE, Kraemer T, Rentsch KM, Hammann F, Liechti ME (2017) Pharmacokinetics and pharmacodynamics of lysergic acid diethylamide in healthy subjects. Clin Pharmacokinet 56:1219–1230

dos Santos RG, Osório FL, Crippa JAS, Riba J, Zuardi AW, Hallak JEC (2016) Antidepressive, anxiolytic, and antiaddictive effects of ayahuasca, psilocybin and lysergic acid diethylamide (LSD): a systematic review of clinical trials published in the last 25 years. Ther Adv Psychopharmacol 6:193–213

Fadiman J (2017) Microdose research: without approvals, control groups, double-blinds, staff or funding by Dr James Fadiman. In.: Psychedelic Press. 2017XV. Available: https://psychedelicpress.co.uk/blogs/psychedelic-press-blog/microdose-research-james-fadiman

Fadiman J, Korb S (2019) Might microdosing psychedelics be safe and beneficial? An initial exploration. J Psychoactive Drugs 51:118–122

Fadiman, James. 2011. The psychedelic explorer's guide: safe, therapeutic, and sacred journeys. Rochester, Vt. Park Street Press

Family N, Maillet EL, Williams LTJ, Krediet E, Carhart-Harris RL, Williams TM, Nichols CD, Goble DJ, Raz S (2019) Safety, tolerability, pharmacokinetics, and pharmacodynamics of low dose lysergic acid diethylamide (LSD) in healthy older volunteers. Psychopharmacology 237(3):841–853

Fuentes JJ, Fonseca F, Elices M, Farré M, Torrens M (2020) Therapeutic use of LSD in psychiatry: a systematic review of randomized-controlled clinical trials. Front Psych 10:943

Garner RC, Lappin G (2006) Commentary. Br J Clin Pharmacol 61:367–370

Gasser P, Holstein D, Michel Y, Doblin R, Yazar-Klosinski B, Passie T, Brenneisen R (2014) Safety and efficacy of lysergic acid diethylamide-assisted psychotherapy for anxiety associated with life-threatening diseases. J Nerv Ment Dis 202:513–520

Gasser P, Kirchner K, Passie T (2015) LSD-assisted psychotherapy for anxiety associated with a life-threatening disease: a qualitative study of acute and sustained subjective effects. J Psychopharmacol 29:57–68

Ghose T (2015) Short trip? More people 'microdosing' on psychedelics drugs. In Live science. https://www.nbcnews.com/science/weird-science/short-trip-more-people-microdosing-psychedelic-drugs-n390791

Greiner T, Burch NR, Edelberg R (1958) Psychopathology and psychophysiology of minimal LSD-25 dosage; a preliminary dosage-response spectrum. AMA Arch Neurol Psychiatry 79:208–210

Griffiths RR, Johnson MW, Richards WA, Richards BD, McCann U, Jesse R (2011) Psilocybin occasioned mystical-type experiences: immediate and persisting dose-related effects. Psychopharmacology 218:649–665

Griffiths RR, Johnson MW, Carducci MA, Umbricht A, Richards WA, Richards BD, Cosimano MP, Klinedinst MA (2016) Psilocybin produces substantial and sustained decreases in depression and anxiety in patients with life-threatening cancer: a randomized double-blind trial. J Psychopharmacol 30:1181–1197

Hasler F, Grimberg U, Benz MA, Huber T, Vollenweider FX (2004) Acute psychological and physiological effects of psilocybin in healthy humans: a double-blind, placebo-controlled dose-effect study. Psychopharmacology 172:145–156

Horowitz M (1976) Interview with Albert Hofmann, High Times

Hutten NRPW, Mason NL, Dolder PC, Kuypers KPC (2019a) Motives and side-effects of microdosing with psychedelics among users. Int J Neuropsychopharmacol 22:426–434

Hutten NRPW, Mason NL, Dolder PC, Kuypers KPC (2019b) Self-rated effectiveness of microdosing with psychedelics for mental and physical health problems among microdosers. Front Psychiatry 10:672

Hutten N, Mason NL, Dolder PC, Kuypers KPC (2019c) Self-rated effectiveness of microdosing with psychedelics for mental and physical health problems among microdosers. Front Psych 10:672

Isbell H, Belleville RE, Fraser HF, Wikler A, Logan CR (1956) Studies on lysergic acid diethylamide (LSD-25): I. Effects in former morphine addicts and development of tolerance during chronic intoxication. AMA Arch Neurol Psychiatry 76:468–478

Johnson MW, Griffiths RR (2017) Potential therapeutic effects of psilocybin. Neurotherapeutics 14: 734–740

Johnstad PG (2018) Powerful substances in tiny amounts: an interview study of psychedelic microdosing. Nordic Stud Alcohol Drugs 35:39–51

Kuypers KPC, Ng L, Erritzoe D, Knudsen GM, Nichols CD, Nichols DE, Pani L, Soula A, Nutt D (2019) Microdosing psychedelics: more questions than answers? An overview and suggestions for future research. J Psychopharmacol 33:1039–1057

Lea T, Amada N, Jungaberle H, Schecke H, Klein M (2020) Microdosing psychedelics: motivations, subjective effects and harm reduction. Int J Drug Policy 75:102600

Lea T, Amada N, Jungaberle H (2019) Psychedelic microdosing: a Subreddit analysis. J Psychoactive Drugs:1–12

Lee MR (2010) The history of ergot of rye (Claviceps purpurea) III: 1940–80. J R Coll Physicians Edinb 40:77–80

Liechti ME (2017) Modern clinical research on LSD. Neuropsychopharmacology 42:2114–2127

Madsen MK, Fisher PM, Burmester D, Dyssegaard A, Stenbaek DS, Kristiansen S, Johansen SS, Lehel S, Linnet K, Svarer C, Erritzoe D, Ozenne B, Knudsen GM (2019) Psychedelic effects of psilocybin correlate with serotonin 2A receptor occupancy and plasma psilocin levels. Neuropsychopharmacology 44:1328–1334

Mason NL, Mischler E, Uthaug MV, Kuypers KPC (2019) Sub-acute effects of psilocybin on empathy, creative thinking, and subjective well-being. J Psychoactive Drugs 51:123–134

McGlothlin W, Cohen S, McGlothlin MS (1967) Long lasting effects of LSD on normals. Arch Gen Psychiatry 17:521–532

Meaux JB, Hester C, Smith B, Shoptaw A (2006) Stimulant medications: a trade-off? The lived experience of adolescents with ADHD. J Spec Pediatr Nurs 11:214–226

Moreno FA, Wiegand CB, Taitano EK, Delgado PL (2006) Safety, tolerability, and efficacy of psilocybin in 9 patients with obsessive-compulsive disorder. J Clin Psychiatry 67:1735–1740

Nichols DE (2016) Psychedelics. Pharmacol Rev 68:264–355

Osmond H (1957) A review of the clinical effects of psychotomimetic agents. Ann N Y Acad Sci 66:418–434

Passie T (2019) The science of microdosing psychedelics. Psychedelic Press. London, UK

Penn E, Tracy DK (2012) The drugs don't work? Antidepressants and the current and future pharmacological management of depression. Ther Adv Psychopharmacol 2:179–188

Pokorny T, Duerler P, Seifritz E, Vollenweider FX, Preller KH (2019) LSD acutely impairs working memory, executive functions, and cognitive flexibility, but not risk-based decision-making. Psychol Med 50(13):2255–2264

Pokorny T, Preller KH, Kometer M, Dziobek I, Vollenweider FX (2017) Effect of psilocybin on empathy and moral decision-making. Int J Neuropsychopharmacol 20:747–757

Polito V, Stevenson RJ (2019) A systematic study of microdosing psychedelics. PLoS One 14:e0211023

Predictable SEAU (2006) Side effects of antidepressants: an overview. Cleveland Clin J Med 73:351

Prochazkova L, Lippelt DP, Colzato LS, Kuchar M, Sjoerds Z, Hommel B (2018) Exploring the effect of microdosing psychedelics on creativity in an open-label natural setting. Psychopharmacology 235:3401–3413

Sakloth F, Leggett E, Moerke MJ, Townsend EA, Banks ML, Negus SS (2019) Effects of acute and repeated treatment with serotonin 5-HT2A receptor agonist hallucinogens on intracranial self-stimulation in rats. Exp Clin Psychopharmacol 27:215–226

Studerus E, Kometer M, Hasler F, Vollenweider FX (2011) Acute, subacute and long-term subjective effects of psilocybin in healthy humans: a pooled analysis of experimental studies. J Psychopharmacol 25:1434–1452

Tylš F, Páleníček T, Horáček J (2014) Psilocybin – summary of knowledge and new perspectives. Eur Neuropsychopharmacol 24:342–356

Vanderhasselt M-A, De Raedt R, De Paepe A, Aarts K, Otte G, Van Dorpe J, Pourtois G (2014) Abnormal proactive and reactive cognitive control during conflict processing in major depression. J Abnorm Psychol 123:68–80

Webb M, Copes H, Hendricks PS (2019) Narrative identity, rationality, and microdosing classic psychedelics. Int J Drug Policy 70:33–39

Yanakieva S, Polychroni N, Family N, Williams LTJ, Luke DP, Terhune DB (2019) The effects of microdose LSD on time perception: a randomised, double-blind, placebo-controlled trial. Psychopharmacology 236:1159–1170

18

Partnering with the FDA

Katie McCarthy and Niki Gallo

Contents

19.1   A Shared Mission – 276
19.1.1   FDA Mission – 276
19.1.2   FDA Infrastructure – 277
19.1.3   Fostering a Collaborative Relationship with FDA – 278

19.2   Building a Relationship with the FDA – 279
19.2.1   The Role of the Regulatory Project Manager – 279
19.2.2   Knowing When to Engage – 279
19.2.3   Effect of Special Designations/Expedited Development
         Pathways on Drug Development – 281
19.2.4   A Successful Approach to Engaging with and Building
         a Relationship – 281

19.3   Current FDA Initiatives – 282

The US Food and Drug Administration (FDA) has existed for more than 150 years, traced back to the creation of the Agricultural Division in 1848. Although the mission and infrastructure of the FDA have evolved to a great extent, the basic principle has remained the same – to protect the public health. This chapter will walk through the structure and mission of the FDA and will examine best practices for sponsor companies who seek collaborative and successful relationships with the agency to aid in the efficient development of products to improve human health. This chapter will also walk through the different types of interactions with FDA and will explore strategies for ensuring optimal FDA engagement in your program(s). Furthermore, this chapter will highlight key opportunities for sponsors to successfully partner with FDA to enable their programs to progress from the bench to the clinic to commercialization with ease. This chapter will also explain the role and benefits of Special Designations and Accelerated Approval Pathways and will highlight the FDA special initiatives and points of interest and will help the reader understand how to keep up in this constantly evolving field.

⬌ Learning objectives

- To understand the structure and mission of FDA and its relationship with drug developers
- To comprehend FDA communications and meetings
- To understand best practices for communicating and collaborating with FDA
- To recognize the effects special designations can have on a drug development program
- To understand current FDA initiatives and/or points of interest

19.1 A Shared Mission

19.1.1 FDA Mission

The US Food and Drug Administration (FDA) is part of the executive branch of US government and falls under the jurisdiction of the US Department of Health and Human Services (DHHS). The FDA has evolved from its humble beginnings in the 1800s as a laboratory that analyzed samples of food, fertilizers, and agricultural products to becoming a regulatory authority overseeing the development of drugs, biologics, and devices designed to impact human health. The mission of FDA today, as well as the infrastructure needed to support the mission, has far exceeded the expectations of the original intent in 1862. Great strides and advances in science and biotechnology have helped to shape the framework within the FDA.

The FDA's current mission states (▶ https://www.fda.gov/about-fda/what-we-do#mission):

> » *The Food and Drug Administration is responsible for protecting the public health by ensuring the safety, efficacy, and security of human and veterinary drugs, biological products, and medical devices; and by ensuring the safety of our nation's food supply, cosmetics, and products that emit radiation.*
>
> *FDA also has responsibility for regulating the manufacturing, marketing, and distribution of tobacco products to protect the public health and to reduce tobacco use by minors.*
>
> *FDA is responsible for advancing the public health by helping to speed innovations that make medical products more effective, safer, and more affordable and by helping the public get the accurate, science-based information they need to use medical products and foods to maintain and improve their health.*
>
> *FDA also plays a significant role in the Nation's counterterrorism capability. FDA fulfills this responsibility by ensuring the security of the food supply and by fostering development of medical products to respond to deliberate and naturally emerging public health threats.*

The FDA's authority covers all 50 states and extends to the District of Columbia, Puerto Rico, Guam, the Virgin Islands, American Samoa, and other US territories and possessions. As noted in the current mission, the FDA is a part of the DHHS which is responsible for protecting the health of all Americans and providing essential human services.

19.1.2 FDA Infrastructure

The FDA consists of the following offices/divisions:

- Office of the Commissioner
- Office of Foods and Veterinary Medicine
- Office of Medical Products and Tobacco
- Office of Operations
- Office of Global Regulatory Operations and Policy

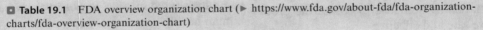 For the purposes of this chapter, we are going to focus on the Office of Medical Products and Tobacco, specifically Center for Drug Evaluation and Research (CDER), Center for Biologics Evaluation and Research (CBER), and Center for Devices and Radiological Health (CDRH), as this is the part of FDA that interfaces with sponsors in developing new drugs.

The Office of Medical Products and Tobacco provides high-level coordination and leadership across the centers for drugs, biologics, medical devices, and tobacco products. The office also oversees the agency's special medical programs. Effective March 31, 2019, FDA begun operational implementation of an agency reorganization. FDA's reorganization reflects the agency's commitment to modernizing its structure to advance its mission to protect and promote public health and to meet the challenges of rapid innovation across the industries regulated by FDA. The FDA's reorganization will realign several entities across the agency to promote strategic priorities and will elevate the role of the centers, offices, and field forces.

In line with the FDA's mission, there is a common thread among the FDA centers that are responsible for the review of new and innovative products that impact human health; this commonality is the review and oversight to ensure that these products are safe and effective for the people.

Within the Office of Medical Products and Tobacco are the following centers/offices:

- Office for Special Medical Programs
- Center for Drug Evaluation and Research (CDER)
- Center for Biologics Evaluation and Research (CBER)
- Center for Devices and Radiological Health (CDRH)
- Center for Tobacco Products (CTP)
- Oncology Center of Excellence

☐ Table 19.1 displays the organizational structure of the Food and Drug Administration

☐ **Table 19.1** FDA overview organization chart (► https://www.fda.gov/about-fda/fda-organization-charts/fda-overview-organization-chart)

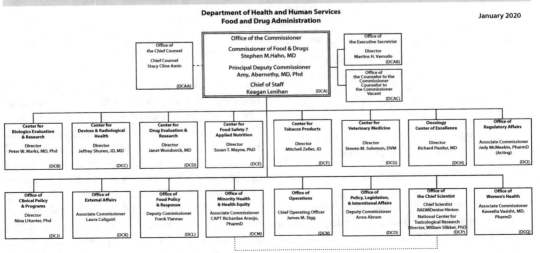

reporting to the Commissioner of Food and Drugs within the Office of the Commissioner.

> **Definition**
>
> CDER regulates prescription and over-the-counter drugs, including therapeutic biologics and generic drugs from the initial Investigational New Drug (IND) application through to the New Drug Application (NDA) and post-marketing activities to ensure the products are safe and effective.

CDER has multiple offices that are responsible for the review of a product's application, specifically the Office of New Drugs (OND) is composed of numerous review divisions that review new drugs by body system.

> **Definition**
>
> CBER regulates biological products including vaccines, blood and blood products, allergenics, tissue and tissue products, xenotransplantation, cellular and gene therapies, and biosimilars.

Like CDER, CBER regulates biologics from the initial (IND) application through to the Biologics Licensing Application (BLA) and post-marketing activities to ensure the products are safe and effective.

> **Definition**
>
> CDRH is responsible for regulating firms who manufacture, repackage, relabel, and/or import medical devices sold in the United States. In addition, CDRH regulates radiation-emitting electronic products (medical and non-medical) such as lasers, X-ray systems, ultrasound equipment, microwave ovens, and color televisions.

CDRH is tasked by ensuring American's safety by ensuring medical devices are safe and effective. Like CDER and CBER, there are product life-cycle activities (Investigational Device Exemption [IDE], 510K applications,

Premarket Approval Application [PMA]) that CDRH regulates.

19.1.3 Fostering a Collaborative Relationship with FDA

Academic and industry sponsors are dedicated to developing innovative, impactful, curative therapies for patients in need. The foundation of a sponsor organization is built upon the goal to generate therapies or devices that will make a positive impact on human health. As sponsor organizations create infrastructure to support the various facets and stages of drug development, the FDA has also created infrastructure to allow for efficient regulation of these products.

> ❯ Analogous to the infrastructure needs, both sponsor organizations and the FDA require talented scientists, clinicians, medical professionals, regulatory affairs specialists, quality assurance specialists, and pharmacovigilance professionals, along with information technology specialists, to shepherd the product through the development product life-cycle from concept to market access.

There are several commonalities between sponsors and FDA with the most notable being the shared mission to positively impact human health.

> ❯ Recognizing the shared mission is a key component to successful development of new drugs.

Some sponsors view the FDA as an obstacle in bringing new and innovative therapies to patients and thus have a complicated relationship with the FDA, which makes for a challenging road. However, sponsors who respect and understand FDA's authority, experience, and perspectives on bringing new, safe, and effective products to market will have a collaborative relationship with the FDA and thus a clear understanding of the data needed to support a marketing application. Under these circumstances, product manufacturing plans, clinical study designs, content and organization of an NDA/BLA, and labeling claims are discussed and vetted with the agency prior to implementing these plans, which mitigates the risk of delays to key studies and activities and impact product approval.

19.2 Building a Relationship with the FDA

When it comes to building relationships, Dale Carnegie's bestseller "How to Win Friends and Influence People" published in 1936 is as relevant today as it was nearly 100 years ago. It is a universal guidebook on how to interact with peers which can also be applied to building a relationship with the FDA. "If there is any one secret of success," he states, "it lies in the ability to get the other person's point of view and see things from that person's angle as well as from your own." This passage is essential to building a collaborative and successful relationship with the FDA.

> In addition to recognizing the shared mission to impact human health and appreciate the FDA's perspective and authority over investigational products, the approach to engaging with the FDA and knowing when to engage in a dialogue with the FDA is key to success in developing drugs.

The FDA must regulate and review numerous investigational products at a given time and it is important to remember that your product is one of many that is typically under review by the review team. Displaying awareness of this and being respectful of FDA's time will help to further strengthen the relationship.

19.2.1 The Role of the Regulatory Project Manager

Typically, all communication from a sponsor to the FDA funnels through the Sponsor's Authorized Representative to the assigned FDA regulatory project manager (RPM) for your application (i.e., IND/IDEs, NDA/BLAs, etc.). Within CDER, CBER, and CDRH and their respective reviewing divisions, for example, sponsors developing drugs for central nervous system (CNS) indications would submit their applications to CDER to the Office of Neuroscience which encompasses the following review divisions: Division of Anesthesiology, Addiction Medicine, and Pain Medicine (DAAP), Division of Neurology I (DN I), Division of Neurology II (DN II) and Division of Psychiatry (DP) . Upon receipt by the FDA, an RPM is assigned to that application (CDER [▶ https://www.fda.gov/about-fda/center-drug-evaluation-and-research-cder/cder-offices-and-divisions] and CBER [▶ https://www.fda.gov/about-fda/center-biologics-evaluation-and-research-cber/center-biologics-evaluation-and-research]). From that point forward, all communications are funneled through the RPM to the appropriate FDA review team staff.

> RPMs generally oversee multiple applications at one time and are not fully dedicated to one sponsor; therefore, it is of great importance to establish a great relationship with the RPM by providing organized, thoughtful, and prompt responses to requests.

By doing so demonstrates the sponsors willingness to collaborate with the FDA on the development of a new therapy and in some cases helps the FDA respond faster than anticipated since they appreciate the sponsors willingness to comply and collaborate with the FDA's requests.

19.2.2 Knowing When to Engage

Understanding the appropriate time to engage with the FDA can make or break a drug path toward reaching the end goal, which is affording patients in need access to a new therapy.

> Ultimately, the FDA has the final say in the approvability of your new drug, and thus having a collaborative relationship through the drug's development is essential.

Knowing when it is appropriate to engage and seek feedback from the FDA versus proceeding with your development plans without input can speak volumes to the FDA.

The FDA has prepared and published guidance documents on the types of formal meetings a sponsor can have with the agency. For example, there is FDA guidance for sponsors developing PDUFA products entitled, "*Formal Meetings Between the FDA and Sponsors or Applicants of PDUFA* Products Guidance for Industry*," (December 2017). The types of meetings, meeting timelines, and expectations for content to support these meetings are outlined in this FDA guidance. The agency often views these meetings as critical points in the regulatory process and therefore it is important that there are efficient, consistent procedures for the timely and effective conduct of such meetings. Outlined below are the four types of formal meetings under the Prescription Drug User Fee Action (PDUFA[1]) that can occur between sponsors and FDA staff: type A, type B, type B EOP (end-of-phase), and type C.

> Type A meetings are those that are necessary for an otherwise-stalled product development program to proceed or to address an important safety issue. Before submitting a type A meeting request, requesters should contact the review division or office to discuss the appropriateness of the request.

> Type B meetings are as follows: pre-investigational new drug (IND) meetings, pre-emergency use authorization meetings, pre-NDA meeting, post-action meetings requested 3 or more months after an FDA regulatory action other than an approval (i.e., complete response letter), meetings regarding risk evaluation and mitigation strategies (REMS) or post-marketing requirements that occur outside the context of the review of a marketing application, meetings held to discuss the overall development program for products granted breakthrough therapy designation.

> Type B EOP meetings are certain EOP1 meetings (i.e., for products that will be considered for marketing approval under 21 CFR part 312, subpart E, or 21 CFR part 314, subpart H, or similar products) or EOP2 or pre-phase 3 meetings typically used to garner feedback on the pivotal registration clinical trial design or other key development aspects that need to be in place to support that clinical study.

> Type C meeting is any meeting other than a type A, type B, or type B (EOP) meeting regarding the development and review of a product, including meetings to facilitate early consultations on the use of a biomarker as a new surrogate endpoint that has never been previously used as the primary basis for product approval in the proposed context of use.

There are three meeting formats for formal FDA meetings: face-to-face, teleconference/videoconference, and written response only (WRO).

> Face to face – traditional face-to-face meetings are those in which the majority of attendees participate in person at the FDA.

> Teleconference/videoconference – teleconferences/videoconferences are meetings in which the attendees participate from various remote locations via an audio (e.g., telephone) and/or video connection.

> Written response only – WRO responses are sent to requesters in lieu of meetings conducted in one of the other two formats described above.

Usually, sponsors will have multiple formal meetings with the FDA. These typically include a pre-IND meeting, end-of-phase (EOP) meeting, pre-NDA meeting, and potentially other type C meetings that are focused on a specific development milestone. The first interaction a sponsor has with the FDA is usually the type B pre-IND meeting. It cannot be emphasized enough that the quality

---

1   The Prescription Drug User Fee Act (PDUFA), a law passed in 1992 by the US Congress to collect fees from Sponsors to fund the FDA drug approval process.

of the documents provided and the outcomes of the meeting influence the relationship between a sponsor and the FDA. More on how to build a relationship with the FDA is discussed below.

Although the guidance noted above speaks to formal meetings, there are opportunities for informal meetings, correspondence, and feedback once a relationship is established with your assigned RPM. The willingness of the RPM and the reviewing division to provide informal feedback stems from the sponsor's preparedness and quality of previously submitted documentation. The quality of the IND application and the sponsor's preparedness and representation at an FDA meeting are likely key metrics the FDA will use to measure the adequacy of the sponsor's approach to developing a drug and thus likely influence their willingness to provide informal feedback or recommend a formal avenue to allow for official documentation of the advice provided by the FDA. This informal correspondence is not mandated by PDUFA timelines and, therefore, you are subject to the RPM and the review team's availability since formal feedback and PDUFA deliverables take priority over non-PDUFA activities.

19.2.3 Effect of Special Designations/Expedited Development Pathways on Drug Development

Receiving a special designation brings additional benefits and incentives to the sponsor when working with the FDA.

> For drug development programs that have a special designation, i.e., Orphan Drug Designation (ODD), Fast Track Designation (FTD), Breakthrough Therapy Designation (BTD), Regenerative Medicine Advanced Therapy (RMAT), the agency sees the drug has the potential to address an unmet medical need for a serious disease.

This often results in frequent dialogue with the FDA, both informal and formal routes of communication, as the FDA wants to closely collaborate with the sponsor to bring this new drug to market under expedited pathways.

19.2.4 A Successful Approach to Engaging with and Building a Relationship

The approach to building a successful relationship with the FDA is analogous to the relationships one builds in their work and home lives. Successful relationships are often built on trust, and the same can be said for the sponsor/FDA relationship. A real-world example is the press surrounding the FDA, Novartis, and Novartis's drug Zolgensma. There are claims that Novartis delayed submitting preclinical safety information until the marketing application review of Zolgensma indicated for Spinal Muscular atrophy (SMA), which caused the FDA to change their approach to working with Novartis. This lack of trust as result of not sharing nonclinical safety data in a timely manner, nor altering the FDA of these findings in real-time, has cast a shadow on the drug industry, and steps should be taken to gain the agency's trust again.

Key steps to building a collaborative working relationship with the FDA are as follows:

- View the FDA review team members as your peers rather than authority figures
- Participate in a scientist to scientist discussion
- Engage with the FDA in a collaborative and transparent nature
- Always put your best foot forward by being prepared and organized
- Don't waste their time on trivial details or questions that could be answered by referring to FDA guidance and/or regulatory precedent

Sharing data and/or concerns with the FDA should not be a game of cat and mouse. Sponsors often incorrectly fear that by sharing certain data points and/or asking the agency certain questions will raise red flags with the FDA causing them to scrutinize a dataset and request additional information. It is very important to remain transparent and ask the questions that are keeping you awake at night as it will benefit the FDA's review of your IND and/or marketing application. There will be little to no surprises with future submissions if these points are discussed with the FDA. These discussions will likely result in an agreed-upon plan to put these issues to rest.

Treating the FDA review team members as your peers and engaging in scientist-to-scientist dialogue are instrumental to a successful relationship. A drug sponsor must realize that the FDA staff are also scientists and share the same mission to bring new and innovative therapies to patients in need; however, they are also responsible for determining that the therapy is both safe and efficacious. One must also realize that FDA has seen multiple drug applications and is aware of ongoing clinical investigations and thus they often make recommendations or ask for additional safety measures, nonclinical studies, or changes to manufacturing procedures because they may have seen something in another drug development program. Remember that it is the duty of the FDA to make recommendations that protect humans and prevent harm from happening to clinical trial participants and patients. As a sponsor, don't take the point of view that FDA is singling you out; that is not the case, they are trying to prevent any harm to humans.

19.3 Current FDA Initiatives

As science and new drug modalities are evolving, so are the guidelines the FDA is providing in order to support and keep up with these new therapeutic modalities. Specifically, in neurological disorders, where the progress for new disease-modifying treatments of well-studied neurodegenerative diseases like Alzheimer's and Parkinson's has been slow, the FDA is eager to collaborate with sponsors to find new treatment approaches.

To accelerate innovation and provide safe patient access to new therapeutics tools (e.g., gene and cell therapies and most recently also digital therapies), the FDA has issued statements on how they are internally evolving to address the arising questions. The FDA encourages early dialogue with sponsors on the development of new bio markers which may help to define surrogate endpoints beyond clinical symptoms which includes new digital endpoints (i.e., AI-enhanced imaging tools).

As former Commissioner Scott Gottlieb outlined in a statement, the FDA is moving toward a more team-based approach as part of the CDER Office of New Drugs modernization, with interdisciplinary teams across different review divisions. Plans include issuing additional guidance documents more frequently (cite). Even more so, working groups, such as the Center of Excellence for Digital Health, are created to help shape these new guidelines in close collaboration with industry leaders and scientific experts.

So how to effectively keep up with the constantly evolving field? In order to best collaborate with sponsors while being completely transparent with other stakeholders, especially consumers, the FDA set itself very high standards. Besides a very comprehensive website on already approved drugs (Drugs@ FDA), the FDA provides a vast tool box to keep up to date on regulator intelligence right from the source. One can subscribe to daily news updates on FDA Voices: Perspective from FDA Leadership and Experts, policy, consumer safety and enforcement, medical products, food, and tobacco (cite). In addition, the FDA offers an educational platform for all stakeholders: healthcare professionals, industry, consumers, and academia (cite) and holds regular informative meetings, workshops, and seminars (cite); some of which are free to the public.

Independent sites (i.e., FDAnews) can be used as alternative sources to find information. Professional organizations, such as the Regulatory Affairs Professional Society (RAPS) or the American Society for Quality (ASQ) can be referenced to examine the guidelines of the evolving regulatory landscape. Even more so, many regional bio-clusters offer events to update their local life science ecosystems on regulatory strategies, many of them organized by independent regulatory consultancies.

> ### Conclusion

FDA and sponsor companies share a common mission to bring safe and effective therapies to the public. Recognizing this shared goal is a key step in building a collaborative relationship with FDA, which will help sponsors successfully develop their products with limited surprises along the way. It is crucial for sponsors to know when and how to engage with and seek feedback from the FDA. Meetings with the agency are critical points in the regulatory process and the development of a product. Being prepared and thoughtful in your approach during all meetings and interactions will positively influence FDA's view of the adequacy of the sponsor's approach. It is also important for sponsors to be aware of the possible expedited pathways for their product, as this often brings opportunities to collaborate more frequently with FDA. Building a positive and collaborative relationship with FDA can positively impact the development and approval of your product. It is important to build trust, to listen, and to be collaborative and transparent throughout all of your interactions. By working collaboratively together, sponsors and the FDA can transform their mission to reality and bring safe and efficacious treatments through development and to the market for patients in need.

Correction to: Positron Emission Tomography in Drug Development

Frans van den Berg and Eugenii A. (Ilan) Rabiner

Correction to: R. Schreiber (ed.), *Modern CNS Drug Discovery*, https://doi.org/10.1007/978-3-030-62351-7_11

The original version of Chapter 11 was inadvertently published without the first name of the co-author. The name "A. (Ilan) Rabiner" has now been corrected to "Eugenii A. (Ilan) Rabiner".

The updated online version of this chapter can be found at
https://doi.org/10.1007/978-3-030-62351-7_11

Printed in the United States
by Baker & Taylor Publisher Services